Integrated Ecosystem Management Approach and Application

Proceedings of the International Workshop on Integrated Ecosystem Management Approach and Application held in Beijing, November 6-7, 2008

Jiang Zehui

China Forestry Publishing House

Editorial Board

Editor: Jiang Zehui

Associate Editors: Zheng Xiaosong, Zhang Yongli, Wei Diansheng

Members of the Editorial Board:
Wang Yan, Wang Hong, Wang Zhongjing, Wang Zhenjiang, Ran Dongya, Feng Renguo, Feng Baoshan, Ye Bing, Tian Baoguo, Bai Jinming, Liu Shirong, Liu Weihua, Liu Xuhong, Liu Keyong, Sun Xuefeng, Zhuang Guotai, Zhuang Chunyan, Qu Guilin, Wu Jinkang, Zhang Kebin, Zhang Xuejian, Shen Jianzhong, Zou Ciyong, Zhou Zhenhua, Mong Yongqing, Yue Zhongming, Hu Zhangcui, Zhao Tao, Yin Weiping, Tu Fengxin, Guo Wenfang, Guo Aijun, Gao Zhixin, Kou Jianping, Cui Xianghui, Huang Wenhang, Dong Yan, Lu Shengli, Chu Liming, Xue Ping

Foreword

Land is the most essential environmental element in the earth ecosystem and the fundamental human beings are dependent on. Land is called 'the Mother gestating human beings'. Global land degradation has been worsening due to excessive development and unsustainable management of land resource. Land degradation has caused vulnerability of some land ecosystems, dramatic reduction in agricultural and animal husbandry productivity and has directly threatened food security and ecosystem health of the world. As a major environmental issue threatening human beings livelihood and development, land degradation has increasingly received great attention from the international community. Land degradation control and safeguarding of the security of the ecosystem to achieve sustainable development are not only wishes of the local communities but also holy responsibilities of all governments and a common task for all human beings.

China is one of the countries suffering from most severe land degradation which has been a particularly outstanding issue in the arid and semi-arid areas in West China. Therefore, land degradation control in West China is not only one of the fundamental programs and a major measure for China's implementation of its West China Development Strategy and achievement of sustainability in economic and social development, but also an important part and a key step for safeguarding the security of global environment and acceleration of the sustainable development of the internationalization process.

The PRC-GEF Partnership on Land Degradation in Dryland Ecosystems is the first partnership on land degradation control of the Global Environment Facility (GEF). It aims to use the Integrated Ecosystem Management (IEM) concept as direction to develop a new approach that addresses the root of land degradation by setting up a cross-sector, cross-department and cross-region institutional framework for sustainable natural resource management. Positive achievements have been obtained thanks to great attention paid by the Ministry of Finance (MoF) and the State Forestry Administration (SFA) of China and their coordination, strong support from GEF, Asian Development Bank (ADB), national and international consultants and great efforts made by the Central Project Management Office (CPMO), Provincial Project Management Offices (PPMOs) and the communities in the Project areas since the inception of the Partnership. These achievements have strongly enhanced reform of China's traditional land degradation control and ecosystem rehabilitation techniques and models, and are of global value for land degradation for other countries.

To present and extend the achievements of the Partnership, strengthen exchange of experience with other countries in natural resource management, in particular, in land degradation, and to enhance the development of the Partnership and the collaboration with other partnerships, the Ministry of Finance (MoF) and the State Forestry Administration (SFA) of China jointly organized an International Workshop on IEM Approach and Application in Beijing on November 6-7, 2008, which GEF and ADB gave strong support. Over 200 participants including government officials, researchers and representatives from relevant ministries, the 6 Project Provinces, GEF, World Bank, ADB, other international organizations and well-known scholars from national and international institutes attended the Workshop. All discussions were held over relevant topics, resulting in good outcomes.

This collection of the Proceedings of the International Workshop on IEM Approach and Application has been prepared to ensure the outcomes of the Workshop are extensively shared. It collects the main results of various researches that were shared at the Workshop, including presentations and academic papers prepared by national and international experts, which covers theory and practice of various sectors including laws and policies, strategies and plans, monitoring and evaluation and pilot sites for land degradation control as well as land degradation control and climate change. The proceedings have been divided into 5 parts: Part I includes opening addresses of government officials and the keynote speech; Part II presents the achievements of the Partnership on land degradation; Part III presents the achievements of the Capacity Building to Combat Land Degradation Project; Part IV displays the practices and application of the IEM approach; and Part V gives good description on climate change and land degradation.

These proceedings are crystals of wisdom of many national and international decision makers, managers, practitioners and professionals. They are also a large collection of the achievements and practical experience of land degradation control by many countries. It is expected that the publishing of these proceedings will further enhance extension and application of the IEM approach and provide useful experience for land degradation control in other areas in China and other countries.

Compliers
November 19, 2009

Contents

Foreword

CHAPTER I : Opening Addresses and Keynote Speech

1. **Opening Address by Ministry of Finance**
   Mr. Zhang Tong 2

2. **Opening Address by State Forestry Administration**
   Mr. Li Yucai 6

3. **Opening Address by GEF**
   Madam Monique Barbut 10

4. **Opening Address by ADB**
   Mr. Kunhamboo Kannan 11

5. **Opening Address by Kyrgyz Republic**
   Mr. Kambarali Kasymov 13

6. **Opening Address by World Bank**
   Ms. Sari Söderström 15

7. **Join Hands to Promote Sustainable and Healthy Development of the PRC -GEF Partnership**
   Professor Jiang Zehui 17

CHAPTER II: Global GEF Partnerships on Land Degradation

8. **Application of IEM in Land Degradation Control in West China: Successful Stories of PRC-GEF Partnership on Land Degradation in Dryland Ecosystem**
   Ms. Hu Zhangcui 25

9. **Central Asian Countries Initiative for Land Management**
   Mr. Kambarali Kasymov 32

10. **Sustainable Land and Water Management: An African Initiative to Develop Agriculture under Pressures of Land Degradation**
    Phiri Elijah, Bwalya Martin 35

11. The role of the Partnership Initiative on Sustainable Land Management in Combating Land Degradation in Caribbean SIDS

  Leandra K. Sebastien ··· 46

12. China Biodiversity Partnership Framework

  Sun Xuefeng ··· 59

CHAPTER III: Outcomes of Capacity Building to Combat Land Degradation Project

13. Application of IEM Concepts & Methods in Land Degradation Control in Western China

  Li Sandan ·· 62

14. Capacity Building in Combating Land Degradation: International Application of Legal Aspects of PRC-GEF Partnership Project

  Dr Ian Hannam ·· 68

15. Executive Summary and Overview of Strategy and Action Plan for Combating Land Degradation in Northwest Six Provinces of China

  Zhang Kebin ·· 85

16. Land Degradation Status and Control Measures in Xinjiang Autonomous Region

  Cui Peiyi, Gao Yaqi and Liu Xiaofang ·· 97

17. Participatory IEM Approach Used in Land Degradation Control in the Community: a Review on Mandulahu Gacha (Village) Pilot Site in Naiman Banner Inner Mongolia

  Gao Guiying ··· 106

18. Challenges Facing Legislation for Land Degradation Control and Their Countermeasures in China

  Wang Canfa, Feng Jia ··· 115

19. Practice and Application of IEM Concept in Legislation for Combating Land Degradation in Gansu Province

  Wan Zongcheng ·· 123

20. Achievements of Hundan Watershed Pilot Site in Huangyuan County Qinghai Province and Lessons Learned

  Cai Chengyong ··· 133

21. Building Information Platform Serving for Ecological Development

  Wang Zepeng ··· 138

22. Community Capacity Building and Achievements in GEF Pilot Sites, Shaanxi Province
    Wen Zhen 142
23. Study on Evaluating Forest-Related S&T Innovation Capacity in Western Areas Stricken by Land Degradation of China
    Liu Xilin, Hu Zhangcui 149
24. Combating Land Degradation in Mediterranean Coastal Areas: Using Erosion Mapping for Assessment of Land Degradation in O. Rmel Watershed
    R. Attia, S. Agrébaoui and H. Hamrouni 158
25. Legislation on Prevention and Combating Desertification Caused by Land Use in China
    Yu Wenxuan, Zhou Chong 166
26. Improving China's Legislative Framework for Land Degradation Control
    Zhou Ke, Cao Xia and Tan Baiping 169

CHAPTER IV: IEM Approach and Application

27. Developing a Global Strategy to Combat Land Degradation and Promote Sustainable Land Management
    Michael Stocking 186
28. Using the Integrated Ecosystem Management Principles, Implementing Practice to Combat Desertification
    Liu Tuo 201
29. Land Degradation and Sustainable Land Management in the Central Asia
    Mr. Umid Abdullaev 206
30. National Mechanism for Sharing Land Degradation Monitoring and Evaluation Information
    Wu Bo 208
31. A Legal and Policy Framework for IEM of Soil and Water—the New Zealand Model
    David P Grinlinton, Kenneth A Palmer 215
32. Application and Practice of Integrated Ecosystem Management (IEM) in Gansu GEF Pastoral Development Project
    Hua Limin 227
33. An Introduction of Ningxia Integrated Ecosystem and Agricultural Development Project of Asian Development Bank
    Ma Minxia 239

34. Implementing GEF Objectives in a Systems Framework in Western China
    Victor R. Squires 243
35. An IEM Approach to the Land Degradation Control and Conservation of Biodiversity in Dryland Ecosystems
    Zheng Bo 248
36. CPRWRP's Approach and Challenges to Slow down Degradation in Mountainous Areas of the Upper Yangtze River Basin.
    Piet van der Poel 254
37. Application and Promotion of Integrated Ecosystem Management in China
    Cai Shouqiu 265

CHAPTER V: Climate Change and Land Degradation

38. Brief Introduction on Climate Change Impact and China's Action
    Gao Yun 274
39. IEM and the Future Opportunities for Growth in China from 2010 onwards; the Ultra Green approach.
    Ian R. Swingland 280
40. Climate Change and Carbon Forestry
    Liu Shirong, Jiang Youxu and Shi Zuomin 300
41. Mitigating Climate Change and Better Ensuring Agriculture's Adaptation for impending Climate Change through Conservation Agriculture
    Des McGarry 309
42. Working Together to Combat Rangeland Degradation
    Brant Kirychuk 326

CHAPTER I
Opening Addresses and Keynote Speech

1. Opening Address by Ministry of Finance

Mr. Zhang Tong
Assistant Minister of Ministry of Finance

Distinguished delegates, Ladies and Gentlemen,
Good Morning!

It's my great pleasure to participate in the International Symposium on Concept and Practice of Integrated Ecosystem Management (IEM). First of all, on behalf of the Ministry of Finance of the People's Republic of China, I would like to offer my sincere congratulations on the opening of the symposium and the successful implementation of the GEF project and the achievements made so far, and express my heartfelt gratitude to the Global Environment Facility, the Asia Development Bank, the World Bank as well as other international organizations, governments of relevant countries and friends from home and abroad who have been continuously caring for and supporting the ecological development and environmental protection of China.

As is known to all, in the past over one century, with the global population boom and improved productivity, while making great achievements in economic development, the mankind has also destroyed and used the natural resources without restrain to the point that the natural resources are running out rapidly and the ecosystem is deteriorating and the survival and development of human beings are hence threatened by such disastrous consequences as energy crisis, environmental pollution, water resource shortage, global warming, desertification, and the extinction of a large number of species of wild fauna and flora. Therefore, it has drawn great attention from the countries all over the world to strengthen ecological improvement, enhance environmental protection and establish a harmonious society, which has become significant to realize the sustainable social and economic development of a country or a region.

China is the biggest developing country, and its resources and environment have always been key factors that constrain its sustainable social and economic development. The Chinese government has been attaching great importance to the protection of ecosystem and the sustainable social and economic development. Since the beginning of the 21st century, following the scientific outlook on development, China has made a series of strategic decisions, including building up resource-conserving and environment-friendly society, building an innovating country, establishing a harmonious socialist society and constructing new socialist rural areas. Especially at the 17th National Congress of the CPC in 2008, China, viewing from the strategic perspective of China and the world, made a strategic decision to build ecological civilization and determined that China shall be turned into a country with sound ecology by the year of 2020. Ecological civilization is a new type of civilization along the history of human civilizations. Based on the respect in nature and maintaining the nature, it aims to realize the harmonious co-existence of human beings, the nature and the society, pursuits the establishment of sustainable mode of production and consumption, and leads to sustainable and harmonious development. To build ecological civilization is a strategic measure to carry out the scientific outlook on development, which is one of the actions China takes to improve the global ecosystem, facilitate mankind progress and human civilization and to implement the international conventions and fulfill the millennium goals of the United Nations.

Among all the ecological problems confronting China, land degradation in arid zones especially in Western China is one of the most serious problems. According to the 3rd national survey on land desertification, there are currently a total of 2.6362 million m^2 of desertified land and 1.7397 million m^2 of sandified land, accounting for 1/3 and 1/5 of the national land respectively. The sandification in some areas is still growing. The direct economic loss caused by land sandification amounts to as high as over 50 million yuan. Nearly 400 million people across country are threatened by land desertification and sandification. Half of the poverty population is living in those areas. Land desertification has become China's another great concern after flood and drought.

The Chinese government has attached great importance to the control and management of land degradation and desertification. As early as the end of 1970s, China launched the "Three-North Shelterbelt System Development Program". Since the late 1990s, China has initiated and implemented such national key ecological programs as the Natural Forest Protection Program, Program for Converting Farm Land to Forest, Program for Combating Sandification in Areas in Vicinity of Beijing and Tianjin, Grassland Development and Protection Program, Protective Farming Program, and Water-Saving Draught-Resistant Agriculture Program. The Chinese government has input huge amount of money in this regard. Take the Program for Converting Farm Land to Forest as an example, by the end of 2007, an accumulation of over 160 billion yuan from the national budget has been injected in this project, of which 62.5% has gone to Western China. Of the over 23 billion yuan of funding that has been allocated this year in particular, 61% goes to Western China. Another example is the arrangement of the national poverty alleviation project, in which the Chinese government also favors its support to the western region. Of over 44 billion yuan of poverty alleviation funds allocated from 2006 to 2008 alone, over 62% has gone to the western region. The prominent achievements made in these key programs and poverty alleviation projects have effectively kept the key sandy areas from being further sandified, greatly improved the ecology and enhanced the local economic and social development. However, as land degradation in China is widely distributed with large area affected, plus the influence of global warming and human activities, China is still confronted with severe land degradation problem and has a long way to go to combat land degradation and desertification.

The causes that lead to land degradation in dry ecosystem are rather complicated, which involve both the natural factors and human factors. To combat land degradation is therefore a long-term and complex systematic project. It needs highly advanced science and technology as well as improved laws and policies; it also needs the concerted efforts and close cooperation of domestic departments of agriculture, forestry, environmental protection, water resources and land resources as well as the full support of the international community. The Chinese government attaches great importance to the international cooperation and exchange in combating land degradation and has established the cooperative partnership to combat land degradation with GEF in 2002, namely PRC-GEF Partnership on Land Degradation in Dryland Ecosystems.

PRC-GEF Partnership on Land Degradation in Dryland Ecosystems is the first partnership that GEF established with a government in the field of ecology. The Partnership aims to establish an integrated management mechanism across sectors and regions, and to coordinate the policies, laws, programs and actions through the concerted efforts of relevant departments so as to realize an integrated management of land degradation in Western China especially in ecologically vulnerable areas and to achieve the goals of reducing poverty, maintaining ecological sustainability and facilitating the sustainable social and economic development.

The Capacity Building to Combat Land Degradation Project is the core project of the Partnership. Since the four-year implementation of the project, with the guidance of the steering committee and the joint efforts of the Ministry of Finance, the State Forestry Administration, the central project member departments and the provinces involved in the project, the project has made great achievements, playing a positive role in establishing and improving the framework of technology, policies, laws, and organizations in combating land degradation in the arid zones of Western China, furthering the Partnership on Land Degradation, and setting a good example for the global land degradation combating.

The Partnership marks a new change for GEF's traditional project-based practice. This large-scale cooperation with detailed steps and plans also sets a landmark in the history of the exchange and cooperation between the Chinese government and the GEF. The successful implementation of the Partnership has promoted the application and development of the concept and approaches of IEM in China. Through enhancing the understanding and application of the approaches of IEM by the participating departments and provinces, the Partnership improved the framework of combating land degradation from the perspectives of laws, policies, technologies, information and demonstration, accelerated the progress of land degradation combating in arid zones of Western China, effectively promoted the rehabilitation of the vulnerable ecosystem and the protection of biodiversity in arid zones of Western China, and contributed to the improvement of the global ecology.

Currently, the IEM is still a new concept and approach in China, which needs to be improved by more practice and innovation based on the achievements made in the last phase of capacity building project so that it could be further extended and applied in the global land degradation combating. Based on the new strategic goals and direction of GEF, the Ministry of Finance is trying to make a better coordination between land degradation combating and biodiversity protection and addressing climate change, in a hope to promote and strengthen the China-GEF Partnership through the deepened cooperation between the Chinese government and the GEF.

This symposium provides us a good opportunity for the international exchange and a good platform to explore the practice of combating land degradation. I sincerely hope that through this symposium we can exchange views with each other, learn from each other, and share our knowledge, information and experience so that our capability to combat land degradation can be improved and the IEM corresponds to and fits better the reality of China be explored.

Ladies and gentlemen,

The successful implementation of PRC-GEF Partnership on Land Degradation in Dryland Ecosystems and the achievements made so far marks a new step for the co-operation between the Chinese government and the GEF. I truly believe that with the great support and cooperation of the international organizations including the GEF, ADB and WB, the relevant departments of the Chinese government and the provinces that implement the project, the Partnership will develop in depth in more fields and make a greater contribution to land degradation combating in China and the whole world.

To conclude, I would like to wish a great success of the symposium!

Thank you!

2. Opening Address by State Forestry Administration

Mr. Li Yucai
Deputy Administrator of State Forestry Administration

Distinguished delegates, Ladies and Gentlemen,
Good morning!

With the concerted efforts and preparation of all parties concerned, the International Symposium on the Concept and Practice of Integrated Ecosystem Management jointly hosted by the State Forestry Administration and the Ministry of Finance now opens at this beautiful autumn. First of all, on behalf of the State Forestry Administration, I would like to confer my congratulations on the opening of the Symposium, extend my warm welcome to the guests and friends present today, and express my heartfelt gratitude to the international organizations, foreign governments and friends from all walks of life for your continuous care for and support to the ecological improvement and forestry development in China.

This Symposium is held under the context that the PRC-GEF Partnership on Land Degradation in Dryland Ecosystems has been implemented for four years and achieved significant success for the current stage. It is an important meeting to break new ground for the future implementation of the Partnership. The experience and achievements obtained from the four-year implementation of the Partnership prove that the Partnership guided by the concept of Integrated Ecosystem Management (IEM) is one of the most efficient approaches to combat land degradation.

The combination of national planning framework and long-term national planning established by the Partnership ensures that the complicated problems of land degradation could be managed in a long-term, sustainable and integrated way. The improvement of the laws, regulations and policies has solved the land degradation problems fundamentally caused by human factors. The integrated strategies to combat land degradation and the action plans have enhanced the coordination of relevant departments and the effective use of funds. The ameliorated public participation mechanism and incentive policies, and the community based participatory land use planning has mobilized the rural inhabitants and private sectors to invest in land degradation combating.

This Symposium aims to provide a platform for us to learn from each other, to exchange views with each other and to draw lessons and to reflect on the practice, in a bid to further summarize the successful examples and experiences in combating land degradation with the IEM concept and to further explore the models and approaches to combat land degradation with the IEM concept, so as to contribute to the development of the concept and innovation of the approaches of the IEM and to play bigger role in combating land degradation in western arid zone and in continuous improvement of the ecological status in western region.

Ladies and Gentlemen,
The past 30 years since the reform and opening-up has witnessed a rapid economic

growth in China, on the other hand, China confronted with huge ecological pressure. Due to the climate change and human activities especially, China has always been one of the countries in the world that has a large and widely-distributed area of desertified land and suffers from desertification most seriously. At present, the area of desertified land in China totals 2.6362 million km^2, amounting to one third of the total land area, and the area of sandified land totals 1.7397 million km^2, accounting for one fifth of the total land area. Land sandification not only threatens directly the existence and development of over 100 million people in China and influences the living environment and quality of over 400 million people, but also causes a direct economic loss up to 54 billion yuan, severely constrains the all-round, balanced, and sustainable economic development of China, threatens the safety of national land and ecological security, and has become another great concerns of China after flood and drought.

The Chinese government has always attached great importance to land degradation combating and desertification control, and has always set land degradation combating especially desertification control as an important strategic task. Centered on the task of forest and grass vegetation rehabilitation in arid and semi-arid areas, China started to develop farmland shelterbelt forest and sand-fixing and wind-breaking forest as early as in the 1950s, and implemented the "Three-north" Shelterbelt Development Program and the National Desertification Combating Program in the 1980s and the 1990s respectively. Since the beginning of the 21st century, a number of key ecological programs have been launched, including the Land Conversion Program, Natural Forest Protection Program, Program for Sandification Control in Areas in Vicinity of Beijing and Tianjin, Graze Land Conversion Program and Program for Comprehensive Micro-basin Management. In the mean time, the Chinese government promulgated and implemented the Desertification Combating Law and the National Program on Desertification Combating, issued the State Council's Resolution on Further Enhancing Desertification Combating, and convened the National Conference on Desertification Combating. Entrusted by the State Council, the State Forestry Administration has signed Duty Agreements on Desertification Combating Target with provinces and autonomous regions with heavy tasks of desertification combating. Through the above-mentioned significant measures, land degradation and desertification combating conduction has yield great results, playing very important role for the improvement of China's ecological status.

However, due to natural factors and the impact of human activities, land degradation, even sandification, remains severe, which is demonstrated by serious expanding of sandified land in some areas, the difficulties for controlling sandification remains great, the achievements of desertification combating is still weak, and the hidden threat caused by human activities is still large. Therefore, to combat land degradation and accelerate the control of desertification is still an arduous task, which requires our continuous and strong efforts.

To combat land degradation and accelerate desertification control is a long-term task and a complex systematic project as well. It depends not only on projects implementation with sufficient funds as guarantee, but also on scientific concepts and advanced technologies as guidance and support. It requires both the coordinated strength from relevant domestic departments to double their efforts and the great supports from international organizations and foreign governments to enhance cooperation and exchange. In recent years, the Chinese government has made remarkable achievements in terms of conducting cooperation with the international community. First, China has earnestly implemented the UNCCD. As the contracting party of the UNCCD, China has attached great importance to the implementation of the Convention. On the one hand, China has organized and implemented a series of key projects to strengthen the efforts to combat desertification; one the other hand, China has played a positive role in taking on its international obligation by promoting regional cooperation and conducting bilateral cooperation in the field related to desertification combating with Germany, Japan, the Netherlands, Australia, Canada and Sweden, which contributed significantly to facilitating UNCCD process. Second, China established the first Partnership with the GEF on Land Degradation Combating. Since its initiation in 2004, the Partnership Project established effective project implementation and coordination mechanism in line with the IEM method. The Project formulated the Legislative and Policy Framework on Provincial Land Degradation Combating, compiled the provincial Integrated Land Degradation Combating Strategy and Action Plan, conducted the study on National Land Degradation Data Sharing and Coordination Mechanism, and developed provincial IEM information Center. The Project also participated in the modification of the National Action Plan on UNCCD and compiled the community based participatory planning program. Through these activities, the dissemination of IEM concept and application of IEM method were effectively promoted, the capacity of national and local authorities in combating land degradation was improved, and a number of talented personnel equipped with the knowledge of IEM were fostered, playing significant role for land degradation combating and desertification control in arid and semi-arid areas in western China and laying sound foundations for the further development of the Partnership. On behalf of the State Forestry Administration, I would like to avail myself of this opportunity to extend my sincere thanks to such international organizations as the GEF, ADB, WB and other domestic departments who have been offering continuous support to the implementation of the Partnership Project in the past years, and to express my high respects to all staff participated in the organization and implementation of the Project.

Ladies and Gentlemen,

Forestry is the main body for ecological improvement, serving such significant functions as building forest ecosystem, protecting wetland ecosystem, improving desert ecosystem and maintaining biodiversity, and shouldering prominent historical task to build ecological civilization. The position of forestry in China's national economic development is more and more important, the role of forest is more and more prominent, and the task more and more arduous. Currently, we are making our all efforts to promote the development of modern forestry in line with the decision made by the Chinese government on enhancing ecological improvement and accelerating forestry development with the Scientific Outlook on Development as the guidance and forestry reform as the momentum. To build modern forestry is to make every effort to develop a complete forestry ecological system, advanced forestry industry system and prosperous ecological culture system, to realize a sound and rapid forestry development, to improve the three main functions of forest and to bring the three effects of forestry into full play so as to meet the diversified demands of the society from the forestry to the largest extent.

To combat land degradation and accelerate desertification control is an important content for building a complete forestry ecological system. We will further enhance our cooperation with the GEF, ADB and WB to expand application of IEM concept, strengthen land degradation combating and accelerate the speed of desertification control based on the summarization of the achievements and experiences obtained by applying IEM concept in China's land degradation combating. By the year 2010, we try to basically restrain the ecological environment deterioration trend in desertified region and remarkably improve the ecological status of key areas. By the year 2020, we try to further improve the ecological protecting system, put more than half of treatable desertified land under basic control, and greatly improve the ecological status of desertified region. And by the middle of this century, we try to put all treatable desertified land under basic control, establish a relatively complete ecological protecting system, relatively advanced desert-related industry system and relatively prosperous ecological culture system, so as to achieve a remarkable improvement of the ecosystem in the desertified region.

Enhance sandification control and accelerate desertification combating can not only stop and mitigate land degradation, but also make great contribution to address climate change problems. The degradation of land is a process of carbon emission, and the control of land degradation and rehabilitation of vegetation is increasing the absorption of carbon, decreasing the content of CO_2 in the atmosphere and thus mitigating the green house effect. Furthermore, the rehabilitation of forest and grass vegetation will enrich the biodiversity, improve ecological status and maintain ecological balance.

Ladies and Gentlemen,

Autumn is a season for harvest. I hope we can cherish this precious opportunity to make good use of the platform for mutual learning and exchange, enhance communication, inspire with each other and make joint contribution for further promoting the application of IEM concept in the field of land degradation combating, accelerating land degradation combating and facilitating the improvement of global ecological environment.

Finally, I wish the Symposium a great success!

Thank you all!

3. Opening Address by GEF

Madam Monique Barbut
CEO/ Chairperson of GEF

Distinguished colleagues and supporters of the PRC-GEF Partnership on Land Degradation in Dryland Ecosystems Program;

In April 2008, the GEF Council endorsed the second phase of the PRC-GEF Partnership on Land Degradation in Dryland Ecosystems Program. Following a successful and experience-rich first phase of this partnership program, I am very pleased and proud to provide my continuous support to this important joint investment by the GEF and the Government of China.

Your meeting brings together a wide range of stakeholders and partners of this program. I wish you a highly successful meeting and a fruitful dialogue on lessons learned, experiences and new ideas that hopefully will enhance the expected results of the program. I am proud that the GEF has engaged with the Government of China in this ambitious undertaking which links the sustainable development of China's rural areas with achievements benefiting the global environment. China's dedication to manage its natural assets in a sustainable way to ensure a healthy production basis for future generations is by now known to many. It is my intention to also in future closely work with the Government of China to make this dream come true.

Next week, the GEF Council will meet for its November 2008 meeting. One of the most important discussion points will be the launch of the GEF-5 replenishment period. We will work with the GEF donor countries to develop an attractive strategy for the GEF that defines an attractive vision for the GEF in the context of a very dynamic change of the international financing architecture for global environmental concerns. I will propose essential reforms to the GEF as a network institution and strive for the largest ever replenishment of the GEF. It is my hope that the Land Degradation focal area will be replenished at least double of its current size in order to better meeting the demands of countries that are affected by the impacts of land degradation. I will promote an integrated approach to natural resources management that will enable countries to address the multiple threats to their natural resources and create synergetic livelihood and global environmental benefits. 1 am sure that for China, any future GEF investment will further and greatly benefit from this integrated approach which has been piloted through the partnership since 2002.

China has been at the forefront of piloting this challenging program which has served as pilot for the GEF to support an integrated approach to natural resources management. I would like to thank you for your innovative ideas, your courage to pilot it with support by the GEF and congratulate you to the first impressive results this program can present today.

Again, I wish you a successful meeting. I apologize that the GEF Secretariat will not be able to participate but I am sure that you understand the importance of the upcoming Council meeting.

4. Opening Address by ADB

Mr. Kunhamboo Kannan
Director of Agriculture, Environment and Natural Resources Division, East Asia Department, ADB

Dear Representatives, Ladies and Gentlemen,

On behalf of the Asian Development Bank, I would like to express my sincere thanks to the Chinese Government for organizing this important international conference, and also for providing the opportunity to make a few opening remarks.

First of all, it is a pleasure to be part of a gathering of so many distinguished experts from different organizations and regions to share knowledge and experiences on combating land degradation. Recognizing the increasing pressure on dryland ecosystems worldwide, there is a continuous need to strengthen regional and international cooperation to most effectively address seemingly local environmental concerns that actually have global significance and impacts.

The ADB, with the support from GEF, has actively contributed to the development and implementation of the PRC–GEF Partnership on Land Degradation in Dryland Ecosystems (the Partnership). The Partnership, initiated in 2002, aims to facilitate the cooperation between national and international organizations to introduce and support integrated ecosystem management (IEM) approaches in combating land degradation, reducing poverty, and restoring dryland ecosystems in the western region of China. It also recognizes the need for a long-term approach to address land degradation and associated global environmental concerns such as loss of biodiversity, climate change, and desertification.

As further presented during the Conference, we have been pleased to see that the implementation of the project has effectively promoted the application of the IEM concept and approach over the last few years. It enhanced national and local capacities to combat land degradation and raised the national and local capacity in combating land degradation. Multi-level and multi-agency coordination mechanisms have effectively enhanced coordination among the central and provincial agencies, opened channels of cooperation from the central to the county level natural resources management agencies, and improved coordination between the sectoral plans and programs and between central and provincial budgets. Coordination of laws and regulations has been facilitated through the formulation of the legal framework for land degradation at provincial and regional level and the revision of relevant national laws and provincial policies. Land degradation issues have been integrated into the provincial 11th five year plans, the provincial strategies and action plans for combating land degradation, and the participatory community development plans. The mechanism for land degradation data sharing has been established and has integrated scattered data resources, resulting in data sharing across the sectors and the provinces/autonomous regions. Implementation of the pilot sites activities has improved the rural infrastructure and empowered community members to address local land degradation. In addition, the lessons and experiences have been widely disseminated.

The Partnership is now in a critical phase of its development. While it contributed significantly to the introduction of the IEM approach in China, as we will be further hearing during the conference, there are still new challenges ahead. The PRC Government expressed its intention to focus future work under the Partnership on scaling up activities to deepen understanding of the IEM approach, disseminating experiences with associated policy and institutional reforms, and seek further cooperation and integration with other ongoing programs in and outside of China.

The ADB, with GEF support, is keen to remain an active partner and is currently discussing further support for the Partnership through a new Capacity Building project, to start in 2009. Strengthening its cooperation with the Partnership is in line with the recently approved new ADB long-term strategic framework for 2008–2020 (Strategy 2020), that serves as ADB's corporate-wide planning document. ADB pursues its vision and mission by focusing on three complementary strategic agendas: inclusive growth, environmentally sustainable growth, and regional integration. All of these are most relevant for the activities of the Partnership.

As such, ADB sees this conference is a most timely opportunity to further discuss ways of cooperation with and support for the Partnership. The Government has already prepared draft documents outlining the future cooperation, which will also be presented later during the conference. Obviously, we like to benefit from the wealth of knowledge and experience of experts gathered for this conference and would welcome all discussions and suggestions on how to strengthen ADBs cooperation with the Partnership.

On behalf of ADB I again want to thank the Chinese Government for organizing this important conference, and like to wish all participants successful and interesting discussions.

Thank you very much for your attention.

5. Opening Address by Kyrgyz Republic

Mr. Kambarali Kasymov
State Secretary of Ministry of Agriculture, Water Resource and Process Industry, Kyrgyz Republic

Dear Representatives, Ladies and Gentlemen,

I am pleased to address this conference on behalf of the Central Asian countries Kazakhstan, Kyrgyz Republic, Tajikistan, Turkmenistan and Uzbekistan, and our program, the Central Asian Countries Initiative for Land Management, which is also known as CACILM.

The Central Asian region is one of the most ancient agricultural regions of the world, and is populated with more than 50 million people. We have a diverse agrarian culture, of sophisticated irrigation systems, rainfed lands providing crops of grain, forage, fruits, nuts and vegetables, and huge areas of pasturelands in the mountains, hills, steppe and deserts for grazing our cattle, sheep, goats, horses and even camels.

Although agriculture has been practiced in Central Asia for more than two thousand years, some of our current framing and grazing methods are not sustainable, and these practices are degrading our land and soil, plants and water.

In current times, the Central Asian countries are facing great challenges to change our practices so that our agriculture to be successful in terms of food security, and for our agriculture to be environmentally sustainable.

For sustainable and successful agriculture in Central Asia we must address very challenging issues: due to population growth, due to the migration of rural peoples to our cities; due to the increased costs of production for labor, fuel, equipment, fertilizers and seeds; Also, we have agricultural methods which are not sustainable, for example: poor irrigation and drainage practices which cause soil salinity, erosion and water logging; inefficient agronomic practices such as the excessive tillage of dry lands, and the over application of herbicides, pesticides and fertilizers. We have much old and inefficient machinery and equipment; and the overgrazing of pasturelands is widespread.

In addition to these challenges, the people of Central Asian face immense challenges due the forces of nature, particularly changes to our agro-ecosystems that are related to changes of climate – lower rainfall and snowfall, longer and colder winters, and from hotter and windy summers.

For agriculture to be successful and sustainable we know that we must manage our land and water resources wisely.

So what actions have we undertaken in Central Asia to improve our land and water management?

Over the past 10 years, we have made some progress on resolving and rationalizing transboundary water use issues through the frameworks of cooperation of the Central Asian countries, notably: the International Fund for Saving the Aral Sea (IFAS) and the Interstate Commission for Water Coordination (ICWC) and other with other inter-state and inter-governmental committees.

An Intergovernmental Committee on Sustainable Central Asia Development was established to develop and initiate a "Subregional Strategy on Central Asian Countries Sustainable Development" specifying priorities for rational use of natural resources and related economic development planning.

The Central Asian countries participate in frameworks to address issues beyond our region through the Shanghai Cooperation Organization and the Eurasian Economic Community.

Issues related to the development of the member countries' agro industrial complex is included in the Agenda of the Business Council of the Shanghai Cooperation Organization and includes connection to integration of processes and cooperation on tourism and for hydro-electric engineering.

The People's Republic of China (PRC) and Central Asian Countries (CAC) cooperate within the framework of UNCCD - United Nations Convention to Combat Desertification. The PRC and CACs are affiliated with the Asian Group and cooperate in the context of single Convention Addendums.

A Treaty of Friendship, Neighborliness and Cooperation was concluded on bilateral level between the PRC and the Kyrgyz Republic in 2002. A Program on Cooperation between the PRC and Kyrgyz Republic for 2004-2014 was ratified by the KR law of 2006. A Kyrgyz-Chinese joint group on agrarian cooperation was established within the Program on the basis of a signed Memorandum.

There are similar cooperation agreements between the People's Republic of China and other Central Asian countries.

We believe there is an excellent opportunity for cooperation with the People's Republic of China through our Central Asian Countries Initiative on Sustainable Land Management (CACILM).

The Central Asian Countries Initiative for Land Management (CACILM) is a partnership of Central Asian countries (CACs) and ten development cooperation partners dedicated to combating land degradation and improving rural livelihoods.

The Global Environment Facility (GEF) is one main partners and supporters of CACILM. We would like to actively cooperate with other GEF initiatives in Asia and particularly in China.

Our CACILM activities are guided by the National Program Framework and by National Coordination Councils established in each Central Asian Country.

Through CACILM, the Central Asian countries are implementing multicountry projects for sustainable land management information systems, knowledge management and dissemination systems, for applied research and for capacity building.

CACILM has sustainable land management demonstration projects in agro-ecological systems of five Central Asian Countries including mountain, desert and steppe pasturelands; in irrigated and rainfed farmlands. We have large scale sustainable land management investment projects in Uzbekistan, Kyrgyz Republic and Tajikistan that are co-financed by ADB and the GEF.

We are most interesting in the land management experience of our colleagues in the Peoples Republic of China, and would welcome opportunities to collaborate and cooperate with you in the near future.

On behalf of Central Asian Countries I wish success to the Chinese Government and all participants in achieving the goals and objectives of this Conference.

6. Opening Address by World Bank

Ms. Sari Söderström
Rural Sector Coordinator and Chief Operations Officer, China and Mongolia Sustainable Development, World Bank

Ladies and Gentlemen.

I would like to extend my warm greetings to Mr. Li, Madame Jiang and Mr. Zhang; our partners in the line ministries and agencies (MOF, Ministry of Environment Protection); our foreign development partners (ADB, GEF, EU etc) and in particular to all participants from the provinces, counties, and academic community.

The World Bank is delighted to be part of this important workshop on integrated eco-system management, because as you know – we are also part of the PRC-GEF partnership through the Gansu and Xinjiang Pastoral Development Project, which we are going to learn more about later in this workshop, and also through watershed management activities jointly financed with the Ministry of Water Resources and the EU in the Changjiang River Basin.

I am glad to see so many old friends here today in the audience. This bids for productive discussions.

Four years ago I attended a similar workshop to this one on IEM in China. During my opening remarks at that workshop, I pointed out both the importance of IEM approaches and challenges in its application. Integrated eco-system management - by definition - can address multi-focal problems. At its best – it not only facilitates sustainable economic development, but also functions as a mechanism to optimize scarce ecological, social, and economic resources promoting sustainable development at all levels.

Four years has passed quickly! Today we are here to take stock of what has happened with IEM in China and elsewhere, and take stock of the changes that are taking place at the local levels; changes that arise from people's initiatives; and to learn where integrated eco-system management has worked, and where not.

The past year (2008), we witnessed the 30th anniversary of China's reform and opening-up policy. During these three decades, China has transformed from an agriculture-based economy to an industrialized, urban-based economy, growing by double digit numbers. More importantly, China has successfully lifted more than 400 million rural people out of absolute poverty, and it is well on the way to achieve the other Millennium Development Goals.

During the same period, the Chinese government has been gradually incorporating integrated natural resource management approaches into its development strategies by focusing on scientific approaches and balanced development. Guided by the objective of building a harmonious society and an energy efficient and environmentally friendly socialist market economy - a large number of programs and projects have been launched on water and soil conservation, afforestation, combating desertification, renewable energy, grassland management, eco-system restoration, etc. Among which the PRC-GEF partnership on land degradation has been able to contribute to the success of these programs.

On the other hand, China's rapid economic growth is having a significant impact on both the quality and quantity of the natural resources base, with both domestic and international environmental consequences, including water scarcity, water pollution, desertification and land degradation, air pollution, and loss of biodiversity. Nearly 40% of China's land is eroded; 10% of its arable land is polluted; 60% of all monitored rivers are too polluted to be used for human consumption. China is also the world's largest producer and consumer of fertilizer, the second largest producer and consumer of pesticides. And - China has become one of the world's largest source of greenhouse gases.

With this as a background, application of integrated eco-system approaches is even more important. To solve this set of complex and multi-faceted problems, we need to firmly embrace integrated, system-wide, and cross-sectoral approaches that take into consideration ecology, people, and economics. We need to make concerted efforts to draw on past experiences, distill useful practices and scale them up in new programs; - further promoting integrated eco-system management in China and globally.

We see that the Chinese Government is determined to achieve sustainable economic development and is taking significant steps to achieve this. The 17th Party Congress meeting last month reaffirmed that the Government considers harmony between man and nature as one of the most important aspects of a harmonious society. To support this - decisions passed at the Party Congress, promise changes in the way that farmers can transfer land rights which will further liberate the rural economy and promote sustainable land use. (The Decision on Major Issues Concerning the Advancement of Rural Reform and Development was approved by the CPC Central Committee). Similarly, recently the State Council Information Office issued a white paper (China's policies and actions on climate change) which proposes a series of policies, principles and programs to address climate change issues.

In this context we hope that the PRC-GEF partnership on land degradation will continue to play an important role in the protection and management of ecosystems. In the beginning of my speech, I said that integrated eco-system management, by definition, can address multifocal problems. However, for this to happen, close and effective inter-agency coordination is a must. In reality, this still poses a big challenge at all levels. Integration requires collaboration, cooperation, participation, and search for synergies. Integration requires new ways of thinking – a change of mindset - and new ways of working at each level. In particular, integration requires active involvement by local leaders and officials and communities, as managing natural resources in a sustainable way is a process to balance the productive use of land, water and vegetation with the needs of people whose livelihood depends on them. We are here today to learn what new ways of thinking are surfacing around us.

Ji Wang Kai Lai - now it is time to build on past successes and look forward to the future -

Ladies and gentlemen, I believe this workshop will provide an excellent opportunity for all participants to discuss the experiences and lessons-learnt from the past practice of application of integrated eco-system management and put forward your ideas for practical ways to further promote it in China as well as China passing on its experiences to other developing countries.

Finally, I wish the workshop a big success!

7. Join Hands to Promote Sustainable and Healthy Development of the PRC -GEF Partnership

Professor Jiang Zehui
Vice Director, Committee of Population, Resources and Environment of CPPCC
Co-Chair of Board of Trustees, International Network for Bamboo and Rattan (INBAR)
Director, Chinese Society of Forestry
Director, Steering Committee of PRC-GEF Partnership on land Degradation in Dryland Ecosystems

Distinguished guests, dear friends,
Ladies and gentlemen,

Today, the "international conference on IEM approach and application" is opened ceremoniously here in Beijing. First of all, on behalf of the Steering Committee of the PRC-GEF Partnership Program, please allow me to extend my warm welcome to all guests and friends! I would like to express my sincere thanks to the Ministry of Finance, the Committee of Legislation of the NPC, the State Development and Reform Commission, the Ministry of Science and Technology and other ministries for their great attention and support. I would like to thank all project provinces/regions for their full cooperation. Thanks are also given to GEF, World Bank, ADB and other international organizations as well as friends from various institutions for their long-term commitment and support to land degradation control in China.

Nowadays, land degradation dominated by desertification has become a key ecological issue threatening human's life and development. Combating Land degradation, maintaining ecological security and realizing sustainable development are therefore the major ecological and environmental issues that the human are facing with. I would like to take this opportunity to share with you my understandings and views on application of IEM concepts in land degradation control in China.

1. Opportunities and challenges for China's land degradation control

"Reduction of land degradation, alleviation of poverty and rehabilitation of dryland ecosystems" are not only the goal of the PRC-GEF partnership, but also China's national objective of combating land degradation. With continuous efforts by generations of people for tens of years, China's land degradation control has made great achievements which have drawn worldwide attention and made great contribution to safeguarding sustainable socioeconomic development in China. Meanwhile, we should be clearly aware that while China's land degradation control is facing with unprecedented opportunities it is also facing with the first-ever challenges. There is still a long way to go.

On one hand, China's land degradation control has made great achievements that have drawn worldwide attention. For a long time, Chinese government has always given high priority to addressing land degradation problem as one of its strategic tasks, and effective measures have been taken. Since the entry into the new century, China has developed and implemented the "Desertification combating law", the "Environment impacts assessment law", the "Regulations on implementation of the forest law" and other laws and regulations as well as the 6 major forestry programs, grassland protection and development program, water and soil conservation program, integrated management of inland rivers program which are related to land degradation control. Since 2001, the annual area of treated desertification land reached 1.92 million ha, playing an important role in reversing overall trend of desertification. Currently, 20% of China's desertification land has been treated to various extents, the vegetation cover in priority areas have increased by more than 20 percent, ecological conditions have been significantly improved in some areas. With continual efforts by generations, China's efforts in combating land degradation including degradation of farmland, forestland and grassland have made great achievements which draw worldwide, making significant contribution to safeguarding sustainable social and economic development.

On the other hand, land degradation is another serious threat to China's sustainable development besides flood and drought disasters. China is the most populous country in the world, and one of the countries subject to severest erosion and desertification. China's total area of sandy desertification land is as high as 1.743 million square kilometers, accounting for 18.2% of the total land area, and still expanding at an average of 3,436 square kilometers annually. The western region is the worst-hit area of land degradation, the population under poverty in the western region exceeds 30 million. Desertification not only directly threatens the life and development of more than 100 million people, but also affects the living conditions and standards of 400 million people. At present, China's land degradation is still a serious problem, not only the priority and difficulty of ecological development, but also a major constraint to sustainable economic and social development.

It needs to be highlighted that China's land degradation control has got valuable opportunities and favorable advantages such as worldwide attention, nationwide concern, increased investment and firm foundation of previous work. For land degradation control in China or even in the world, to grasp the opportunities, meet the challenges and steadily move forward, it needs to emancipate our mind, renew our ideas and renovate the way of thinking. Encouragingly, integrated ecosystem management (IEM) advocated and promoted by GEF provided new concept and approach to addressing land degradation for China and the world, setting up an effective platform for action.

2. Practices of IEM in China's land degradation control and innovation

During learning and application of IEM concepts, considering China's realistic situation and characteristics of economic and social development in China, China-GEF partnership has made creative additions to IEM concepts during practices of IEM application. While making significant achievements in land degradation control, it has accumulated valuable knowledge and experiences in land degradation control for China and the world.

2.1 Significant achievements have been made in IEM application in China

In recent years, with implementation of the "PRC-GEF partnership on land degradation in dryland ecosystems", China has incorporated IEM concept and approach in the development of land degradation control plan, through integrating all interrelated ecosystem components including human, holistic design, systematic planning and multiple levels of implementation. Encouraging progress has been made in optimization of allocation of resources and funds, innovative management system, improvement of operational mechanism.

On one hand, the partnership has shoveled many problems during project implementation through innovations in coordination and information sharing mechanisms, and also promoted communication and coordination among agencies which reflects appropriate organizational structure and efficient operation, not only conforms to Chinese realistic situation but also innovation of GEF project management. Meanwhile, the partnership program supported each province (regions) to set up an IEM information center, largely improving the use efficiency of existing data, and providing important evidences for administrative agencies to make science-based decisions. On the other hand, by assessing current policies and regulations on land degradation control in each province/region during the development of legal and policy framework and strategic action plans, the partnership program has made recommendations for revisions and improvement, meanwhile provided guidance for designing and implementing short-, medium- and long-term land degradation control actions for project provinces/regions by identifying priority areas and activities for policies, investments and other actions. Important achievement is also reflected in capacity building and demonstration through which a large number of government decision makers and technical specialists were trained with IEM concepts and approach. Significant achievements have been made in practices of land degradation control in the 22 pilot sites in the 6 project provinces/regions.

Through continuous exploration and practices, PRC-GEF partnership program has developed a set of implementation methods such as combination of ecological protection and poverty alleviation, diversification of project activities aiming at both ecological protection and poverty alleviation, establishment of long-term and stable financing mechanism, promotion of bottom-up and participatory management, multi-agency participation, cooperation and management as well as introduction of competition, and promotion of ecological protection and sustainable use of resources in project areas. The partnership also accumulated a lot of knowledge and management experiences, becoming a model for other partnerships of GEF.

2.2 PRC-GEF partnership contributed to IEM development

The smooth implementation of the PRC-GEF partnership and the significant achievements were attributable not only to the application of IEM concepts in project planning and implementation, but also to the contribution of Chinese traditional ecological culture, the new concept of harmonious development and the new approach of ecological civilization to the development of IEM concept and approach.

1) The implementation of PRC-GEF partnership added the idea of "harmonious development" to IEM. Building a harmonious society and realizing harmonious development are the major goal of China's sustainable development. Harmonious development, to a large extent, depends on the level of development of the social productive forces,

depends on the level of protection of resources and environment and depends on the coordination during development. During the implementation of GEF projects in China, the IEM concept of balancing protection and development was effectively integrated with the ideal of harmonious development of economy, society and environment. During the process of land degradation control, attentions were given to the harmony of human and nature, harmony of human and society, harmony among agencies and between higher and lower levels, harmony between resources and environment protection and economic and social development, further enriching IEM concept while the partnership program is moving forward and making achievement. IEM integrated with the idea of harmonious development will more actively cope with the human-nature relation, draw greater attention from the public, encourage public participation, promote cross agency coordination and push forward rapid development of land degradation control in China and the world.

2) The implementation of PRC-GEF partnership injected the concept of "ecological civilization" into IEM. Human civilization has gone through primitive, agricultural and industrial civilizations, currently at an important stage transition from industrial civilization to ecological civilization. Development of ecological civilization is a major strategic decision of the Chinese government based on complete awareness of the rules of economic and social development and on profound self-examination in ecological and environmental protection. It has significant implications to China's sustainable development and far-reaching impacts on maintaining global ecological security, promoting human civilization. A major task of developing ecological civilization is in the development of a resource-saving and environment-friendly society to strengthen the control of water, air and soil pollutions to improve the living conditions in both city and rural areas on one hand, and to emphasize on development of water resources, forestry and grassland, strengthen desertification and rocky desertification control and promote ecological rehabilitation on the other hand. Needless to say, the ecological civilization vigorously pushed by the Chinese government is in consistence with the direction of IEM development. The PRC-GEF partnership as a platform further enriched the content of IEM. I deeply believe, in IEM practices, active building up the concept of ecological civilization will certainly make the "resource saving" and "environment friendly" gradually the common sense of worth and conscious effort of the people in regions subjected to land degradation, and by making unremitting endeavor, eventually leave a beautiful ecological homestead for future generations.

3) The implementation of PRC-GEF partnership enriched the implication of "ecological culture" to IEM. The famous ecologist Donald Vorster once pointed out that "Today the causes for the global ecological crisis we are facing with are not the ecosystem itself but our cultural system". Carrying forward ecological culture and strengthening development is an important strategic choice to counteract the global ecological crisis including land degradation. What is "ecological culture"? Our studies indicated that ecological culture is a culture that human and nature coexist and develop in harmony. Specifically, ecological culture is a culture to explore and address the complex relation between human and nature, a culture based on ecosystem and respects ecological rules, a culture

aiming to meet diverse needs of human by realizing the multiple ecosystem values, and a culture penetrating into material culture, institutional culture and spiritual culture and reflecting the ecological views of harmonious coexistence of human and nature. It is particularly worth to mention that on October 8, 2008, one month ago, China Ecological Culture Association was formally founded. Following the commitment of "carry forward ecological culture, promote green life and co-develop ecological civilization", we are looking forward to a flourished ecological culture in China. Within a good atmosphere of ecological culture, PRC-GEF partnership settled in China and received wide concern and participation from the public. At the platform of Chinese ecological culture system, IEM has been further enriched in its cultural context. I believe that the integration of China's ecological culture and IEM concepts will certainly play a greater role in and make a greater contribution to land degradation control in China and the world.

3. Promote sustainable and healthy development of the partnership in China

The worsening ecological environment and poverty caused by land degradation will be the main constraint for economic and social development in the arid and semiarid regions for a relative long time, and this requires us to join hands to push forward sustainable and healthy development of the partnership, and to persistently counteract the long-term formidable challenge.

3.1 The partnership has a broad space for development in China

The implementation of the 11th 5-year plan creates a more favorable environment for the development of the partnership. To balance urban and rural development, balance the development among different regions, balance economic and social development, balance harmonious development of human and nature, and to balance domestic development and opening up to the outside world are the basic requirements by the scientific development view advocated by Chinese government, and also the basic principle of the partnership. According to the overall requirements of holistic, harmonious and sustainable scientific development view, the 11th 5-year plan has set the development goals of building harmonious society, building new socialist rural areas and pushing forward western development, defined tasks for ecological development including continual efforts in combating desertification and stony desertification. These series major ecological development programs will provide a broader space for development of the partnership. The implementation of phase II country program framework of the partnership will definitely provide support and services to the major national ecological programs. Meanwhile, the implementation of phase II partnership will have significant implications to achieving the strategic goal of "promoting harmonious regional development" of the 11th 5-year plan, and this is also the basic requirement of GEF principle of national development as priority.

The efforts by international community in mitigation of climate change, conservation of biodiversity and alleviation of poverty provide a broader space for development of the partnership. The ecological and environmental problems such as Climate change, land desertification and extinction of species have drawn increasing worldwide attention, among these the climate change becomes the focus of the problems and the key issue of global political agenda. Climate change worsens drought and desertification, and land degradation, deforestation and declining water resources directly threaten biodiversity, further worsening the global climate change. Poverty alleviation, environmental sustainability and global cooperation are the three major ones of the UN millennium development objectives, and these are also in consistence with the partnership objectives of land degradation control, biodiversity conservation and poverty alleviation. These require us to actively respond to the new international and domestic situations during the implementation of phase II partnership, and to combine the implementation of partnership program with the international focus such as prevention of climate change and conservation of biodiversity, combine with the UN millennium development objectives and combine with China's national development objectives of building ecological civilization and new socialist rural areas. This would further improve the capacity of relevant countries and regions in the world such as China and African countries to combat land degradation, making even greater contribution to the mitigation of climate change and realization of UN millennium development goals while pushing forward regional social and economic development.

The implementation of GEF projects in its focal areas laid a firm foundation for development of the partnership. With assistance from the ADB, the Chinese government and the GEF established the first Partnership in 2002 at country level to explore effective approaches to addressing land degradation problem which has obtained satisfactory results. In October 2007, the 4th assembly of GEF and its 4th funding plan identified climate change, biodiversity, international water body, Ozone depletion, land degradation and persistent organic pollutants as the focal areas. PRC-GEF partnership as the first one of the GEF established at country level in land degradation is an important attempt towards its future development strategy, and it will continue to provide demonstration for global land degradation control. The implementation of phase II partnership will further intensify the investment in integrated land degradation control in China through its seed fund. The achievements obtained during the phase I implementation will be further extended, more favorable for pushing forward all activities carried out in GEF focal areas.

3.2 Pushing forward development of the Partnership in China

• Looking forward into the future, the effective mechanisms established by the partnership, the personnel trained by the partnership and the dissemination and extension of IEM concept have paved a solid foundation for future implementation of the partnership in China. At the same time, the joint efforts in counteracting the ecological crisis by the international community and the strategy and actions of China's sustainable economic and social development will also provide a favorable environment for the practices of the partnership in China. In the follow-up activities, we will further deepen the PRC-GEF partnership in the following 4 aspects:

• Actively counteracting the new challenge of global environment protection, integrating the partnership implementation with the burning issues of international concern such as mitigation of climate change, conservation of biodiversity and alleviation of poverty, further elaborate the active roles of land degradation control in global environment protection.

• Further integrating the partnership development with national development priorities, integrating with national development objectives of building ecologically civilized society and new socialist rural areas, providing effective services for development, improvement and implementation of regional land degradation control plan, and consolidating and developing synergic and cooperative mechanism under the partnership.

• Continual summarizing practical experiences, enriching and developing IEM theory and method, further improving policy and legislative environment, promoting integration of provincial/regional land degradation strategy and action plan into medium and long-term regional development and enforcing in decision-making of the government at various levels.

• Continual to build demonstration sites. Effective approaches will be employed to disseminate the successful experiences and models of IEM application at 22 demonstration sites. Continual to push forward and expand the investment projects for demonstration, continual to improve the capacity of land degradation control.

Ladies and gentlemen,

Adaption global environment change and combating land degradation are the common challenges, common responsibilities and common commitment of us. Let's join hands and work together, further strengthen the partnership, apply and develop IEM concepts, build ecological civilization, promote harmonious development, make greater contribution to land degradation control in the world.

Finally, I wish you all a very happy time and good health in Beijing.

Thank you!

CHAPTER II
Global GEF Partnerships on Land Degradation

8. Application of IEM in Land Degradation Control in West China: Successful Stories of PRC-GEF Partnership on Land Degradation in Dryland Ecosystem

Ms. Hu Zhangcui
Deputy Director General, Department of Science and Technology, State Forestry Administration
Director, Central Project Management Office

Distinguished Delegates, Ladies and Gentlemen,

As a critical environmental issue threatening human beings' livelihood and development, land degradation (LD) has increasingly gained extensive concern from the international community. There is still a need to continuously seek how to eliminate the root of LD and improve effectiveness and efficiency of LD control. PRC-GEF Partnership on Land Degradation in Dryland Ecosystems (hereafter the Partnership) has made good trials in this field. The project has employed the IEM approach in practice in LD control, and has gained remarkable achievements.

1. Background of the Partnership

The PRC-GEF Partnership on Land Degradation in Dryland Ecosystems is GEF's first Partnership on land degradation. Invested by the Chinese government and GEF, the Partnership is administered under a long-term programming framework in which integrated ecosystem management (IEM) is brought into West China's land degradation control. Such initiative is the first of its type for both China and GEF.

The Partnership aims to establish a cross-sector, cross-industry and cross-region framework for sustainable natural resource management and develop a plan for land degradation control which takes all interactive elements of the ecosystem into consideration and incorporates funds of the government at all levels and from international organizations to realize an optimized allocation of funds and resources. At the same time, the Partnership pursues an innovative management system and an improved operational mechanism to identify a new approach to address the root of LD.

To effectively reach the goal of Partnership, the Chinese government and GEF have established a 10-year Country Programming Framework (CPF). The framework includes the following 4 components: i) improving enabling environment for improving laws and policies for land management in dryland ecosystems; ii) improving operational mechanisms to increase IEM capacity; iii) establishing an LD monitoring and evaluation system; and iv) initiating demonstration activities to trial and demonstrate the IEM approach.

Four projects are being implemented under the CPF. They are: i) Capacity Building to Combat Land Degradation Project (CBCLDP); ii) Xinjiang/Gansu Pastarol Development Project; iii) Helanshan IEM Project in Ningxia; and iv) Conservation and Rehabilitation of Dryland Ecosystems Project.

The core project under the 1st phase of the Partnerhsip is the Capacity Building to Combat Land Degradation Project for which I would like to provide a summary of its achievements.

2. Results and effectiveness of CBCLDP

Capacity Building to Combat Land Degradation Project (CBCLDP) is the first project implemented under the partnership. It aims to use IEM to strengthen the enabling environment and develop institutional capacity to combat LD, reduce poverty and rehabilitate the dryland ecosystems in selected provinces and autonomous regions including Inner Mongolia, Shaanxi, Gansu, Qinghai, Ningxia and Xinjiang which suffer most from severe land degradation. A series of activities were designed to support the Partnership. Notable achievements have been realized after 4 years' implementation.

(1) The organizing management system innovated by the project has set a good example for government sectors' co-ordination mechanism

A Steering Committee led by the Ministry of Finance (MoF) and involving 11 member agencies including Legislative Working Committee of the NPC, the National Development and Reform Commission (NDRC), the Ministry of Science and Technology (MOST), the Ministry of Land Resource (MLR), the Ministry of Water Resource (MOW), the Ministry of Agriculture (MoA), the Ministry of Environmental Protection, the Stat Forestry Administration (SFA), the Legislative Affairs Office of the State Council and the Chinese Academy of Science was established. A Project Coordinating Office (PCO) was established in MoF and a Project Management Office (PMO) was established in SFA. Accordingly, six project provinces/regions established Leading groups, PCOs and PMOs at provincial level involved by relevant departments and agencies. Vice-governors (vice-chair) in charge of agriculture act as leaders of the leading groups. Such a high-level and multi-sector organizing institution is unprecedented in China, even in the world. It indicates the Chinese government has attached great attention to partnership and is a way of using IEM with China's situation taken into consideration, which has innovated the GEF management model. Regular or irregular meetings have been organized for the Steering Committee, the PCOs, PMOs and member agencies to address problems raised from implementation and enhance communication and cooperation among relevant agencies

To date, the mechanism has not only played an active role in project implementation but also formed a general long-term mechanism in project provinces (regions). For example, the newly revised *The Implementation Measures on 'Desertification Prevention and Control Law of P.R.C.' of Xinjiang Uygur Autonomous Region* promulgated that governments above county level, is responsible for organizing instructing or coordinating sand prevention and sand control in their administrative region, forestry administrative sector is responsible for practical work, departments of forestry, agriculture, husbandry, hydropower, finance, development reform commission and national resource, environmental protection the competent administrative authorities and meteorological competent authority will take charge of sand prevention and sand control in line with their own authority. *Regulation on Wetland Protection of Shaanxi Province* promulgated Forestry Administrative agency is responsible for provincial wetland protection work, hydro-power, agriculture, national land resource, environmental protection, Development and Reform Committee and Construction sectors will take charge of wetland protection work in their due responsibility. In the above two cases, the mutual coordination among the sectors have been facilitated.

(2) Incorporating IEM into legislation for LDC, effectively promoting the coordination among laws, regulations and Policies

To incorporate the IEM approach and principles into the provincial (regional) policy and regulation framework formulation is one of the major activities of Capacity Building to Combat Land Degradation Project (CBCLDP). 19 legal assessment indexes which reflex the IEM idea were proposed, with international references and Chinese legislation actual condition taken into consideration. We take the laws and regulations in national resource protection, sand prevention and sand control, soil and water conservation and protection of plain resource, forest resource and water resource, agriculture, wild flora and fauna and environmental as the assessment objects, complying with corresponding process and method, conducted provincial (region) regulation and policy capacity assessment. Through the assessment, we have defined the existing issues in the current administrative regulations, especially those in contradict with IEM idea; we have proposed the amending and refining suggestions accordingly. To date, 33 local regulations or government statutes in 6 provinces (regions) have been formulated or mended. Gunsu's experience and approaches of applying IEM in legislation has been written into Gansu Legislation, a series edited and publicized by Gansu People's Congress.

Focus on the ecological system structure, function and the inter-relation between them is the basic requirement of IEM. Great attention has been attached to the internal correlation between plants, animals, micro-organism and the whole ecological system in the local laws and regulations of the six provinces (regions) concerning LDC. For example, *The Implementation Measures on 'Grassland Law of P.R.C.' of Qinghai Province* has promulgated obligations and prohibiting stipulations for plain ecosystem as well as wild life and plants at different levels. This reflects our full attention to the integrity of the ecological system and the inter-relationship of all ecological factors and therefore, it cemented a legal foundation for carrying out ecological priority ideas.

(3) IEM strategy and action plan for Land degradation has become guidelines for LDC actions at provincial level

The formulation of provincial (regional) Strategy and Action Plan for LD control is a major approach to use IEM in practice. According to the project design, each province would establish a strategy and action plan working group consisting of experts from agriculture, forestry, hydro-power and land resource sectors. The compiling of Provincial Land Degradation IEM Strategy and Action Plan is based on the results of LDC cost efficiency assessment, with the IEM idea and approach fully introduced and human activities taken as one of the main factors.. At the same time, the bottom-to-top working style was adopted. In this approach, stakeholders are the essential and ecological system management aim takes economic and social factors into account. Meanwhile, priority areas and actions in terms of needed policy, investment and other series of activities were defined.

The formulation of the provincial (regional) land degraded IEM Strategic Plan and

Action Program has converted the situation in which each project was solely independent, separated and overlapping. The provincial governments (regions) attached great importance to the Plan and positively coordinated forestry, agriculture, hydro-power, transportation sectors and integrated the project with other government investment construction projects in forestry, hydro-power, transportation, poverty alleviation and education, which fully reflects the integrity of the LDC project. To date, many contents of the provincial (regional) LD IEM strategic plan and action program have been incorporated into provincial eleventh five-year plans, becoming an important target and task in local people' economic and social development. Qinghai Lake integrated control plan is a good example in this field.

(4) The establishment of provincial (regional) IEM Information center has made LD basic data and information sharing come true

Establishing a "National Land Degradation Data Sharing and Coordinating Mechanism" is an important component of the Partnership and the IEM approach. By setting up a comprehensive and coordinated land degradation monitoring and assessment system, the situation of sector separation and information inconsistency has been changed. It offers opportunities for scientific decision making to support LDC. To improve provincial (regional) capacity on land degradation monitoring, assessment, integrated analysis, decision making and to facilitate the integrated management of degraded land data, the project supports each province to set up an IEM information center, giving technical support for promoting the IEM approach in addressing land degraded issues.

In each Project province/region, an LD control metadatabase was established. Agencies related to LD signed on an LD control data sharing agreement. This data sharing mechanism is the first of its kind in China, covering several sectors, it largely improves the current data's utilization and perfectly satisfies the needs of LD data and information. Take Ningxia as an example, the establishment of the IEM information center which is supported by geological information system and documentary information system has enabled relevant sectors to share the information on LDC. In setting up the metadatabase, all relevant bureaus publish information required by the Center on their government websites, providing data on LD trend data to support national and provincial decision makers in terms of ecological system change and land degradation.

(5) Development of IEM pilot sites increases overall LDC capacity of the community

One of the important activities of Capacity Building to Combat Land Degradation Project (CBCLDP) is setting up pilot sites. 22 sites in 6 provinces/regions were selected to implement IEM practices, with factors including the type of the ecological system, land degradation type, land use and the local economic and social development level as well as the ethnic group differences taken into consideration. The activities undertaken in the pilot site included (i) undertaking participatory community planning. The Project, guided the community to formulate and implement its own sustainable livelihood and LDC plan; (ii) establishing farmers field school to integrate theory into practice and modern technology into indegenous knowledge so as to improve farmers' ability of applying IEM; (iii) conducting participatory LD monitoring and assessment, involving farmers and herders in the monitoring assessment activity to raise their awareness and enthusiasm; (iv) identifying best field practices through conservation farming, water-saving irrigation, grassland protection and vegetation closure, cash forest cultivation, farmland belt forest, barren hill afforestation, machinery sand fixation, plant sand fixation, clean energy utilization, increase of agricultural production improvement of infrastructure, salinization management and technical demonstration for installation cultivation, eliminate LD inducing factors in both production model and living style, offering new channels for farmers and herders to increase their wealth.

Community's capacity to control desertification has been dramatically improved by establishing demonstration sites and experience has been accumulated for other regions. From the following cases, it is easy to see the community has benefited from capacity building project.

- Annual vegetable income per person has reached 4500RMB in demonstration site in Wuda district in Wuhai City, Inner Mongolia, due to development of installation agriculture using sunlight green house to produce organic vegetables, introducing new varieties and extending new techniques;
- Integrated development of biogas tanks, warm shelters and toilets, a trinity system, was initiated in Naiman, Inner Mongolia Demonstration site has established a new energy model. Manure is flowing into the field in a more environmentally friendly way. It increases both production and efficiency;
- Kongtong demonstration site in Gansu has combined small watershed management with cycling economy. In this system, water-saving techniques are used in agriculture, loess fruit and vegetable cultivation and conversion from farmland to grassland. At the same time, livestock is raised with straw and stock to produce manure which is then flowed back to farmland. This technique has changed the single planning structure, reduced the over-use of land, enriched the soil fertility and effectively mitigated soil and water losses.
- Jingtai Demonstration site, Gansu Province has developed more than 600 mu wolfberry in abandoned saline land, with output of dry fruit for 50 kilo gram/mu, which is an effective approach to control the land secondary salinization and increase economic benefits at the same time.

(6) The implementation of capacity building to combat land degradation project (CBCLDP) Has trained a large number of professionals equipped with IEM knowledge

The implementation of CBCLDP has cultivated a large number of personnel equipped with IEM knowledge and LDC expertise, almost covering all the main strength engaging in natural resource managements from governments, research institutes, universities and the agricultural sector. All these people will be the mainstay of China's LDC in dryland ecosystems in the future. According to preliminary statistics, more than 1000 government officials and well-known experts in different fields have directly taken part in the project activities. Local experts have been intensively involved. They developed good understanding of IEM approach in the course of project implementation. This is a unique feature of the project and as well as the most successful practice. By demonstration and training, many farmers and herders have internalized the IEM approach. In addition, project experts have initially introduced the IEM into environment laws or natural resource law courses or related courses in the university. For example, Wuhan University, China Renmin University, Qinghua University, Northwest Political and Law

University and Lanzhou University have made such attempts. It is understood that people are the critical factor for every success. The dissemination of the IEM approach is of large and profound significance for LDC in the future.

(7) Establishment of the partnership enhanced the collaboration among International Partners and Increased project's cost effectiveness

The establishment of the partnership has enhanced the communication among international cooperation partners and government sectors It provides a platform for sharing LDC information.and dredged the channel for collaboration of different departments. These departments are now able to take their own advantages and functions to contribute to the Project. Therefore, funds supporting the LDC project have been well coordinated and used, which have enabled cost effectiveness to be improved notably.

Since the implementation of the project, ADB and CPMO have held several coordinating meetings for international donors including WB, FAO, UNEP, UNDP, KFW, IFAD, UNIDO, EU, AusAID, CIDA, CI, TNC, WWF and JICA, providing opportunities for extensive cooperation and communication, which fully reflects the coordinating function of partner relationship identified in the Country Programming Framework (CPF). It is well welcomed and accepted by international cooperation partners. Through these activities, the influence of partnership has been expanded. Project management experience has been enriched and fund allocation has resulted in improved efficiency. Impelled by the partnership, more funds have been pooled to support LDC efforts in West China.

3. Experience from Implementation of the Partnership

We consider the following experience obtained from Project implementation could be shared with every one involved:

(1) Establishing a multi-level and multi-sector coordinating mechanism is significantly helpful in achieving coordination of policies, incorporation of funds, sharing of information and increasing efficiency in resources use. It is a mechanism mobilizing all relevant aspects to enhance land degradation control.

(2) Integrated Ecosystem Management (IEM) is an effective approach to combat land degradation. However, it creates best practical effects only when it is used in legislation, policy making, planning and technical support.

(3) Building IEM concept and approach among professionals especially local managers, technicians and land users is essential to ensure the IEM concept is incorporated into practices and to ensure policies, technology and funds play effective roles in land degradation control.

(4) Land degradation control is a long term process which requires that a strategy and action plan must be adopted into social development plans to achieve the goal of land degradation under a regional sustainable development framework.

(5) Land degradation control must be incorporated into farmers' and herders' production and livelihood. Therefore, farmers and herders must be involved in planning and implementation of LDC projects. The sustainability of the project is ensured only when the enthusiasm of farmers and herders is mobilized and maintained.

4. Prospect for the Partnership

The achievements obtained in the 1st phase of the Partnership have demonstrated that IEM is an effective approach to combat land degradation. It is in our plan that the achievements should be scaled up and international cooperation must be strengthened to enhance the development of the Partnership.

The goal of the 2nd phase of the Partnership under the Country Programming Framework (CPF) is to strengthen IEM incorporation in China's 11th 5-Year Plan, Socialist Rural Development Program and West-China Development Program., Further improving and expanding IEM pilot sites, promoting poverty reduction in dryland ecosystems in West China, enhancing biodiversity conservation, increasing carbon sequestration, mitigating climate change will bring better global benefit.

Proposed ideas for the 2nd phase of the the Partnership under the CPF

(1) Make good use of the achievements obtained in the Capacity Building to Combat Land Degradation Project (CBCLDP); further improve policies and legislation for LDC; and promote the implementation of ecological conservation development plans involved multi sectors in Western Regions of China.

(2) Continue to promote and expand the polite site demonstrations at the community level; undertake further training to increase farmers and herders' acceptance of new ideas and new approaches, to improve their participatory decision-making skills, increase their awareness of the importance of ecological conservation and LDC while improving their livelihood.

(3) Use the current LDC experiences, develop and extend new technology and measure to effectively control LD in dryland areas.

(4) Strengthen the capacity of monitoring and assessing the influence of LDC on ecology, society and economy; Expand the channels of data collecting so as to forecast and mitigate LD influence.

(5) Use IEM approach in coping with global environmental issues, accelerate design and implementation of follow-up demonstration projects .

(6) Extend the experience of the partnership in China and world wide to address poverty and LD issues.

A greater half of the Partnership life has been completed, the IEM approach initiated by the partnership has been widely accepted. The achievements of capacity building are remarkable and the preliminary achievements of LDC in west part of China have been gained. However, the LD issue in west part of China is still severe and the progress made is but only a benign beginning. Dry-land area is also the key area concerning global climate change and bio-diversity. We will further summarize experiences and address the issues through designing and implementing the management supporting project under the PRC-GEF Partnership on Land Degradation in Dryland Ecosystems to make due contribution to global environmental improvement.

Thank you!

9. Central Asian Countries Initiative for Land Management

Mr. Kambarali Kasymov

State Secretary of Ministry of Agriculture, Water Resource and Process Industry, Kyrgyz Republic

Abstract: This paper illustrated the status of land degradation in Central Asia and the world. It also illustrated CACILM (Central Asian Countries Initiative for Land Management), including its background of establishment, mission, organizational structure, project funding, action plans, anticipated environmental benefits, and the cooperation between China and Central Asian Countries within the framework of CACILM.

1. Global scale of land degradation

Land degradation is one of the most important challenges to global environment, making productive capacity of lands catastrophically worse and destroying ecosystems integrity, especially in drought-affected countries. By UNEP assessment 73% of pastures are degraded in dry areas at present time. Due to salinization 3 ha of arable lands are lost each minute on the Earth that equal 1.6 million ha of arable lands per year (FAO).

2. Basic types of land degradation

Basic types of land degradation include: Erosion, salinization and waterlogging; Decreasing of pasture fertility; Decreasing of arable land productivity; Reduction of forest area and forest production; Internal and external impact of mining operations; Increasing of risks of landslides and floods; Reduction of ecosystem stability.

3. Ecological problems of Central Asia

Ecological problems of Central Asia include: water supply has been reduced; water pollution by agrochemicals, municipal wastes; dust and mineralization of precipitation, melting of glaciers; land/water erosion, salinity, water logged soils, desertification; biodiversity, collapse of fishing industry in Aral Sea; loss of indigenous plants, loss of phenotypes that are adaptable to climate change.

Current status of the region: Agricultural Yields are declining by 20-30%

In Kazakhstan, 60% of top-soil is affected by degradation; in Kyrgyzstan, 40% of arable lands are degraded; in Tajikistan, mountains which are 85% of the total territory have been catastrophically deforested; in Turkmenistan, 70% of the total land area has become desert; in Uzbekistan, 50% of irrigated lands are affected by salinity.

4. CACILM (Central Asian Countries Initiative for Land Management)

CACILM is the first multicountry program aimed to combat land degradation and to raise living standard of rural people. It is started by a Strategic Partnership Agreement in 2004. It is a single partnership that is achieving the goals of United Nations Convention to Combat Desertification (UNCCD) at multicountry level. CACILM is designed to address the issues identified in the National Programming Frameworks to for sustainable land management in each Central Asian Country.

CACILM Program Goals

The goals of CACILM include restoration, maintenance and enhancement of the productive functions of land in Central Asia, leading to improved economic and social well-being, or more simply stated: combating land degradation and improving livelihoods.

CACILM Partnership :

CACILM involves 5 Central Asian Countries (Kazakhstan, Kyrgyzstan, Tajikistan, Turkmenistan and Uzbekistan) and 12 international development partners of the Strategic Partnership Agreement(ADB, CIDA, GTZ CCD, GEF, GM, ICARDA, IFAD, SDC, UNDP, UNEP, World Bank, FAO).

CACILM Implementation

CACILM was launched in November 2006, and 10 years program of activities at national level in each Central Asian Country has been developed at multicountry level. The total program financing is to up to $1.4 billion over the period 2006-2016.

CACILM Organizational Structure

CACILM National Programming Frameworks: Programs Areas

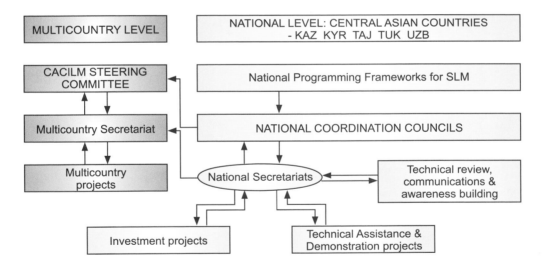

The programs areas include: Capacity building for strengthening of the enabling environment; Capacity building for integrated land use planning and management; Sustainable agriculture in rainfed lands; Sustainable agriculture in irrigated lands; Sustainable forest management; Sustainable pasture management; Integrated resource management; Protected area management and biodiversity conservation; Rehabilitation of the former Aral Sea ecosystem.

CACILM Multicountry activities: Kazakhstan, Kyrgyzstan, Tajikistan, Turkmenistan, Uzbekistan

The CACILM Muticountry activities include SLM Research (ICARDA + National Research Institutes), SLM Information System (FAO/ADB + NSIUs), SLM Knowledge Management (ADB + NSECs), SLM Capacity Building (UNDP + SLM Stakeholders).

CACILM activities at national level: GEF Funded SLM Projects: UNDP

UNDP's CACILM activities at national level include: Rangeland Ecosystem Management in Kazakhstan; Demonstrating Sustainable Mountain Pasture Management in Suusamyr Valley in Kyrgyzstan; Rural Development and Demonstrating Local Responses to Combating Land Degradation and Improving Sustainable Land Management in SW Tajikistan; Capacity Building and On-the-ground Investment for Integrated and Sustainable Land Management in Turkmenistan; Achieving Ecosystem Stability on the Exposed Aral Seabed and the Kyzylkum Desert and Land Improvement in Uzbekistan.

CACILM activities at national level: ADB SLM Investment Projects with GEF co-financing

ADB's CACILM activities at national level include: Southern Agricultural Areas Development Project in Osh, Jalalabad and Batken in Kyrgyzstan; Rural Development Project in Tajikistan; Land Improvement in Bukhara, Navoi and Kashkadarya in Uzbekistan.

Central Asian Countries Benefits from the CACILM Program over the 10-year CACILM program period:

CACILM's benefits for Central Asian Countries include: More efficient investments in the maintenance and improvement of land assets; Use of an integrated approach to land use planning and management and appropriate adoption of sustainable practices; Increasing of productivity and profitability of lands and pasture, decreasing of waterlogged and saline lands and increasing of forest cover; Sustainable funding for SLM activities.

Global Environmental Benefits from the CACILM Program

Global environmental benefits include: Significant reduction in loss of vital soil in dust storms; Reduction in soil and pesticide runoff into rivers; Improvements in water availability; Reduction in the loss of carbon stocks sequestered in soils or forests, and hence a reduction in greenhouse gas emission; Reduction of the loss of biodiversity.

Cooperation between PRC and CAC within the framework of CACILM

The People's Republic of China and Central Asian Countries cooperate within the frameworks of UNCCD, and had signed many cooperation agreements. There is an excellent opportunity for Central Asian Countries to cooperate with the People's Republic of China through CACILM.

10. Sustainable Land and Water Management: An African Initiative to Develop Agriculture under Pressures of Land Degradation

Phiri Elijah, Bwalya Martin

University of Zambia, School of Agricultural Sciences, Department of Soil Science, P. O. Box 32379, Lusaka, Zambia. T

Abstract: A major challenge facing governments and society worldwide on the threshold of the twenty-first century is to achieve sustainable economic growth by means which alleviate poverty without jeopardizing the quality of the environment. While this is a task of global significance it presents particular problems to the agricultural sector in Africa especially Sub-Saharan Africa because of the direct links between production and the natural resource base, and high dependence on agriculture for income and employment. Assessment of arable land in Africa suggests that large areas of countries are vulnerable to land degradation, a resource environmentally fragile and easily degraded. Land resources lay the heart of social, cultural, spiritual, political and economic life of the most of African continent. Africa's ability to conserve and manage her land resources is key to sustainable development. Major trends to land degradation and agricultural productivity include loss of water for agriculture to other uses, reduction of soil and water quality, and loss of farm land. Through the AU/NEPAD initiative under the Comprehensive Africa Agriculture Development Programme (CAADP) a framework is being developed at continental level to help national governments manage their land and water resources by paying particular attention to issues related to soil fertility, agricultural water and land policy/administration. It is envisaged that with political commitment and common shared vision at the highest level, Africa can implement strategies that addresses issues of land degradation in an agriculture-led economic development.

Keywords: poverty alleviation, policy initiatives, food security

Background

The African Union has identified the failure of African agriculture to provide adequately for Africa's people as a major impediment to economic growth and improved human welfare. Whilst previously the development philosophies of many African nations and of the international development assistance community centred on creating increased purchasing power through non-agricultural investments, there is to-day a more general realization among African leaders that agriculture is an important engine for economic growth and poverty alleviation, not the least for rural people that have the lowest purchasing power and the highest level of food insecurity on the continent. Food is produced on the land, from soil resources, water resources and biological resources. The management of these natural resources is a key element in any quest for increased food production. The African continent is not generally well endowed in all these resources. In particular its soils are often strongly weathered and therefore low in plant nutrients, its water resources – both in the form of precipitation, rivers and lakes and underground resources – are spatially unevenly distributed and subject to weather vulgarities, whilst the continent does possess rich biological resources, both indigenous genetic resources and access to the global crop and livestock pool. Good management of land and water resources is a prerequisite for increased production.

Agricultural production as well as non-agricultural activities provide the basic means for rural livelihoods and also form the driving force for the promotion of the rural economies of the countries in Sub-Saharan Africa. Agriculture alone contributes between 30% and 40% to the GDP and provides employment for 80% of the population. Most economies in Sub-Saharan Africa (SSA) are agriculturally-based and about two-thirds of Africans depend on agriculture for their livelihoods. In this region, most farmers are smallholders with 0.5 to 2ha, earn less than US$1 a day, face 3-5 hunger months, have large families and are generally malnourished. The fate of the agricultural sector, therefore, directly affects economic growth, poverty alleviation and social welfare in Africa. As the region's population continues to grow rapidly (3% per annum), outpacing the growth rate in other regions of the world, the carrying capacity of its agricultural land is becoming lower, bringing closer the land frontier.

Africa suffers from geologically induced and inherently low soil fertility as the bedrock consists of mostly granites and gneiss. African rocks are among the oldest in the world. The relationship between the parent materials and the soil forming factors are very complex because the land surface has undergone a series of shifts in vegetation and climate. Nearly one-third of the central plateau of Africa is of Pre-Cambrian age (over 600 million years old). The rest of the surface is covered with sand and alluvial deposits of Pleistocene age (less than 2 million years old). A recent volcanic activity occurred mainly in the eastern and southern parts of the continent, principally between Ethiopia and Lake Victoria. For this reason, most of the soils in Africa are characterized by a low proportion of clay, making them easy to work, but also easy to lose. Not only is Africa geologically old and afflicted with a harsh climate, but also large parts of the continent have been occupied by human beings much longer than in other continents, human-induced bush fires during hunting and wild-food gathering were early evidence causes of land degradation. Human activities in obtaining food, fibre, fuel and shelter have, therefore, significantly altered the soil. Though degradation is largely man-made, and hence its pace is governed primarily by the speed at which population pressure mounts, irregular natural events, such as droughts, exacerbate the situation. Exploitation of marginal and quality-poor lands has reached upper limits and, when farmers do intensify land use to meet increasing food and fiber needs, they do it without proper management practices and with little or no external inputs. Resulting consequences are a lowering of soil organic matter in already poor soils, a depletion of nutrients that have contributed to a stagnation or decline of crop production in many African countries. In some cases, the rate of nutrient depletion is so high that even drastic measures, such as doubling the application of fertilizer or manure or halving erosion losses, would not be enough to offset nutrient deficits. Unless African governments, supported by the international community, take the lead in confronting the factors that cause nutrient depletion and land degradation, deteriorating agricultural productivity will seriously undermine efforts to bring about food security and to strengthen the foundations of sustainable economic growth in SSA.

Natural Resource Endowment

Africa is a vast continent with a tremendous resource endowment and offers great potential for increased agricultural productivity (FAO, 1993). Despite this endowment, Africa with total land mass of about 30.7 million kilometers and a population exceeding 746 million people at growth rate of 3.3% has generally lagged behind in agricultural development and continues to register declining per capita food production (Figure 1) in the last few decades (WRI, 1994). Recurrent droughts are also a major factor in the degradation of cultivated land and rangelands in many parts of Africa. The two problems are often interlinked. While drought increases soil degradation problems, soil degradation also magnifies the effect of drought (Ben Mohamed, 1998). Hunger remains a major threat to many people, particularly in Sub-Saharan Africa where about 200 million are chronically hungry; 30 million requiring emergency food and agricultural assistance in any given year. Sub-Saharan Africa (excluding South Africa) is the poorest developing region, with 29 out of 34 countries being some of the poorest in the world (FAO, 1993).

In spite of the inherent fragility of Africa's soils, the continent's climatic variability, and the uneven distribution and availability of both surface and subsurface water resources, there is substantial untapped potential for the development of the continent's water and land resources for increasing agricultural production.

FAO estimates that the current area under managed water and land development

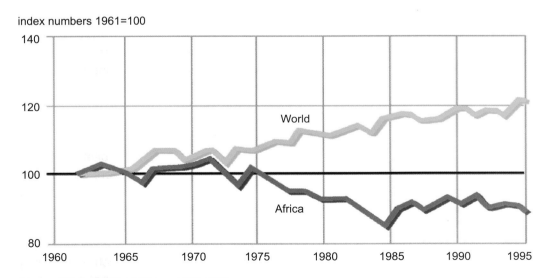

Source: WRI, UNEP, UNDP and WB 1998
Figure 1: Per capita food production

totals some 12.6 million ha[1] equivalent to only some 8 percent of the total arable land. Substantial public and private investments in developing and improving the management of these land and water resources will be essential to enable African countries reach the levels of agricultural production required to meet the targets for poverty alleviation, food production and economic recovery by 2015. Building up soil fertility and the moisture holding capacity of agricultural soils, and rapidly increasing the area equipped with irrigation, will not only provide farmers with opportunities to raise output on a sustainable basis but will also contribute to income and food security at household level.

Land Degradation

Land degradation is a major challenge that cuts across issues of poverty, health, environment and economic growth in an African setup. Up to two-thirds of Africa's productive land area is affected by land degradation, while close to 100% is vulnerable to it. As African farming is essentially a low-input low-output system (Badiane and Delgado, 1995), land degradation is rampant and studies show a progressive decrease in the productivity of the land. In the context of productivity, land degradation results from a mismatch between land quality and land use (Beinroth et al., 1994). The extent and rate of land degradation show a large variation in the available statistics on global estimates; however Africa shares quite a substantial area under degraded condition (Table 1). The semi-arid and weakly arid areas of Africa are particularly vulnerable to land degradation as they have fragile soil, localized high population densities, and generally practice low-input agriculture. About 55% of Africa's land area is unsuitable for agriculture and only 11% of the continent has high-quality soil that can be effectively managed to sustain more than double its current population (Eswaran et al., 1997). Most of the remaining usable land is of medium or low potential, with at least one major constraint for agriculture. This land is at high risk of degradation under low input systems. By 1990 soil degradation was estimated to have affected 500 million hectares, or 17% of Africa's land (UNEP, 1997). Susceptible drylands (arid, semi-arid, and sub-humid aridity zones), covering 43% of Africa, are the worst-affected areas, impacting 485 million people (Reich et al., 2001). Approximately 65% of agricultural land, 31% of permanent pastures and 19% of forest and woodland in Africa were estimated to be affected by some form of degrada¬tion in 1990 (Oldeman, 1994). The current situation is undoubtedly worse. Soil moisture stress inherently constrains land productivity on 86% of soils in Africa (Eswaran et al., 1997), but soil fertility degradation now places an additional serious human-induced limitation on productivity.

Status of Land Degradation in Sub-Saharan Africa

The InterAcademy Council (2004) estimated that land degradation in SSA is caused by soil erosion (46%), wind erosion (36%), loss of nutrients (9%), physical deterioration (4%), and salinization (3%). Overgrazing (49%) followed by agricultural activities (24%), deforestation (14%) and overexploitation of vegetation cover constitutes the primary

---

[1] http://www.fao.org/docrep/005/y6831e/y6831e-03.htm#P458_107823#P458_107823

Table 1: Estimates of degraded lands (in million km^2) in dry area of the world

Continent	Total Area	Degraded area*	% degraded
Africa	14.326	10.458	73
Asia	18.814	13.417	71
Australia and the Pacific	7.012	3.759	54
Europe	1.456	0.943	65
North America	5.782	4.286	74
South America	4.207	3.058	73
Total	51.597	35.922	70

Source: Dregne and Chou, (1994) *comprises land and vegetation

causes of land degradation in rural areas. As a result of the inherent low soil fertility of African soils worsened by subsequent land degradation, only sixteen (16) % of the land has soil of high quality and about thirteen (13) % has soil of medium. About nine (9) million km^2 of high and medium quality soils support about four hundred (400) million or about 45% of the African population (Bationo at al., 2006). Indeed, up to two-third of African's productive land area is affected by land degradation, while about 100% is vulnerable (Bationo at al., 2006). The situation is worsened by the continuous demands for more productive land at a time when land degradation and/or poor land management continues to be a major challenge for SSA. These increasing demands for land, coupled with continuing land degradation and heightened competition for water mean that two-thirds of Africa's croplands could become effectively unproductive by 2025 (MEA, 2005). Up to 75 % of Africa's poor still live in rural areas and the most vulnerable among them are often relegated to the highly marginal lands where productivity is most at risk. If no remedial actions are taken, further land degradation will continue to adversely affect rural livelihoods and ecosystems.

Restoring and sustaining the productivity and ecosystem service functions of the soils will need to solve the development bottleneck on the degraded status of the African soils that include: (i) problem diagnosis and impact assessment; (ii) integrated soil management; (iii) management of soil ecosystem services; and (iv) processes and policies for adoption of integrated soil management.

> **BOX 1: SOME OF THE KEY CAUSES OF LAND DEGRADATION IN SUB-SAHARAN AFRICA** [2]
> **The most important natural factors relate to the risk of:**
> water erosion – steep slopes, high intensity rainstorms, erodible soils;
> wind erosion – strong winds, semi-arid/arid climatic zones with sparse vegetative cover;
> soil fertility decline – strong leaching of soil nutrients, rapid decay and mineralization of soil organic matter, weathered acidic soils low in organic matter and soil nutrients;

---

[2] Extracted from "The SLWM Pillar Framework, NEPAD Secretariat, June 2008"

degradation in soil physical properties – weak structured soils low in organic matter; salinization – semi-arid/arid climates with high evaporation rates and low leaching intensity; vegetation degradation – low and erratic rainfall limits vegetative recovery following disturbance; and climate variability - decline in water quality and quantity –alternating abundance and scarcity according to the season (wet or dry), or natural climatic cycle (El Niño/La Niña).

The direct (human) causes, or pressures on the land include:
inappropriate management (shorter fallows, exposed soil, etc.) of the land for the cultivation of annual rainfed, irrigated and/or perennial crops
poor management of natural forest and tree plantation/woodlot areas;
removal and degradation of natural vegetation through deforestation and/or overexploitation of local species;
overgrazing of natural and planted pastures
poor management and over use of surface and groundwater resources; and
poorly planned and managed urban and industrial development (resulting in the physical loss of good farm land, pastures and forest areas as well as on- and off-site pollution).
Forest Fires
Population growth

The key root causes or driving forces of particular importance in SSA are:
poverty/economic disadvantage (poor people cannot afford to forgo short term production/resource exploitation to take care of immediate income needs for the sake of long term sustainability);
lack of awareness of the consequences of land degradation which happens progressively but for which the symptoms are not immediately evident
population pressure leads to small land holding size, in high potential areas, with traditional fallowing practices abandoned as individual plots are of necessity cultivated on a continuous basis;
high input costs, low produce prices, and other market failures are disincentives to investing in improved land management practices;
under nourishment and ill health are interlinked, rural households with food shortages are more susceptible to the ravages of malaria, HIV-AIDS and tuberculosis, which in turn reduces their ability to produce their own food, or earn their livelihoods in off-farm employment;
rural households with insecure user rights, for their farm plots, pasture and forest resources, are less willing to invest in ensuring future productivity, being unsure as to whether they will be the ones to benefit
inappropriate development policies driven by short term output targets that ignore long term sustainability; and
Weak/none existent advisory support services limiting land users' access to improved farm inputs and information on alternative land use enterprises and improved land management practices.

BOX 2: SUMMARY OF LAND DEGRADATION STATUS AND ILLUSTRATIVE CONSEQUENCES

Status of Land Degradation
• Land degradation affects 67 percent of the total land area of SSA with 25 percent characterized as severe to very severely degraded and some 4 to 7 percent as non-reclaimable.
• Africa is currently exporting 1.7 billion tons per year of sediment causing productivity losses and contaminated water sources.

- The productivity loss in Africa from soil degradation since World War II has been estimated at 25 percent for cropland and 8 to 14 percent for cropland and pasture together.
- There is a negative nutrient balance in SSA's croplands with at least 4 million tons of nutrients removed in harvested products compared to the 1 million tons returned in the form of manure and fertilizer. Soil fertility degradation is considered the single most important food security constraint in SSA.
- Some 86 percent of African soils are under soil moisture stress.

Illustrative Consequences
- Over 3 percent of Africa's agricultural GDP is lost annually - equivalent to US$ 9 billion per yet - as a direct result of soil and nutrient loss.
- By 2015, SSA will host half of the world's poor.
- The World Food Programme has spent US$12.5 billion (45 percent of its total investment since its establishment) in Africa and 50 percent in 2001.
- Africa spent US$18.7 billion on food imports in 2000 alone.
- In 2000 Africa received 2.8 million tons of food aid, over a quarter of the world total.
- In 2001 28 million people in Africa faced food emergencies due to droughts, floods and strife, with 25 million needing emergency food and agricultural assistance.
- Hunger and malnutrition in SSA and degradation of water resources has increased susceptibility to life threatening diseases.
- In sub-Saharan Africa, 15 percent of the population or 183 million people will still be undernourished by 2030 – by far the highest total for any region and only 11 million less than in 1997-99. Malnutrition is expected to increase by an average of 32 percent.
- Land degradation has led to forced migration of individuals, rural households and whole communities.
- Conflicts (between settled farmers, herders and forest dwellers) over access to land resources have increased as households and communities search for productive land for their crops and/or livestock.

BOX 3: THE DOMINANT TYPES OF LAND DEGRADATION WITHIN SUB-SAHARAN AFRICA (AFTER DOUGLAS, 1994)

Soil degradation – decline in the productive capacity of the soil resources as a result of adverse changes in their biological, chemical, physical and hydrological properties, which in turn increase the vulnerability of erosion prone areas to accelerated soil loss through both water and wind erosion.

Vegetation degradation – decline in the quantity and quality of the grasses, herbs and woody species found in grasslands, woodlands and forest, combined with a decrease in the ground cover provided by such plants.

Biodiversity degradation – loss of wildlife habitats and decline in genetic resources, species and ecosystem diversity.

Water degradation – decline in the quantity and quality of both surface and ground water resources and increased risk of downstream flood damage.

Climate deterioration – adverse changes in the micro and/or macro climatic conditions that increase risk of failure of crop and livestock systems and impact negatively on plant growth in rangelands, woodlands and forests.

Land conversion – decline in the total area of land used, or with potential to be used, for crop, livestock and/or forestry as a result of land being converted to urban, industrial, mineral extraction and infrastructure purposes.

Comprehensive Africa Agriculture Development Programme (CAADP) Initiative

New Partnership for Agriculture Development (NEPAD), the mechanism constituted by the African Union in support of the agriculture sector, has identified natural resources management as a field in need of major improvement if the ambitious aim of 6% annual growth in the agricultural sector can be reached through the allocation of 10% of national budgets to the agricultural sector. The sustainable development of land and water resources in Africa in general, and in Sub-Saharan Africa in particular, is a vast subject, interwoven between many sciences and subject to social, political and economic frameworks beyond the natural sciences. Indeed, it may be argued that it has been the failure of more narrow approaches to specific natural science topics (e.g. soils, hydrology, and agronomy) that has led these sciences to have less than optimal impact on practical implementations for sustainable development. This is further exemplified in the difficulties experienced in up-scaling successful pilot projects to community level and beyond. TerrAfrica is a most recent concerted effort to place agriculture as an engine of growth in poverty alleviation in Africa in general and as a contribution to achieving the Millennium Development Goals by Year 2015 (or – realistically – later).

Under Comprehensive Africa Agriculture Development Programme (CAADP), Africa's governments have identified four continent-wide entry points (Pillars) for investment and action in pursuing increased and sustainable productivity in agriculture, forestry, fisheries and livestock management. These entry points include: (i) Extending the area under sustainable land and water management (Pillar 1); (ii) Improving Market access through improved rural infrastructure and trade-related interventions (Pillar 2); (iii) Increasing food supply and reducing hunger across the region by increasing small holder productivity and improving response to food emergencies (Pillar 3); and (iv) Improving agricultural research and systems to disseminate appropriate new technologies, and increasing the support to help farmers adopt them (Pillar 4). Each of these pillars incorporates policy, institutional reform and capacity building and has a framework through which the challenges prioritized by CAADP might effectively and efficiently be achieved. The primary targets are: (i) agriculture-led growth as a main strategy for food security and poverty reduction (MGDs); (ii) pursuit of a 6% average annual agricultural sector growth rate at the national level; and (iii) allocation of 10% of national budgets to the agricultural sector;

CAADP Pillar 1 Framework: African Sustainable Land and Water Management (FASLWM).

The SLWM Vision Paper for Africa and the corresponding SLWM Country Support Tool provide the foundation for the FASLWM: These documents elaborate both the strategic vision for scaling up the area under sustainable land and water management in Africa and the practical tools and modalities for pursuing this vision at the national level. The Country Support Tool has been developed with regard to the need to provide a clear instrument in ensuring clear and concrete linkages between the CAADP agenda at large with the unfolding demands of the country roundtable processes. In addition, the paper on Investment in Agricultural Water for Poverty Reduction and Economic Growth in Sub-Saharan Africa identifies key priorities and entry points for approaching the agricultural water agenda through sustainable land and water management.

The Pillar 1 framework aim to bring the CAADP principles, values and targets into the development agenda presented as a set of tools which development players would use in order to: (i) guide country strategies and investment programmes; (ii) allow regional peer learning and review; and (iii) facilitate greater alignment and harmonization of development efforts.

The Comprehensive Africa Agriculture Development Programme (CAADP) Pillar 1 embraces three mutually related aspects of sustainable natural resource management. These are: (i) Soil fertility and sustainable land management; (ii) agricultural water underline the critical nature within which water presents a key entry point in address sustainable natural resource management); and (iii) land policy/land administration within the scope of sustainable land management underlining the critical significance land policy/land administration on the ability to achieve sustainable land and water management objectives .

Sustainable land and Water Management (SLWM) is seen as the foundation of sustainable agriculture and a strategic component of sustainable development, food security, poverty alleviation and ecosystem health. SLWM can be defined as 'the use of land resources, including soils, water, animals and plants, for the production of goods to meet changing human needs, while simultaneously ensuring the long-term productive potential of these resources and the maintenance of their environmental functions' (UN Earth Summit, 1992). Sustainable Land and Water Management (SLWM) is a knowledge-based procedure that helps integrate land, water, biodiversity, and environmental management including input and output externalities) to meet rising food and fibre demands while sustaining ecosystem services and livelihood (World Bank 2006) Sustainable Land and Water Management is considered an imperative for sustainable development and plays a key role in harmonizing the complementary, yet historically conflicting goals of production and environment. Thus one of the most important aspects of sustainable land and water management is this critical merger of agriculture and environment through twin objectives: i) maintaining long term productivity of the ecosystem functions (land, water, biodiversity) and ii) increasing productivity (quality, quantity and diversity) of goods and services, and particularly safe and healthy food.

SLWM encompasses or contributes to other established approaches such as sustainable agriculture and rural development, integrated natural resources management, and ecosystem management (as noted above) and involves a holistic approach to achieving productive and healthy ecosystems by integrating social, economic, physical and biological needs and values. Thus it requires an understanding of:

the natural resource characteristics of individual ecosystems and ecosystem processes (climate, soils, water, plants and animals);

the socio-economic and cultural characteristics of those who live in, and/or depend on the natural resources of, individual ecosystems (population, household composition, cultural beliefs, livelihood strategies, income, education levels etc);

the environmental functions and services provided by healthy ecosystems (watershed protection, maintenance of soil fertility, carbon sequestration, micro-climate amelioration, bio-diversity preservation etc); and the myriad of constraints to, and opportunities for, the sustainable utilization of an ecosystem's natural resources to meet peoples' welfare and economic needs (e.g. for food, water, fuel, shelter, medicine, income, recreation).

SLWM recognizes that people (the human resources) and the natural resources on which they depend, directly or indirectly, are inextricably linked. Rather than treating each in isolation, all ecosystem elements are considered together, in order to obtain multiple ecological and socio-economic benefits. Sustainable Land Management is considered an imperative for sustainable development and plays a key role in harmonizing the complementary, yet historically conflicting goals of production and environment. Thus one of the most important aspects of sustainable land management is this critical merger of agriculture and environment through twin objectives: (i) maintaining long term productivity of the ecosystem functions (land, water, biodiversity) and (ii) increasing productivity (quality, quantity and diversity) of goods and services, and particularly safe and healthy food. To operationalize the sustained combination of these twin objectives, sustainable land management must also take into account issues of current and emerging risks

Conclusion

Several global and regional and national efforts have been put in place to address land degradation in Africa, these include:

- The Comprehensive Africa Agriculture Development Program (CAADP) launched in 2002 by the New Partnership for Africa's Development (NEPAD) as an African-led commitment to address issues of growth in the agricultural sector, rural development and food security;
- The Action Plan of the Environment also launched by NEPAD, in 2003, as an integrated action plan designed to address environment challenges whilst also combating poverty and promoting socio-economic development;
- Regional Economic Communities of the Africa Union (RECs) –which by 2005, along with their member countries took ownership of the implementation process within CAADP, identified priority investment programs and immediate actions, and agreed upon basic principles and procedures for implementation and governance involving also farming and agribusiness stakeholders;
- The Alliance for a Green Revolution in Africa - launched in 2006 as an alliance through the Bill and Melinda Gates and Rockefeller foundations to build a prosperous agricultural system focused on economic, social, and environmental aspects required to double or triple farmers' yields;
- The Great Green wall initiative (GGWI): launched in 2005 in Ouagadougou with the goal to promote a socio-economic development of the target zones vulnerable to desertification by the implementation of projects of conservation and restoration of the natural resources and promotion of economic activities (agricultural, livestock, fishing, handcraft industry)
- The Soil Fertility Initiative (SFI) launched during the 1996 World Food Summit, the first regional level concerted attempt toward reversing the detrimental effects of soil

degradation and nutrient depletion;
• The Global Environment Facility (GEF) designated land degradation as one of its key GEF focal areas at the Second GEF Assembly held in Beijing (October 2002) in response to growing global concern over the issues of desertification and deforestation;

References

Badiane, O. and Delgado, C.L., eds. 1995. A 2020 Vision for Food, Agriculture, and the Environment in Sub-Saharan Africa. Food, Agriculture, and the Environment Discussion Paper 4. Washington, D.C.: IFPRI.

Beinroth, F.H., Eswaran, H., Reich, P.F. and Van den Berg, E. 1994. Land related stresses in agroecosystems. In: Stressed Ecosystems and Sustainable Agriculture, eds. S.M. Virmani, J.C. Katyal, H. Eswaran and I.P. Abrol. New Delhi, India: Oxford and IBH.

Douglas, M.G. 1994. Sustainable Use of Agricultural Soils. A Review of the Prerequisites for Success or Failure. Development and Environment ReportsNo. 11, Group for Development and Environment, Institute for Geography, University of Berne, Switzerland

DREGNE, H.E. and CHOU, N.T. 1994. Global desertification dimensions and costs. In: Degradation and Restoration of Arid Lands, ed. H.E. Dregne. Lubbock: Texas Technical University

Eswaran H, Almaraz R, van den Berg E, Reich P. 1997. An assessment of the soil resources of Africa in relation to productivity. Geoderma 77:1–18

FAO, 1993. Agriculture: Towards 2010. FAO Conference report C93/24. Food and Agriculture Organization of the United Nations, Rome

Hammond, A. L. 1998. Which World? Scenarios for the 21st Century. Island Press, Washington DC, United States

Interacademy Council. 2004. Realizing the promise and potential of African agriculture. Amsterdam: Interacademy Council

MEA (Millennium Ecosystem Assessment), 2005. Living Beyond Our Means: Natural Assessment and Human Well-being. Island Press, Washington, DC

NEPAD, 2002. New Partnership for Africa's Development, Comprehensive African Agriculture Development Programme. Johannesburg: New Partnership for Africa's Development and Food and Agriculture Organization of the United Nations

Oldeman LR. 1994. The global extent of soil degradation. In: Greenland DJ, Szaboles T, eds. Soil Resilience and Sustainable Land Use. Wallingford: CAB International

Reich PF, Numbem ST, Almaraz RA, Eswaran H. 2001. Land resource stresses and desertification in Africa. In: Bridges EM, Hannam ID, Oldeman LR, Pening FWT, de Vries SJ, Scherr SJ, Sompatpanit S, eds. Responses to land degradation. Proceedings of the 2nd International Conference on Land Degradation and Desertification, Kon Kaen,Thailand. New Delhi: Oxford Press

UNEP, 1997. United Nations Environment Programme, World Atlas of Desertification. 2nd ed. London: Arnold

World Bank, 2006. Sustainable Land Management. Washington, DC: World Bank.

WRI, 1994. World Resources 1994-95. Oxford Univ. Press, pp. 400.

11. The role of the Partnership Initiative on Sustainable Land Management in Combating Land Degradation in Caribbean SIDS

Leandra K. Sebastien

PISLM Coordinator, PISLM Support Office,
The Caribbean Network for Integrated Rural Development (CNIRD), Trinidad and Tobago, West Indies

Abstract: The Partnership Initiative on Sustainable Land Management (PISLM) is presently the only innovative initiative taking place in the Caribbean region that addresses land degradation specifically as it applies to the implementation of the United Nations Convention to Combat Desertification (UNCCD). Land degradation is a serious and significant problem in Caribbean SIDS. Based on the differences between the countries there is no regional mechanism for coordinating land management issues at the regional level and this provides a void which the PISLM attempts to fill. Concerted efforts need to be given to the implementation of the UNCCD as one of the major problems identified in Caribbean Small Island Developing States (SIDS) is the difficulty in implementing multi-lateral environmental agreements (MEAs).

The PISLM was developed based on the recognition of the need for a systematic approach to the implementation of the Convention and it aims to promote synergistic approaches with other areas such as biodiversity conservation, climate change, disaster management and sustainable rural/agricultural practices with particular linkages with food security and poverty alleviation. The PISLM consists of various components, one of which is capacity development for sustainable land management. This essentially aims at enhancing the capacity of stakeholders to plan and implement activities and undertake tasks that addresses land degradation and contribute to sustainable land management.

Keywords: Partnership Initiative on Sustainable Land Management, the United Nations Convention to Combat Desertification (UNCCD), land degradation, sustainable land management, desertification, capacity development, Caribbean Small Island Developing States (SIDS), implementation

Background on the PISLM

The importance of integrated land use management in most Small Island Developing States (SIDS) is underscored by competing economic activities for the limited land space of these countries. As clearly outlined by the Programme of Action for the Sustainable Development of SIDS (BPOA), most aspects of environmental management in SIDS are directly dependent on, and influenced by, the planning and utilization of land resources, which in turn is intimately linked to coastal and marine management and protection. On the one hand, the Caribbean SIDS were faced with the challenge of implementing the elements of the BPOA on "Land Resources" and on the other their obligations under the United Nations Convention to Combat Desertification (UNCCD) which provides the global framework for addressing land degradation and sustainable land management issues. Furthermore the official title of the UNCCD which denotes to 'combat desertification' created a question in the minds of the policy maker, with respect to its immediate relevance to the Caribbean, an issue which has been overcome by an emphasis on sustainable land management.

Faced with this dilemma of reconciling these two instruments from an operational standpoint, the decision of the Forum of Ministers of the Environment for Latin America and the Caribbean at their fourteen session held in Panama in November 2003 to establish a Caribbean SIDS Programme for Caribbean SIDS to facilitate the implementation of the BPOA, provided a context. It is against this background that a Caribbean Sub-regional Workshop on Land Degradation, which was held in Port-of- Spain, Trinidad from February 3rd – 6th 2004 decided to initiate the Partnership Initiative on Sustainable Land Management (PISLM). The Partnership Declaration of Caribbean Small Island Developing and Low-lying Coastal States adopted at the meeting called for the establishment of an Interim Sub-Regional Task Force which was later formalized in July 2008 at the Task Force Meeting for the PISLM and the First Meeting of the National Focal Points of the UNCCD in Trinidad and Tobago. To ensure complimentarity and synergy, the Partnership Initiative was incorporated as an integral part of the Technical Programme of the Caribbean SIDS Programme; an initiative adopted in Decision 4, by the Forum of Ministers in 2003[1], to support the sustainable development of Caribbean SIDS.

Another innovative aspect of the PISLM is that it provided a framework for the agencies working on land management issues in the Caribbean to harmonise their approaches and to pool resources to address the land management issues in the Caribbean. The PISLM therefore forged a cooperative relationship between several institutions, including the Global Mechanism of the UNCCD (GM/UNCCD) as the lead organization, the United Nations Environment Programme (UNEP), the Food and Agriculture Organization (FAO) of the UN, the UNCCD Secretariat, CARICOM Secretariat, and University of the West Indies, Civil Society including (RIOD), GTZ and Caribbean SIDS Country Parties of the UNCCD. The PISLM is therefore an expression of the translation of the aims of the UNCCD and the Land Resources Chapter of the BPOA into tangible deliverables.

Following an assessment period in which an analysis was undertaken of the institutional situation for land management in Caribbean SIDS the stage was set for the further consolidation of the PISLM. The first meeting of the Interim Sub-Regional Task Force for the PISLM and the extended Task Force Meeting of participating agencies and Latin American countries for enhancing south-south cooperation between LAC-Caribbean SIDS was held in Bridgetown, Barbados from May 30th to June 1st, 2005. At this meeting the Task Force suggested the establishment of a Support Office to coordinate the activities of the PISLM. It was further recommended that the Government of the Republic of Trinidad and Tobago should be approached to host the Support Office through the

---

[1] At the XIV Forum of Ministers of the Environment for Latin America and the Caribbean meeting held in Panama in November 2003.

Caribbean Network for Integrated Rural Development (CNIRD) located in Trinidad and Tobago. The rational for selecting the CNIRD is because of its mandate[2]. Furthermore, the CNIRD is the only regional institution in Caribbean SIDS that has participated and continues to participate actively in the work of the UNCCD at the national, regional and international levels. It was therefore deemed to be the most suitable regional organisation to be charged with the implementation of the PISLM. It may be considered a novelty for a civil society organization to be given the responsibility for implementing an initiative considered normally as a general rule in the Caribbean as a national responsibility.

Legislative Mandate

The Legislative Mandate of the Partnership Initiative on Sustainable Land Management (PISLM) for Caribbean SIDS is derived from a number of sources as outlined in Table 1.

Table 1: Evolution of the PISLM and its Legislative Mandate

YEAR	EVENT	HIGHLIGHT/DECISIONS/ACTIONS
Aug—Sept 1977	UNCOD, Nairobi, Kenya	Desertification addressed as a worldwide problem for the 1st time and a Plan of Action to Combat Desertification (PACD) adopted
1992	UNCED, Rio de Janeiro, Brazil	Evolution of Agenda 21 and call for the establishment of a Convention to Combat Desertification
1994	Global Conference on the Sustainable Development of SIDS, Barbados	SIDS Programme of Action (SIDS POA) developed as a key global strategy geared towards the implementation of Agenda 21
1996	UNCCD enters into force, Secretariat, Bonn Germany	Enters into force (90 days after 50th ratification was received)
Oct. 2002	GEF Operational Program on Sustainable Land Management (Opt #15), Beijing	Catalyzed a high financial demand for Sustainable Land Management activities
Nov. 2003	14th Session of the Forum of the Ministers of the Environment for LAC— Panama	Caribbean SIDS Programme formulated
Feb. 2004	Caribbean Sub-regional workshop on Land Degradation, Port of Spain, Trinidad	Formulation of the Partnership Initiative on Sustainable Land Management (PISLM) as part of the Caribbean SIDS Programme and the establishment of a Sub-Regional Task Force for the PISLM. (See Box 1 for Partnership Declaration and Decisions)
May-June 2005	Task Force Meeting for the PISLM and Extended Task Force Meeting of Participating Agencies and Latin American Countries for Enhancing South-South Cooperation between LAC-Caribbean SIDS, Barbados	The establishment of a Support Office to coordinate the activities of the PISLM; Formalizing of the Institutional Arrangements and Relations of the PISLM; The definition of the operational modalities of the Task Force; Draft Operational Guidelines for a PISLM Support Office; Review of the Components of the Partnership Initiative; The development of a work programme for the PISLM

---

[2] CNIRD's mandate is, *"To promote sustainable and environmentally sound development, through consultation with and the involvement of communities and their relevant entities, in order to improve the quality of life in rural areas and the well-being of Caribbean people"*.

YEAR	EVENT	HIGHLIGHT/DECISIONS/ACTIONS
Aug. 2005	X Regional Meeting of LAC country Parties of UNCCD, São Luís, Maranhão, Brazil	Decision 1 (b) (2) - recognize that the process within the PISLM in the Caribbean represents the conceptual framework of the Caribbean Sub-Regional Action Programmes (SRAP) for the coordination of new and existing initiatives.
Oct – Nov 2005	Fifteenth meeting of the Forum of Ministers of the Environment, Caracas, Venezuela	Additional guidance was provided, building on Decision 4 of 2003, where the meeting: To urge further development and implementation of the Caribbean SIDS programme and its continued review and assessment so that it might reflect the goals of the MSI and emerging development needs and priorities of the region
Jan-Feb 2008	The Sixteenth Meeting of the Forum of Ministers of the Environment of Latin America and the Caribbean, Santo Domingo, Dominican Republic.	The meeting: Acknowledging also the contribution made by the Government of Trinidad and Tobago, and other regional and international organizations including inter alia, the GM/UNCCD, the UNCCD Secretariat, FAO and UNEP/ROLAC to the development of the Partnership Initiative on Sustainable Land Management (PISLM) as part of the Caribbean SIDS Programme; 9. The request to UNEP, GM/UNCCD, FAO and UNCCD Secretariat to continue supporting the Partnership Initiative on Sustainable Land Management as a main component of the Caribbean SIDS Programme as a vehicle for enhancing synergistic implementation of related MEAs.
April 2008	The Council for Trade and Economic Development (COTED) [Environment], Georgetown, Guyana[3]	The Ministers provided further guidance to the PISLM by: Noted the relevant decisions of the Sixteenth Meeting of the Forum of Ministers of Environment for Latin America and the Caribbean, in particular Decision 5 of the Sustainable Development of SIDS; Acknowledged and thanked the Government of Trinidad and Tobago for hosting the Partnership Initiative on Sustainable Land Management (PISLM) Support Office through the Caribbean Network for Integrated Rural Development (CNIRD), and encouraged their continued institutional and technical support; Agreed that the PISLM should be used as the framework for the implementation of the United Nations Convention to Combat Desertification (UNCCD), and the Land Management components of the Barbados Programme of Action (BPOA) and the MSI/BPOA in Caribbean SIDS, to the extent practicable, and also urged all Member States and relevant regional and international organizations to support and participate actively in this initiative, particularly as it seeks to address issues relating to rural development and poverty alleviation in the rural sector in Caribbean SIDS.
July 2008	Task Force Meeting for the Partnership Initiative on Sustainable Land Management (PISLM) and the First Meeting of the National Focal Points of the UNCCD	The Task Force for the PISLM was formalised at this meeting in Decision 4 to constitute 14 members.

---

[3] The guidance provided by COTED [Environment] must be seen within the broader context of the Revised Treaty of Chaguaramas establishing the Caribbean Community including the Single Market and Economy which provides the legal framework within which COTED functions. The directives given by COTED are therefore of critical importance to the implementation of the Community Agriculture Policy pursuant to Article 56 to 61 of the Revised Treaty.

> **BOX 1: PISLM DECLARATION**
>
> The participants at the meeting which included, the National Focal Points of Caribbean Country Parties to the United Nations Convention to Combat Desertification (UNCCD), representatives of the CARICOM Secretariat, United Nations Multilateral Agencies, Sub-regional Development Partners, Non-Government Organizations, the RIOD Network and Academic Institutions, agreed in the Partnership Declaration of Caribbean SIDS and Low Lying Coastal States that they:
>
> *"see the opportunity and the imperative need to forge a strategic partnership in support of combating land degradation and drought in the Caribbean as part of the process of sustainable development in Caribbean Small Island Development States including Low lying Coastal States."*
>
> In this regard we adhere to the principles of the UNCCD and its fundamental links with the United Nation Framework Convention for Climate Change (UNFCCC) and the United Nations Convention on Biodiversity (UNCBD) and are in keeping with The Bonn Declaration as regards the elaboration of National Action Programmes. The Comprehensive Review of the Programme of Action for Small Island Developing States Decision 4 of the LAC XIV Forum of Environment Ministers convened in November 2003 in Panama.
>
> *Towards this end, the meeting agreed to the following:*
> 1. The establishment of an Interim Sub-Regional Task Force
> 2. Partnership among the following agencies/institutions: - The UNCCD Secretariat, The Global Mechanism, the CARICOM Secretariat, The FAO, UNEP, UWI, NGOs (RIOD), GTZ, Caribbean UNCCD Country Parties.
> 3. The establishment of a formal Sub-regional Coordination Platform

Institutional Modalities

The PISLM, as seen from its legislative mandate, receives its policy guidance from a number of diverse sources to which it has to respond as shown in Figure 1.

Ministerial Oversight

Policy directives for the CNIRD/PISLM Support Office are provided from a number of sources including (a) Forum of Ministers of Environment for Latin America and the Caribbean (b) the CARICOM Ministers of Environment (COTED) [Environment] and the Conference of the Parties of the UNCCD through its regional processes such as the Committee to Review the Implementation of the Convention (CRIC); Regional and Sub-Regional meetings of Latin America and the Caribbean, including Technical Programme Networks. The policy guidance from the various Ministerial bodies is usually provided by way of decisions of those bodies.

The Government of Trinidad and Tobago and the CNIRD/PISLM Support Office

The CNIRD/PISLM Support Office is provided by the Government of the Republic of Trinidad and Tobago through CNIRD (Figure 1). This raises the issue of the relationship between the CNIRD/PISLM Support Office and the Ministry of Planning, Housing and

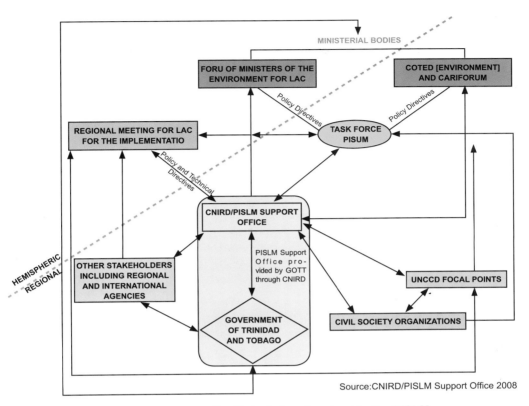

Figure 1: Inter-relationships between the Main Actors involved in the PISLM

the Environment. This relationship takes place at a number of levels, namely, at the Policy and Oversight Level, the assigning of specific projects and the means available to the Support Office to support the implementation of sustainable land management at the national level, participation on various Committees addressing land management at the national level as well as facilitating participation at international meetings. The Government of Trinidad and Tobago also provides financial assistance to the operations of the Support Office.

In terms of the provision of policy oversight, this is provided by the Ministry both in terms of ensuring the operations of the office is consistent with international norms as well as keeping the regional governments abreast of development with respect to the implementation of the PISLM. The policy oversight responsibility of the GOTT includes Support of the PISLM at the Ministerial levels including the Forum of Ministers of the Government and CARICOM Ministers of Environment COTED (Environment).

Sub-Regional Task Force

The main function of the Task Force is to provide operational policy guidance to the PISLM taking into consideration the policy directives which are provide by the various Ministerial bodies (Figure 1).

Network of UNCCD and BPOA Focal Points

This Network is a critical element of the institutional structure of the PISLM. This network comprises the Focal Points (Figure 1) in each participating country of the UNCCD Focal Points and the BPOA Focal Points with specific responsibility for the implementation of the land resources components of the BPOA. These focal points are usually from the Ministries of Environment or the Ministries of Agriculture. Irrespective of where these focal points are located it is of critical importance that there is effective coordination at both the national and regional levels.

Regional Network of Civil Society Organizations

Best practice indicates that the most effective organizations currently undertaking activities in the rural sectors in Caribbean SIDS are the civil society organizations, primarily community based organizations and Non-Governmental Organizations (NGO) (Figure 1). Given this truism, a network of civil society organizations is a necessary addition to the institutional structure of the PISLM, as a vehicle for reaching the most vulnerable in the rural economies of the Caribbean SIDS.

Network of Regional and International Agencies and Organizations

This Network comprises the regional and international institutions which are participating in the PISLM (Figure 1). These are mainly regional institutions and agencies of the United Nations System. A key role of the agencies of the United Nations System is the provision of technical assistance, including the provision of financial resources to support the implementation of the PISLM.

Implementation of the PISLM

Operationalisation of the PISLM Support Office

CNIRD facilitated the necessary discussions with the Ministry of Public Utilities and the Environment re the hosting of the PISLM Secretariat at the offices of CNIRD, under the auspices of the Government of the Republic of Trinidad and Tobago. The hosting of the Secretariat was approved by Cabinet. The PISLM Support Office became operational on December 5th, 2007 with Ms. Leandra Sebastien accepting the post of Coordinator.

Preparation of the Five Year Business Plan

In preparation for the Task Force Meeting for the Partnership Initiative on Sustainable Land Management (PISLM) and the First Meeting of the National Focal Points of the UNCCD On the implementation of the PISLM in Caribbean SIDS from July 15-17, 2008 in Trinidad and Tobago, the PISLM Secretariat worked in conjunction with an independent consultant to prepare a "Draft Five Year Rolling Business Plan for the PISLM for Caribbean SIDS 2009-2013" which was put forward for consideration at the aforementioned meeting. This business plan is currently under revision for finalisation.

Building of linkages and Exchanging of Experiences

In an effort to build linkages and exchange experiences on combating land degradation, the PISLM Coordinator will be attending the International Workshop on IEM Approaches and Application, November 2008 in Beijing China to develop relationships with PRC-GEF Partnership on Combating Land Degradation in Dryland Ecosystems and other relevant stakeholders in an effort to further consolidate the PISLM and achieve its goals and objectives in Caribbean SIDS.

The PRC-GEF Partnership on Combating Land Degradation in Dryland Ecosystems aims to use the Approaches and methodology of integrated ecosystem management (IEM) to combine the ecological, economic and societal purposes in western China and extend the application and demonstration in this regard, so as to realize poverty alleviation, and land degradation combating and biodiversity conservation and to achieve the multi-layer and cross-sector regional, national and global benefits[4].

Fund Raising for the PISLM

Ensuring the financial sustainability of the PISLM is one of the important elements underpinning the preparation of a Five Year Rolling Business Plan. The financing for the PISLM 'corporate budget' is expected to come from two main sources initially, the contribution from the Government of Trinidad and Tobago and from the participating agencies of the UN System, primarily the UNCCD Secretariat, GM/UNCCD, FAO, UNEP and UNDESA which has the responsibility for the implementation of the BPOA. The contribution of the UN Agencies will be on the basis of a shared commitment.

The Programme Budget will be funded by resources mobilized to carry out specific activities either at the national or regional levels. A target of USD 15 million in project/programme funds is being sought for the period 2009-2013. The mobilization of the resources will require the cooperation of all stakeholders of the PISLM in particular the participating Government and agencies and organizations. To facilitate the mobilization of these resources a financial strategy will be formulated.

Situation in Caribbean SIDS

Land degradation is defined by LADA (2005)[5], as the reduction in the capacity of the land to perform ecosystem functions and services (including those of agro-ecosystems and urban systems) that support society and development. Latin America and the Caribbean are actually about one-quarter desert and drylands. According to the Global Mechanism, it is estimated that the economic cost of desertification related problems in Latin America and the Caribbean countries amounts to some USD $1,000 million per year[6]. The figure soars to USD $4,800 million if drought related losses are included.

The various causes of desertification and land degradation in Caribbean SIDS are listed in Table 2 under five major categories.

---

[4] Source: The Second Announcement paper on the International Workshop on IEM Approaches and Application in Beijing, China 6-7 November, 2008 Sponsored by the State Forestry Administration, P. R. China and Co-sponsored by the Global Environment Facility and the Asian Development Bank.
[5] www.lada.virtualcentre.org LADA Secretariat FAO of the UN, Rome, Italy.
[6] See www.global-mechanism.org

Table 2: Drivers/Causes of Land Degradation

CATEGORIES	DRIVERS/CAUSES
Human Induced	Unsustainable land management practices for cropping, livestock and forests Inadequate exploitation of natural resources and Incompatible land uses Limited resources Population increase and Human settlements Poverty Land tenure security Shortages of food and fuel Mining/quarrying Poor infrastructural maintenance Inadequate waste management
Physical	Soil erosion by wind/water Compaction and Water logging Sealing and crusting
Natural	Natural hazards Climate variability (increase in temperature range, change in rainfall patterns) Fires
Chemical	Acidification Salinization/sodification Contamination (excess fertilizers, industrial/farm wastes) Nutrient depletion
Biological Degradation	Reduction in Biomass and Loss of ecosystem productivity Disruption in variety and quantity of biota (biodiversity) Decline in nutritional value for livestock and wildlife Eutrophication

Source of information: FAO (2007)

In addition to the causes outlined in Table 2, other indirect divers/causes of land degradation include: insufficient institutional capacities, poor legislation, lack of monitoring and enforcement of existing laws and limited collaboration among the State, private sector and civil society. Land degradation is a serious and significant problem in Caribbean Small Island Developing States (SIDS). The types and causes of land degradation are essentially the same throughout the islands. However, the extent and severity differs throughout the region. An insight into the land degradation in Caribbean SIDS is provided by highlighting examples in Caribbean SIDS.

The eastern region of Cuba has been feeling the effects of climate change during the recent years coupled with the inadequate use of the land which contributes to desertification. Climatic variations such as Hurricane Ivan devastated Grenada in 2004 leaving in its wake the land severely degraded. Additional activities such as deforestation, road construction, sand mining, quarrying, dumping of garbage, overgrazing, fires and inappropriate agricultural practices also contribute to Grenada's land degradation woes.

In Herberts, Antigua, the clearing of land for housing and sand mining in Barbuda contribute to the problem of land degradation in these islands. In Dominica, land degradation results from changes to both soil and vegetation conditions. Infrastructure develop-

Grenada after Hurricane Ivan 2004
Source of photo: LADA Regional Workshop (2007)

Sand Mining in Barbuda.
Source of photo: LADA Regional Workshop (2007)

ment, population growth, industrial activities such as mining and quarrying also leads to land degradation. The issue in the Dominican Republic is soil degradation which occurs in the upper river basins and watering districts. In the case of the upper river basins, the main contributing factor is deforestation which has produced a drastic forest reduction in the country during the last three decades with a decent of 41.5% of the forests during 1940 to 1998.

In Jamaica, the issue of land degradation in relation to mining remains a serious concern, where inappropriate practices have lead to the creation of numerous 'pits' while the quarrying of limestone and gypsum has resulted in the levelling of sections of hills or entire hills. Research done on the island in 2003 shows that there are nearly 600 squatter settlements island wide and this number has increased since. A large number of these settlements are on hillsides, slopes, coastal and watershed areas leading to severe cases of land degradation.

The UNCCD secretariat notes that poverty and pressure on land resources are causing land degradation in many of these areas. In a region with 465 million inhabitants, around 110 million live below the poverty line. The Convention to Combat Desertification has strong political support. All countries in the region have already joined the Convention and its issues are becoming part and parcel of national agendas on sustainable development and combating poverty. At the 25th Special Meeting of the Council for Trade and Economic Development (COTED) [Environment] from 14th – 18th April, 2008 in Guyana, the Partnership Initiative on Sustainable Land Management (PISLM) was given the additional mandate to address issues relating to rural development and poverty alleviation in the rural sector in Caribbean SIDS. The literature available all stress on the magnitude of these problems and the need to address these issues in a strategic manner.

In St. Lucia, visual evidence suggests a reduction in dry season base flows of rivers. Furthermore, there is a decline in the quantity and quality of both surface and groundwater resources. The problem of land degradation in St. Vincent and the Grenadines are similar to the other islands but with the potential for both earthquake and volcanic changes. In the Dominican Republic, Cuba, Haiti and Jamaica, there are arid zones, as erosion and water shortages are noticeably intensifying in the Eastern Caribbean.

In Haiti, the principal problems underlying the desertification process and drought are: (i) Conversion of forests, recovering secondary forest, grasslands and shade coffee to other land uses. Using forest cover as an indicator of ecosystem stability, the country's forested areas have been reduced to about 1.4%, down from 60% eighty years ago.; (ii) Inappropriate farming and grazing methods on steep lands (burning, hillside tillage, and weeding); (iii) Use of fuel wood as the primary source of energy in rural settlements. For more than 70% of the population, the main source of energy is fuel wood/charcoal (yearly use between 4.5 and 6 million m^3); (iv) Unplanned settlement patterns; (v) Limited access of farmers to appropriate technologies and insufficient awareness at all levels of sustainable land management processes; and (vi) High poverty leading to a dependency on natural resource extraction and unsustainable activities for survival.

Role of the PISLM in Caribbean SIDS

Based on the differences between the countries there is no regional mechanism for coordinating land management issues at the regional level and this provides a void which the PISLM attempts to fill. In addition, there is no regional mechanism which addresses the three Rio Conventions – UNFCCC, UNCBD and the UNCCD – far less to examine the synergies between these conventions and the implications for Caribbean SIDS. Given the fundamental nature of land resources, the PISLM provides a framework for the examination of the synergies between these conventions. Table 3 outlines the challenges in the region and the role of the PISLM for Caribbean SIDS.

The Partnership Initiative on Sustainable Land Management (PISLM) is therefore a platform for collaboration to increase coordination and facilitate the development and implementation of sustainable land management related activities in the Caribbean. The PISLM aims to promote synergistic approaches with other areas such as biodiversity conservation, climate change, disaster management and sustainable rural/agricultural practices with particular linkages with food security and poverty alleviation. It also provides a framework for Caribbean SIDS to meet their obligations to the UNCCD and the land resources section of the BPOA in a holistic fashion. This will be achieved through the execution of the various programme areas contained in the Draft Five Year Rolling Business Plan for the PISLM.

Sourcing funding for projects is always problematic and it is usually much more difficult for SIDS to obtain substantial funds to assist in meeting the various obligations to MEAs. The PISLM aims to provide a more coordinated approach to resource mobilization for the Caribbean SIDS in addressing sustainable land management. In addition, the PISLM has the cooperation from a number of UN agencies and the pooling of resources to facilitate its implementation, which from an agency perspective, forces the agencies to work together to ensure complimentarity between the various programmes in the Region.

The partnership has established a model that encompasses Government-Civil Society-International Agencies which can be produced and replicated in other programmes to assist in building capacities.

In conclusion, the region has been provided with a unique opportunity to address the

Table 3: Challenges in sustainable land management in the Caribbean and the role of the PISLM

Challenges	Role of the PISLM	Actions to be taken
Synergies between MEAs and the UNCCD	Regional /Sub-Regional Stakeholder Platform for the Implementation of the UNCCD, including building complementarities with other MEAs	Facilitate the establishment of a Regional Stakeholder Platform consisting of all relevant stakeholders in rural development for the purpose of identifying needs, priorities for investment based on the obligations contained in the various MEAs, which impact on sustainable land management Explore and develop areas of synergies between the various MEAs and the UNCCD and formulate initiatives to build on those synergies Establish a mechanism to facilitate the participation of stakeholders in the PISLM and its complementarities with other MEAs Establish mechanisms to promote and protect traditional knowledge and culture, in particular of indigenous peoples Development of SLM as a hazard mitigation mechanism tool
Research and Development, Assessment and Monitoring of Land Degradation	Development of Methods and Tools	Development of benchmarks and indicators for SLM to assess the status of land degradation in Caribbean SIDS and to develop a regional database for the storage and retrieval of data on SLM Promote the use of tools and methods for measuring land use change
Assessment of land management issues	Sub-Regional Action Programme for Caribbean SIDS	Contribute to the strategic planning, design and implementation of initiatives on Sustainable Land Management in Caribbean SIDS Develop a Sub-Regional Action Plan based on the NAPs To address land degradation/sustainable land management issues at a macro (Sub Regional) level
Building Capacity	Capacity Development	To develop, strengthen and share capabilities for the management of land and provide opportunities for alternative livelihoods To build the institutional capacities of drought and land management in Caribbean SIDS To facilitate access to appropriate technologies, knowledge and know how To foster the development of partnerships and strengthen the capacities of stakeholder to implement the UNCCD To promote education and public awareness
Technical Co-operation	South-South Cooperation	To facilitate South-South Cooperation between Caribbean SIDS To facilitate South-South Cooperation between Caribbean SIDS and Latin American Countries in SLM To facilitate the establishment of linkages between the PISLM within SIDS and other SLM countries globally
Lack of Harmonization	Harmonization of Public Policies	Mainstreaming of the UNCCD into Public Policies Contribute to the improvement of public policies dealing with and influencing the process of sustainable land management in Caribbean SIDS
Rural Development and Poverty	Rural Development and Poverty Alleviation in the Rural Sector	Design of Interventions in the Rural Sector to address Poverty Alleviation Development of alternative livelihood options as a means of relieving pressure on the environmental resources in the rural economy Promotion of Sustainable Land Management practices in the rural sector Enhancing the environmental goods and services from the environmental resource base through investing in environmental rehabilitation and restoration Strengthen food security potential of rural communities

Challenges	Role of the PISLM	Actions to be taken
Involvement of Youth and the Environment	Youth, Land Management and the Environment	Using the existing Youth infrastructure in Caribbean SIDS, develop specific initiatives to increase involvement in land and natural resources management Encourage change of attitudes among Youth with respect to the environment Target unemployed Youth to increase their involvement in sustainable land use Development of alternative livelihood options Increase awareness amongst school children about the importance of sustainable land management and the environment

Source of information: PISLM Support Office (2008) Draft Five Year Rolling Business Plan for the PISLM

woes of land degradation by working together to achieve sustainable land management. It is an opportunity which should be entirely supported, especially by building capacities among the various stakeholders, employing a systematic approach in order to achieve effective results throughout the region, particularly at the community level.

References

FAO (2007) LADA Brochure and information from the Regional Workshop on Land Degradation Assessment Methodologies in the Caribbean Barbados, 9 – 12 Oct 2007

CNIRD/PISLM Support Office (2008), "Draft Five Year Rolling Business Plan for the PISLM for Caribbean SIDS 2009-2013' for consideration by the Interim Sub-Regional Task Force/PISLM."

www.global-mechanism.org

www.lada.virtualcentre.org

www.pislmcnird.org

12. China Biodiversity Partnership Framework

Sun Xuefeng
Foreign Economic Cooperation Office (FECO)
Ministry of Environmental Protection of the People's Republic of China

1. Why do we need CBPF?

China is one of the countries with rich biodiversity. Ecosystem and biodiversity in China has contributed significantly to domestic and international economic and sustainable development with great resources and ecosystem service. However, due to over development, unsustainable land management, and climate change, natural resources have been increasingly threatened in China. With rapid development of eco-society, the biodiversity conservation in China will face not only opportunity but also threaten and various challenge, although Chinese government has taken many steps yet to protect biodiversity by establishing legal framework for natural resources protection, establishing Chinese natural reserves system, conducting research on biodiversity and fulfilling "The Convention on Biological Diversity (CBD)". Lots of national and international partners are taking action for biodiversity conservation in China. The roles and responsibilities are dispersed, sometimes overlapping and duplicating. The government commitment does not always lead to action in the field. The result has not been satisfied and many gaps, overlapping and duplicating still exist due to insufficient coordination and cooperation. An innovative concept -China Biodiversity Partnership Framework for Action, CBPF, emerged.

2. What is CBPF?

Considering all the challenges and opportunity of biodiversity conservation, a new innovation project– "China Biodiversity Partnership Framework (CBPF)" has been promoted since 2003 by the Chinese government. The project aims to establish a coherent, efficient, integrated biodiversity conservation framework by establishing a platform for efficient communication, coordination, and cooperation for all partners. The project will focus on giving coordination and strategic guidance for different biodiversity conservation activities implemented by government sectors, domestic and international donors, research institutes, local government policy makers, biodiversity managers, rural people, international partners and Non-governmental organizations (NGOs).

CBPF includes two parts, one is Partnership of Participation for Stakeholders, the other is Activity Framework for Results. CBPF will give guidance to partners' activity and investment and make them pay attention to the issues prioritized in Chinese biodiversity conservation by organizing co-participation and consultation of all partners. The purpose of CBPF is "obviously reduce the speed of biodiversity loss and effectively contribute to Chinese sustainable development".

In September, 2007, an action framework for CBPF was established after great efforts were made and an agreement was reached by specialists. These specialists come from National Development and Reform Commission, Ministry of Finance, P.R.C., the Ministry of Land and Resources, P.R.C., Ministry of Environmental Protection, P.R.C.,

Ministry of Housing and Urban-Rural Development P.R.C., Ministry of Agriculture, P.R.C., State Forestry Administration, P.R.C., State Oceanic Administration, P.R.C., and UNDP, GEF, EU, Italy, TNC. The action framework was separated to five separate topics including 27 research areas. The five topics are as follows: Improving biodiversity governance; Mainstreaming Biodiversity into Socio-Economic Sectors, Plans and Investment Decision-Making; Investing Effectively in Reducing Biodiversity loss in Protected Areas; Investing Effectively in Reducing Biodiversity loss outside Protected Areas; Cross-Cutting and CBD Emerging Issues.

3. The Goal of CBPF
• To build momentum around the programmes of the many partners, leading to a shift in the national approach to biodiversity conservation and sustainable use,
• All the partners will agree to use the Results Framework to program their biodiversity-related initiatives in China.
• Better planning, reduces duplication, increases transparency, provides a means for different agencies to coordinate actions, co-finance projects, share lessons learnt, replicate innovative techniques, etc.
• A Significant Reduction of the Rate of Biodiversity Loss as a Contribution to Sustainable Development

4. CBPF's current work and major partners
• Framework and several demo projects under Framework/GEF have been approved by Government, key international partners and the 32nd GEF council meeting in Nov. 2007
• Chinese Gov and GEF committed to finance demo actions within the framework through the GEF RAF allocation.
• The CBPF Results Framework has been adopted by MOF and GEF council meeting as their planning tool for GEF projects on Nov. 2007.
• And, the GEF chose CBPF as their 'model' partnership programme
CBPF has a total of 20 Partners at present:
• 8 from China's key line ministries -MEP, SFA, MOA, MLR, SOA, MOC, NDRC, and MOF
• 5 intergovernmental organizations -GEF, UNDP, World Bank, UNEP and Asian Development Bank (ADB)
• 3 bilateral development program (EU, Gov't of Italy and Norway)
• 3 international NGOs –The Nature Conservancy, Conservation International and WWF
• 1 intergovernmental membership organization -IUCN

5. What is next?
• The GEF Project Prodoct of Priority Institutional Strengthening and Capacity Development to Implement the CBPF will be submitted to GEF for approval in Nov, this year.
• To implement this IS project from the beginning of next year.
• Relevant management, incentive mechanisms and platform, etc. will be established with the IS project implementation.
• As the coherent activities, the other GEF and partners' projects and programs will be implemented at the same time.

CHAPTER III
Outcomes of Capacity Building to Combat Land Degradation Project

13. Application of IEM Concepts & Methods in Land Degradation Control in Western China

Li Sandan
Director General, Qinghai Forestry Bureau

Abstract: Based on ecological situation and severity of land degradation in western China, this paper points out the necessity and importance in improving ecosystem management. It also describes IEM concept and method as well as the accomplishments of GEF/OP12 in Qinghai Province. Thereafter, it summarizes the management practices in land degradation control, using the IEM concept and method. These provide examples for land degradation control to western China and even arid and semi-arid regions of less developed countries worldwide.

Key word: integrated ecosystem management, theory and practice

1. Ecological status and situation of Qinghai Province

1.1 Ecological status

Qinghai Province is located in the northeast of Qinghai-Tibetan Plateau, which is the cradle of the Yangtze River, Yellow River, and Lantcang River. It contributes to 9.25% of the water of the Yellow River, 25% of the Yangtze River, and 15% of the Lantcang come from Qinghai Province, and is famously known as the "Cradle of the Three Rivers" and the "Water Tower of China".

1.2 The situation of land degradation

Over the past decades, due to climate change and effects of human activities, the environment in Qinghai Province has deteriorited and experienced various types of problems as described below.

1.2.1 Serious soil and water loss

At present, the total area of soil and water loss in Qinghai is 33.4 million ha, accounting for 46% of the total area of the province. The sediments exported to the Yellow River reach 88.14 million ton each year, and to the Yangtze River 12.32 million ton each year. It severely affects those rivers in the middle and lower reaches.

1.2.2 Exacerbating desertification

Desertification area in Qinghai Province is 19.17 million ha, accounting for 26.5% of the total land area of the province, of which 12.558 million ha is sandified land, accounting for 17.5% of the total area in the territory.

1.2.3 Shrinking wetlands

Wetland area in Qinghai Province is 4.126 million ha, one of the top 4 provinces that are rich in wetland resource in China. Over the past 20 years, the water area has reduced by 683400 ha, as much as 21.39% of the area in the mid-1980's.

1.2.4 Severely degraded grasslands

The area of degraded grasslands reaches 16.364 million ha, accounting for 51.7% of the available grasslands. Yield of grass production has been reduced by an equivalent of 8.2 million sheep units.

1.2.5 Loss of biodiversity

Qinghai Province has abundant biodiversity. Seventy-four species of wild animals are listed as the national protected species. At present, some 15-20% of the species are threatened by deteriorating living environment. The number of individules for the most endangered animal - pushiyuan antilope - is less than 300.

1.2.6 Frequent natural disasters

Worsening ecological conditions and climate change have caused frequent natural disasters, such as floods, droughts, hails, frosts, and sand storms.

2. The necessity of implementation of the IEM concept and method

The first need is to carry out ecological rehabilitation campaigns in Qinghai Province. Years of practice shows that it is not enough to combat desertification only by increasing inputs, we need new a concept to improve management and to explore new control mode from perspetives of theory, policy, law and regulations, and technical aspects. The second need is to implement the "ecological rehabilitation strategy". Based on the local conditions, the Qinghai Provincial Government adopted the strategy of "protect ecological environment, develop ecological economy and ecological culture". It needs new theory and scientific methods as well as effective ecosystem management practices to implement the strategy. Innovative management mechanism requires introduction of advanced theory and methods.

3. Application of IEM in itemized activities

3.1 Theory and methods

3.1.1 Theory

Application of IEM theory in itemized activities is not only a new concept but also a new management strategy and method. It is also a tool for planning and assessing ecological activities. The first is integration. Based on the goal of reducing poverty and keeping the land degradation on check, different sectors and departments need to work together and cooperate closely, to develop multi-component activities and create multiple benefits. The second is participatory management. It involves community setting down land use programme and demonstration activity and experts providing advices, and community members make a decision following joint discussions, with particular emphasis on women's participation. The third is to combine land degradation control with reduction of county level poverty. We undertiook ecological rehabilitation activities, and at the same time carried out community development in order to help increasing farmers' income. The fourth is to integrate funds; by playing the corresponding roles of provincial and county administrations, to expand on the input channels, absorb more funds, and fully realize funds benefit.

3.1.2 Methods

The first is to bring in competition mechanism, choosing pilot project sites through bidding. According to the standard of the national project office for pilot project site, we chose Baiwu Jia of Minhe County, Hudan watershed of Huangyuan County and Shazhuyu of Gonghe County as pilot project sites out of 11 applications, which respectively represented three different ecosystem types in Qinghai Province. The second is to adopt participatory monitoring and assessment method. The project uses this method to evaluate the effects of activities and public satisfaction, and continuously improves project management and outcomes.

3.2 Organizational development and capability building

3.2.1 Establishment of multi-departmental cooperation and coordination mechanism

The project set up a leading group, with the group leader being the provincial vice-governor in charge of agriculture and members including 13 relavent departments and agencies. The leading group is made up of a project coordination office and a project management office, responsible for organizing, coordinating and implementing the project. According to the work requirement, we established a strategy-planning working group and a policy and law working group, consisting experts of multi-departments and multi-sectors and being responsoible for establishing integrated controlling strategy plan for land degradation at provincial level and carrying out policy and law assessment. Counties and villages of pilot sites also established corresponding groups responsible for organization, coordination, and guidance of project implementation.

3.2.2 Capability building

• Training on theory. Trainings on theory and methods were held on four occassions aiming at the provincial and county ecology and resource management departments. Totally 328 people took part in the training.

• Technological training. Various forms of trainings were held on application of computers and 3S technology, project management, project design, and policy and law in order to improve management skills and technological skills of all levels of government working staff.

• Training on environmental awareness. With the help of Australian Youth Ambassadors, the project villages made efforts to improve the environmental awareness of villagers through environmental education and activities of waste cleaning.

3.3 Project activities

3.3.1 Compilation of *Assessment Report of Land Degradation Policy and Law and Organizational Capacity of Qinghai Province*.

By assessing policy and law on controlling land degradation, measures and suggestions for improving lawmaking and policy environment were brought forward. Priority activities of improving lawmaking and policy environment were studied.

3.3.2 Drafting of *Integrated Strategy and Action Plan for Combating Land Degradation in Qinghai Province*.

Through participatory analysis of the natural environment, economy, policy and law, organization and land degradation in Qinghai Province, we drafted *IEM Strategy and Action Plan for Combating Land Degradation in Qinghai Province*, incorporating integrated ecosystem management theory and methods. It provides guidance for integrated land degradation control in the whole province.

3.3.3 Establishment of Information Center of Land Degradation Control

Affiliated with the Forest Survey and Planning Institute of Qinghai, the provincial Information Center of Land Degradation Control comes to an agreement with all relevant departments and agencies for archieving and sharing land degradation data. By providing special trainings, the center collects and stores land degradation information of the whole province, and builds information database.

3.3.4 Pilot Site Activities

3.3.4.1 For land degradation control activities, we carried out afforestation and planting of grasses on barren mountains, and closed hillsides and sandlands to facilitate vegetation restoration. A total of 49800 mu of degraded land was controlled; the acreage of grassland deratization is 22000 mu. We supported famers to build 87 sheep shelters, and carried out grazing ban and promoted stall feeding. What's more, we developed conservation tillage for demonstration.

3.3.4.2 For alternative energy. we helped formers to build methane pits, and purchased solar cookers, induction cookers, and pressure cookers for them.

3.3.4.3 Community development.

• Community infrastructure — this includes farm road construction, regulation of water discharge, and community-based cultural facilities, etc.

• Vocational technical training — upon farmers' request, we organized some vocational technical training including driving, Tibet blanket knitting, needlecraft learning and pomiculture.. It helped to transfer some surplus rural labors.

• Farmers' field school — field trips were organized to address the questions farmers and herdsmen confronted in their planting and/or breeding activities. We organized farmers and herdsmen to visit field sites and facility agricultural sites and show them how to cultivate plants in greenhouse, control pests, and implement new technology, etc.

4. Project effects

4.1 Extension and application of the integrated ecosystem management approach

Since we implemented the project in 2005, Qinghai Province has enacted six local regulations, specific regulations, and government rules and regulations, all of which fully accepting the advice of assessment reports on amendment of policies and laws. Establishment of land degradation control Strategy and Action plan with IEM concept has been used as reference for making decision by relavent departments. Some departments used to carry out ecological construction with mandatory planning, but now gradually adopt participatory management. In the pilot project villages, "participation" has been fully adopted by villagers and administrators as a decision-making mechanism of major activities.

4.2 Strenthening of coordination capability among departments

Through establishment of integrated Strategy and Action Plan for combating land degradation controlat the provincial level and assessment of policy and law, we promoted multi-department cooperation, established coordination mechanism of multi-department cooperation and data sharing mechanism. All departments at the county level cooperated closely in setting up the pilot project sites.

4.3 Combining land degradation control with improving farmers' living standard

Project activities included not only land degradation control but also community development and increase in farmers' and herdsmen's income. They helped to create multiple benefits, and facilitated a win-win outcome in improving ecological conditions and farmers' income. Establishment of pilot project sites has greatly inspired the farmers and herdsmen to participate in ecological protection and construction.

4.4 Implementation of ecological improvement

In the project area, vegetation is gradually recovering, water and soil losses are reduced, moving dunes are fixed, the damages of sands to farms, grassland and villages are mitigated, and consciousness of farmers and herdsmen on environmental protection is increased. Trainings provided strengthened villagers' environmental protection consciousness and the pride of ownership, which helped with improving the environmental condition of the project pilot sites.

4.5 Increases in farmers' income and improvement in living conditions

Based on survey, the income of farmers and herdsmen in the project area has largely increased. In the period from 2005 to 2007, the net income per capita of Shazhuyu pilot sites increased from 1268 yuan to 1390 yuan, with a net increase of 122 yuan and average annual growth rate of 4.59%, which was 2.1 percentage higher than the average growth rate of the township; the net income per capita of Baiwujia pilot site increased from 1874 yuan to 2465 yuan, with net increase of 591 yuan and average annual growth rate of 16%, which was 2 percentage higher than the average growth rate of the township. Farmers commanded new technology after taking vocational technical training, resulting in an increase in average monthly income from 600 yuan to 1500 yuan. Extension and utilization of alternative energy prevented farmers from destroying vegetation and ecological environment, reduced emission of harmful gases, and improved environmental and living conditions.

4.6 Training of management team for implementing foreign aid projects.

Many project managers and participants with advanced management skills and knowledge were appointed to be in charge of implementation of the project, to help project staff to accumulate experience of implementing external assistance projects, improve their abilities for management and decision-making and their profession, which laid good foundation for successful implemention of external cooperation projects in the future.

5. Project experience

5.1 Support of government and participation of decision-making leaders assures project success.

Implementation of the project changed the way of thinking of governmental decision makers and facilitated transformation from traditional management approach (i.e. planning management, lacking interactions with other departments and agencies.) to integrated management approach (i.e. participatory management, cooperation among multi-departments). The source area of the three major rivers in China, i.e. the Yangtze River, the Yellow River, and the Lantcang River, and Qinghai Lake watershed is currently undergoing ecological rehabilitation under the guidance of provincical ecological protection and construction leading group and the management of relavent resource departments with cooperation with other departments. It is no doubt that the effect of the IEM approach on decision-makers is great.

5.2 Project implementation is based on combining land degradation control with poverty relief.

Human being and natural resources depend on each other. We protect ecological resources and at the same time must also consider human's need. Paying attention to the two issues is an ultimate approach to control land degradation. The project started with alleviating pressures causing land degradation, and with human-oriented concept developed multi-elements activities and made efforts to improve ecological environments and alleviate poverty, The thinking mode has great practicality in our province and even in the western area of China.

5.3 IEM approach urges project fund to exert "seed effect".

Project activities need fund as support, but funding from GEF was limited and not enough for comprehensive pilot activities in project area with IEM theory. However, the advanced theory and methods of IEM have been accepted by the county and township governments of the project area and related departments of the province, and those departments and agencies provided counterpart funds to develop activities. Until the late 2007, total inputs on project activities of the pilot sites in our province were 26.9115 million yuan, of which project funds were 2.3319 million yuan, accounting for 8% of the total costs, inputs of local government and related departments were 21.6833 million yuan, accounting for 80.8% of the total costs, and funds raised by farmers were 2.9963 million yuan, accounting for 11.2% of the total costs.

5.4 Project management approach has radiating and popularizing effects.

Through the project demonstration activities, the management approach of synthetically controlling land degradation with IEM theory has been preliminarily developed in the project area. It has been popularized and applied in the project area and the surrounding areas, has and provided demonstration and reference effects in ecological protection and construction of the whole province.

5.5 Project implementation adopted competition and incentive mechanisms to achieve maximum benefits.

We selected pilot sites by bidding, which made managers better understand the idea of the project and use scientific methods, and made villagers better understand the importance of project and cherish the opportunity. We did not evenly arrange the pilot project activities, but preferentially supported those demonstration villages that have high enthusiasm, good management and good accomplishment. The method promoted effective pilot project activities and sustainable project achievements.

14. Capacity Building in Combating Land Degradation: International Application of Legal Aspects of PRC-GEF Partnership Project

Dr Ian Hannam[1]
Senior Research Fellow, University of New England, Australia

1. INTRODUCTION

This paper describes the contribution Component 1 - Improving Policies, Laws and Regulations for Land Degradation Control (hereafter "the Legal Component") of the PRC-GEF Partnership on Land Degradation in Dryland Ecosystems Capacity Building Project (Capacity Building Project) makes to international environmental law and policy in the areas of ecosystem management and land degradation (LD) control. The legal aspects of LD have in the past been generally neglected at the international level and in many of the world's regions, at the domestic level (Hannam with Boer 2002). Although China started to introduce comprehensive environmental laws in the 1980s it became apparent soon after that more specialized laws were needed to manage the ecological environment and control LD (ADB 2002). As the ADB points out, because the manner in which these early laws were developed, coupled with institutional inadequacies, they have become ineffective in controlling LD and managing natural resources (ADB 2004). The realization of this situation, combined with assistance from international agencies, enabled China to undertake the Legal Component which, on the basis of the number of laws and regulations analyzed and the new alternative frameworks that developed (including training and guidelines), could now generally be regarded as the most comprehensive national environmental law reform project for LD management undertaken anywhere in the world.[2]

1.1 China Project and International Environmental Law

The Legal Component provided China an opportunity to vastly improve the capability of its environmental laws, policies and institutional arrangements to manage LD in the dryland region. The general legislative framework, guidelines and procedures prepared under the Legal Component are transferable to other regions and countries with similar climatic, ecological and socio-economic characteristics to that of dryland China (Millennium Ecosystem Assessment, 2005). The depth and diversity of the Legal Component goes beyond any other integrated environmental law assessment program for

---

[1] Asian Development Bank Consultant, International Environmental Law and Policy Specialist, the PRC-GEF Partnership on Land Degradation in Dryland Ecosystems Capacity Building Project; Senior Research Fellow, Australian Centre for Agriculture and Law, University of New England, Armidale 2351, NSW Australia ian.hannam@ozemail.com.au

[2] Comment is based on comparison of this project with other national environmental law investigations concerning soil and land degradation carried out by the IUCN Commission on Environmental Law through its Specialist Group for Sustainable Use of Soils and Desertification.

LD control attempted in the world to date.[3] In this regard, the benefits from the Legal Component are not only to China, but to the world in general, as the knowledge gained and lessons learned can be utilized to help expand the foundations of an international front against LD control. The legislative findings of the Legal Component make a useful contribution to the objectives of many international environmental law reform initiatives and strategies for developing countries with LD and desertification problems (Hannam with Boer 2003 Section IV), but they are also of considerable value to Western counties and many outcomes of the China project could be adapted to western environmental law and institutional systems (see Chalifour et al 2007). The key aspects of the Legal Component that are relevant to international environmental law implementation and development include:

Approach and methodology – including: the ecosystem approach; role of expert legislative and policy teams; method used in legal and policy framework development; measuring legal capacity, capacity building; a model approach;

Influence on international environmental law and policy strategy – including: Montevideo Program III; World Soils Agenda; multilateral treaties; Plan of Implementation of World Summit on Environment and Development; IUCN Environmental Law Program; UNEP Strategy on Land Use management and Soil Conservation;

2. APPROACH AND METHODOLOGY
2.1 Integrated ecosystem approach

The role of the IEM concept in the Capacity Building Project is explained in Jiang Zehui 2006. An IEM approach as a key strategy and action for land use management is advocated by numerous global environmental strategies where the mandate is sustainable development and poverty reduction by focusing on specific environmental dimensions (Hannam with Boer 2003 Section II, UNEP 2004). The ecosystem approach focuses on the integrated management of land, water and living resources and promotes conservation and sustainable use of resources in an equitable way. Through the use of the ecosystem approach the Capacity Building project makes a direct link between environmental issues associated with LD, sustainable development and poverty reduction. Moreover, the implementation of the IEM approach as the dominant underlying parameter in the assessment of laws, regulations and policy, by using the 12 Principles of IEM advocated by the IUCN (The World Conservation Union) as part of the standard assessment procedure, proved very effective in detecting gaps and weaknesses in the laws, regulations and policies (IUCN 2003). The lessons learned from applying the IEM approach in the Legal Component provides a useful model for the world as its shows that the complex and dynamic nature of ecosystems can be understood more readily when all aspects of the environment are considered within the context of a legal system. It also improves the understanding of ecosystem functioning as well as providing useful

---

[3] Under 9 fields of law, the China project examined hundred's of individual laws, regulations and policies relevant to ecosystem management, land degradation and desertification control.

knowledge and suggestions for a variety of conservation approaches, societal choices and management practices between the local, provincial and national levels (Hannam 2004).

2.2 Specialist legal and policy teams

The central and provincial legal and policy investigation teams assembled for the Legal Component was one of the success stories of the project and provides other countries with useful experiences for an environmental law reform program. These teams managed the selection and assessment of legal and policy materials, prepared interpretative methods, and consulted with government officials and other specialist groups to prepare final recommendations for the provincial legal and policy frameworks. The teams comprised members from government agencies with responsibilities for nine primary law areas examined, as well as legal expertise from universities. In some cases legal expertise was drawn from the private sector. The teams worked in parallel in selecting, analyzing and assessing provincial laws and regulations, and developing the provincial legal and policy frameworks. Joint field investigations were made with scientists and ecologists from other components of the Capacity Building Project to discuss technical aspects of LD before final recommendations were made. Experiences from the team approach in China has already been used in other international environmental law reform programs including projects funded under the Global Environment Facility, United Nations Environment Program and United Nations Development Program, and in developing new foreign donor programs.[4]

2.3 Analytical method

The IEM-based method used for the legislative analysis in the Legal Component is based on the method developed from a worldwide investigation of environmental laws associated with sustainable land management and LD and applied in various jurisdictions in Asia and Europe (Hannam 2003). Lessons from the Legal Component indicate that the basic method which comprises 17 "core" legal elements could also be adapted by other countries undertaking environmental law reform, but modified appropriately (Hannam 2008). The 17 elements comprise the basic essential components of any legal and institutional system (in the form of a principle, a rule of conduct or a power to achieve a particular legal purpose) and can therefore be utilized worldwide. However, for the Legal Component, an additional 3 elements were introduced so that these 20 elements fully represented the procedures in Chinese law and they were developed through an evaluation of legal and ecological principles. It was effectively shown in the project that all 20 elements can be present among a number of individual laws in a legal system.[5] Based on this experience, the "legal elements" approach can be used by other countries to:

---

[4] Close liaison between Chinese environmental lawyers and environmental lawyers of the IUCN Commission on Environmental Law Specialist Group for Sustainable Use of Soils and Desertification passes information through the international environmental law network.

[5] The 20 elements used in the Legal Component were applied to over 200 provincial and central laws and regulations, which comprised the 'legal system' for LD.

To assess the capacity (defined below) of an existing law or regulation to meet prescribed standards of performance for the control of LD, and depending on the assessed capacity of the law to achieve these standards, additional elements can be formulated;

To guide the reform of an existing law, or to develop new legislation for LD control; each legal element must have the capacity to achieve a prescribed level of ecological management or standard for land management;

2.4 Measuring the capacity of legal and policy framework

A main task of the Legal Component, useful to other countries, is measuring the capacity of central and provincial level laws and regulations to manage LD. Capacity is determined by the number and type of essential legal elements present within relevant laws and legal instruments, in a format that enables them to achieve IEM and with the legal, administrative and technical capability within the instruments to take some form of positive action (Boer and Hannam 2003; Hannam 2004). In some instances, capacity for IEM was direct and obvious, but in other places it existed in a format that enabled some form of indirect action. Capacity is represented in the form of legal rights, the type of legal mechanisms, and importantly, the number and comprehensiveness of essential elements (Boer and Hannam 2003). The Legal Component confirms that land management issues are multi-factorial (many have a sociological, legal and technical component), and other countries would have to examine many environmental laws and regulations to effectively determine their role in managing LD issues. The Legal Component also found that many types of legal and institutional elements and mechanisms are required to manage LD, further reinforcing the necessity for other countries to analyze existing environmental laws from all jurisdictions in order to properly identify current management regimes and their interactions. The information generated by the Legal Component proved to be essential as a confident guide to the type of legislative and institutional elements that are necessary to include within a new legislative regime for LD management.

2.4.1 A legal and institutional system

The Legal Component showed that a complex legal and institutional system and organisational and operational regime for LD control can be reliably examined and the weaknesses in an organisational system and its capability to control LD can be confidently detected. The confirmation that several organisations play a role in LD control and that some would need to be partly or wholly re-organised to administer their legal responsibilities in the control of LD is useful for other countries to be aware of in planning a reform program (Hannam with Boer 2003.

2.4.2 Establishing the profile of a legislative system

The Legal Component effectively shows that individual States can adopt a variety of approaches to frame domestic legislation. Comprehensive procedures for land management can be integrated into broader legislation that sets out the responsibilities for protecting and managing the environment, e.g. forests, water, biodiversity, desertification, land management, land administration. In the practical sense, the types of legislation that may fall within the parameters of "land degradation control legislation" will often have a direct role with the management of agricultural land. Agricultural land provides the basic resource for cultivating the food and fibre products essential for humans. It is also valued for its open space, contribution to the natural environment, and for its conservation, landscape, and aesthetic values (Grossman and Brussaard 1992). The Project confirmed that "LD control legislation" plays an important role in the allocation and use of agrarian land so it is expected that other nations might consider the role and benefits of "LD control legislation" in a similar context. The approach adopted for the Legal Component was applied according to the specific administrative and procedural legislative standards of the People's Republic of China to protect and manage land, water, grasslands, and biodiversity and wetland resources. Using the China experience, individual countries would need to adopt an approach that is compatible with their administrative, institutional and legislative characteristics, and apply this as a basis for framing specialised legislation, addressing environmental management matters, or for assisting the process of integrating legislative elements for land management within an existing environmental law, or framing new environmental law (Hannam 2008).

2.4.3 Reform program and priorities

The main outcome of the Legal Component was the program for legislative reform prepared by the six provinces and regions. This amounted to a substantial report which discussed the current status of laws, regulations, policies, and institutional capacity for LD control.[6] For the practical benefit of other countries, the following experiences of the Legal Component could be applied in framing an environmental law reform program:

• Legislative area –legislative system for IEM and LD control; legal and regulatory framework of IEM and LD control at national and provincial level; legal framework for rural land-use rights; deficient elements, inconsistent interpretations, contradictory clauses in current laws and regulations; gaps in legislation; lack of judicial safeguard;

• Policy area - policies for IEM and LD control; policies for economic development, ecological construction, biodiversity protection, agriculture, forestry, grain safety and poverty relief; inadequacies in policies; coordination between policy and regulations;

• Institutional area - functions and responsibilities of institutions involved in LD control and IEM, at city, county, township level; functions and responsibilities of ministries, commissions and central government in LD control and IEM; functions of institutions in drainage areas, coordination mechanism of departments, research institutions, universities, private enterprise, NGOs, community organizations;

2.5 Capacity building experience for law, policies and institutions

For the practical benefit of other countries, the following experiences of the Legal Component could be applied in framing an environmental law reform program:

• Legal Experience - experience in implementation of law; judicial safeguards; prioritizing activities to improve legislation for LD control;

• Policy Experience - coordination of policies with laws; rationality, fairness and efficiency of policy; adherence to democratic and scientific decision-making; introducing reform in IEM; education and popularization of scientific environmental knowledge; im-

---

[6] See the 2007 provincial Legal and Policy Framework reports prepared for Shaanxi, Gansu, Xinjiang, Inner Mongolia, Qinghai, Ningxia

proving LD policy;
• Institutional experience – collective experience of institutions; experience on procedural guarantee of institutions; coordination experiences; prioritizing activities for improving capacity building of institutions with responsibilities in LD control;

2.6 Adaptability by other countries – model approach

The specific lessons learnt from the Legal Component already provide a useful model for other countries and regions to consider when developing an approach to evaluate laws, policies and institutional aspects for LD control (e.g. the Legal Component of UNEP project "Sustainable Land Management in the High Pamir-Alai Mountains" [Kyrgyzstan and Tajikistan, Central Asia] (Hannam 2005), and the UNDP Governance Project in Mongolia, are based on the Legal Component of the China Capacity Building Project, (Hannam 2008). In this regard, the following approach is a guideline for other countries to adopt for the management of LD:

2.6.1 Step 1. Preliminary
• Identify key issues of LD.
• Identify institutions relevant to LD control.
• Identify environmental law relevant to LD control at all levels and jurisdictions.

2.6.2 Step 2. Analysis
• Examine, analyze and interpret the relevant environmental law within an internationally accepted legal and institutional standard for land management.[7]
• For relevant legislation: (i) identify articles, principles or clauses relevant to LD control; (ii) categorize articles and clauses according to which "essential element" they satisfy.
• Determine the legal and institutional profile at each level; includes presence or absence of elements; the most represented elements and least represented elements for each law.

2.6.3 Step 3. Discussion, Results, Outcomes
• Determine characteristics of legal and institutional profiles, summaries and patterns.
• Determine the capacity of the legal and institutional system.
• Document the characteristics, strengths and weaknesses of laws at all levels.
• Prepare recommendations for development of policy, and land management guidelines.
• Identify areas for legislative and institutional improvement, and make suggestions for the legal and institutional reform to improve sustainable use.

3. INTERNATIONAL ENVIRONMENTAL LAW STRATEGY

The Legal Component contributes extensive knowledge to international environmental law on LD and a number of international environmental law initiatives benefit from its outcomes.

---

[7] This "standard" refers to the basic legal and institutional elements considered as essential include within the structure of an individual instrument for its effective implementation within a jurisdiction to achieve land degradation control. For the China project, this was the concept of IEM as developed by IUCN.

3.1 The Montevideo Program III

The UNEP Montevideo Program III concerns the development and review of environmental law for the first decade of the twenty-first century (UNEP 2001). The Program includes specific objectives for soils (Objective 12), forests (Objective 13), biodiversity (Objective 14), and pollution control (Objective 15) as part of a strategic environmental law program. The Program provides for the development of international agreements, international guidelines, principles and standards, and for development of capacity to formulate and implement these actions. In this context, the Legal Component makes a significant contribution to the Program, in particular by its actions to:

- Improve the effectiveness of environmental law for LD control.
- Improve the conservation and management of ecosystems in dryland areas.
- Forging better links between the environmental law for soils, forests, biodiversity, wetland, nature reserves, agriculture, grasslands, land, water and environmental management.

Various areas of the Montevideo Program, and their objectives, strategies and actions provide a list of elements against which to evaluate the Legal Component of the China Project. Moreover, many outcomes of the Legal Component make a direct contribution to the following aspects of the Montevideo Program:

3.1.1 Effectiveness of environmental law[8] – where the objective is to achieve effective implementation of, compliance with, and enforcement of environmental law and the strategy is to promote the effective implementation of environmental law through, inter alia, the widest possible participation in multilateral environmental agreements and the development of relevant strategies, mechanisms and national laws.

The following Montevideo Program III actions were undertaken by the Legal Component:

(i) Commenting on compliance with international environmental law;

(ii) Investigation of effectiveness of domestic environmental law;

(b) Identifying means to address constraints in implementing environmental law for LD control;

(c) Results of assistance being supplied to China Capacity Building Project in:

(i) Establishing and strengthening its domestic law to improve compliance with international environmental standards and enforcement of such obligations through domestic law; [9]

(ii) Developing environmental action plans and strategies to assist in the implementation of international environmental obligations;

(d) Developing advice to competent national authorities, model laws and guidance materials to implement international environmental standards;

(e) Preparing comparative analyses of compliance mechanisms, including reporting and verification mechanisms;

(f) Promoting ways to implement international environmental law standards;

---

[8] UNEP 2001 Montevideo Program III, Program area I-1 'Effectiveness of Environmental Law, Implementation, compliance and enforcement', see Actions (a)-(k) p2.

[9] The Method of analysis developed for the China project takes into consideration the objectives of international environmental law relevant to LD control.

(g) Promoting the use of civil liability mechanisms;

(h) Evaluating and promoting the wider use of criminal and administrative law in the enforcement of domestic environmental laws and standards;

(i) Exploring options for advancing the effective involvement of non-State actors in promoting implementation of, and compliance with, environmental law and its enforcement at the domestic level;

3.1.2 Capacity-building[10] – where the objective is to strengthen the regulatory and institutional capacity of developing countries, in particular the least developed and small island developing States, and countries with economies in transition, to develop and implement environmental law, and the strategy is to provide appropriate technical assistance, education and training to those concerned, based on assessment of needs.

Specific actions undertaken by the Legal Component include:

(a) Assisting the development and strengthening of domestic environmental legislation, regulations, procedures and institutions;

(b) Arranging seminars and workshops for government officials, the judiciary, the legal profession and others concerned, on environmental law and policy, including the implementation of international environmental instruments;[11]

(c) Providing training and support related to environmental law and tools of capacity-building;

(e) Promoting the teaching of domestic, international and comparative environmental law in universities and law schools, and developing teaching materials;

(f) Collaborating with governments and relevant international bodies in facilitating educational programs in environmental law at the provincial and national levels;

(g) Strengthening coordination among international organizations and institutions, including those that provide financing, on educational projects and programs related to environmental law, its implementation and enforcement and the underlying causes of environmental damage.[12]

3.1.3 Harmonization and coordination[13] – where the objective is to promote, where appropriate, harmonized approaches to the development and implementation of environmental law and encourage coordination of relevant institutions, using a strategy that promotes domestic, regional and global actions towards the development and application of appropriate harmonized approaches to environmental law and encourage coherence and coordination of international environmental law and institutions.

Specific actions undertaken by the Legal Component include:

(i) Improving environmental law standards;

---

[10] From UNEP 2001 Montevideo Program III, Program area I-2 'Capacity building', Actions (a)-(g) p3-4.
[11] Training manuals developed under the China project discuss the role of international environmental law in improving national, provincial and local Chinese laws and regulations
[12] During the course of the China project there were many meetings between project management and representatives of foreign donor organizations, Chinese institutions and organizations, to provide information on the project and interchange ideas and experiences.
[13] UNEP 2001 Montevideo Program III, Program area I- 6, 'Harmonization and coordination', see Actions (a)-(c), p7.

(ii) Promoting coherence between environmental law and other laws, both at the domestic level, to ensure they are mutually supportive and complementary;

(iii) Studying the integrated environmental policy and governmental processes;

(b) Conducting studies on the legal aspects of, obstacles to and opportunities for consolidating and rationalizing the implementation of environmental laws, to avoid duplication of their work and functions; [14]

(c) Improving ways of harmonizing and otherwise rationalizing the reporting obligations in environmental laws;

3.1.4 Innovative approaches to environmental law[15] – where the objective is to improve the effectiveness of environmental law through the application of innovative approaches and the strategy is to identify and promote innovative approaches, tools and mechanisms that will improve the effectiveness of environmental law.

Specific actions undertaken by the Legal Component include:

(a) Assessing State practice in utilizing tools such as eco-labeling, certification, pollution fees, natural resource taxes and emissions trading and assist in the use of such tools;

(b) Promoting the development and assess the effectiveness of voluntary codes of conduct and comparable initiatives that promote environmentally and socially responsible corporate and institutional behavior to complement domestic law; [16]

(c) Encouraging consideration of the use of spokesmen for environmental values and concerns, including for the interests of future generations;

(d) Studying the contribution other fields of law can make to environmental protection and sustainable development;

(e) Enhancing, through studies, the relationship of indigenous and local communities embodying traditional lifestyles to the management and protection of the environment;[17]

(f) Promoting ecosystem management in law and practice, including the valuation of services provided by ecosystems, such as environmental benefits;

(g) Encouraging the development of legal and policy frameworks in ways that benefit the environment.

3.2 Implementation of the World Soil Agenda

The Legal Component makes a significant contribution to the implementation of the 9 Agenda of the World Soils Agenda for sustainable land management (Hurni and Meyer 2002; Hannam 2006), in particular Agenda 6, in providing guidance to develop and implement national soil and water conservation policies, and Agenda 9 in providing guidance

---

[14] The China project consistently calls for improvements in China's ability to effectively coordinate major activities concerned with control of land degradation.

[15] From UNEP 2001 Montevideo Program III, Program area I- 9, 'Innovate approaches to environmental law', see Actions (a)-(g), p9.

[16] The China project consistently calls for improvements in environmental law to eliminate corruption and administrative interference in decision-making, where development interests are placed over environmental management interests.

[17] The China project calls for improvements in national and provincial environmental laws and regulations to improve the economic and social situation of disadvantaged people.

for reform of national laws and regulations for sustainable use of soil. The Agenda were prepared using international experience in all aspects of sustainable land management and are viewed as a basis for framing national natural resource management legislation (Hurni and Meyer 2002; Hurni, Giger and Meyer 2006). Comments on each Agenda are based on the experiences, outcomes and lessons learned from the Legal Component:

3.2.1 Tasks for science, monitoring and evaluation:

Agenda 1: Assessment of the status and trends of soil degradation:
- Recommended that provincial laws and regulations be amended to provide for monitoring and evaluation procedures for LD;
- Improving sustainable land management legislation to ensure problems are based on scientific opinions;
- Improving the research capabilities of legislation;

Agenda 2: Defining impact indicators and tools for monitoring and evaluation:
- Recommended that legislation be improved for research agencies to develop indicators and install monitoring systems to assess ecological sustainability;

Agenda 3: Developing principles, technologies and approaches and enabling frameworks:
- Recommending that ministries and agencies with sustainable land management responsibilities should have adequate legislative responsibility;
- Recommending that provincial legislation be reformed to enable research monitoring and evaluation to work towards developing and testing sustainable technologies, their ecological suitability, economic viability, social acceptability and institutional feasibility;

3.2.2 Tasks for policy:

Agenda 4: Identifying a multidisciplinary network;
- Recommending that legislation be improved to raise awareness among policy makers on the need to develop integrated policy and institutional structures;
- Improving legislative mechanisms to obtain advice from multidisciplinary specialists and specialist institutions and ensuring policy is compatible;

Agenda 5: Establishing specialist panels:
- To synthesize relevant information at provincial and local levels;
- To provide information on the impacts of land degradation and desertification;
- To improve policy-making at all levels to achieve sustainable land management;

Agenda 6: Providing guidance to develop and implement national integrated land management policies:
- Recommending development of national sustainable land management policies;
- Using special task forces with multiple disciplines;
- Developing a stronger legislative bases to policy;

3.2.3 Tasks for support of implementation:

Agenda 7: Promoting initiatives for sustainable land management:
- Encouraging government agencies and private enterprise to invest in sustainable land management technologies and approaches;
- Replacing incentive-based conservation projects with economically focused investment programs for sustainable agriculture;

• Assisting disadvantaged people including farmers and herders, women, minorities;

Agenda 8: Ensuring inclusion of sustainable land management in development programs:

• Recommending development cooperation agencies evaluate the impacts of their programs on natural resources;

• Developing legislation for watershed protection, mitigation of site impacts, protecting biodiversity, improving environmental education;

Agenda 9: Providing guidance for national and local action:

• Recommending that policies, projects and programs be improved at local to national levels in all stages of implementation, planning, stakeholder involvement, field activities, monitoring and impact assessment;

• Improving national research institutes to provide expertise and capacity backup;

3.3 Implementation of Multilateral Treaties

The Legal Component makes a significant contribution to the practical implementation of various multilateral environmental treaties that have a role in LD control, in particular the Convention to Combat Desertification; the Convention on Biological Diversity and the Convention on Wetlands of International Importance as Waterfowl Habitat. Each of the six provincial and regional legal and policy frameworks evaluated the provisions of these conventions in regard to their role in provincial legislation and policy reform and formulating actions to control LD. While it is regarded that these instruments have some limitations in protecting the natural environment, they do assist in promoting the management of activities that can control LD and desertification (p152-154, Boer and Hannam 2003).

3.3.1 Convention to Combat Desertification – the Legal Component indicated that the following aspects of this Convention be implemented at the provincial level (United Nations 1992):

• To implement laws, regulations and policies stipulated by the State, formulate local decrees, special decrees, government regulations and policies with the desertification conditions of the province taken into consideration;

• To implement China's State's strategy and action plan on desertification prevention, and integrate the strategy and action plan into the local ecological plans;

• To implement a monitoring system for desertification, conduct scientific research on prevention and control of desertification and introduce relevant controls;

• To control the cutting of wind breaks and sand fixing forests, and manage the closure of desertified land and protect the closed lands;

• To implement the Environment Impact Assessment Law for construction projects in sand areas, and rationally utilize and develop resources in the sand area; [18]

• To undertake publicity and educational activities, raise public awareness of the importance and necessity of preventing and controlling desertification;

---

[18] Environmental Impact Assessment Law came into force 1 September 2003

3.3.2 Convention on Biological Diversity – the Legal Component indicated that the following aspects of the Convention should be implemented at the provincial level (UNEP 1992):
• To implement the laws, regulations and policies of the State concerning protection of biological diversity, and formulate local decrees, regulations and policies;
• To implement the State strategy and action plan for protection of biological diversity, and integrate these into local economy, social development and ecological protection;
• To establish biological diversity protection zones;
• To cooperate with local citizens in the repair and restoration of ecosystems;
• To control risks caused by live organisms transformed by modern biological technology, and to avoid invasion of foreign species that may endanger existing ecosystems, populations and species, and adopt control measures;
• To encourage public participation in biodiversity activities, and evaluate the environmental impact caused by developments that endanger biodiversity;
• To undertake publicity and education, and raise public awareness of the importance of biological diversity, and respect traditional knowledge regarding biological diversity;

3.3.3 Convention on Wetlands – the Legal Component indicated that the following aspects of the Convention should be implemented at the provincial level (United Nations 1971):
• To implement the laws, regulations and policies of the State concerning protection and development of wetlands, and formulate relevant local decrees, special decrees and government regulations and policies;
• To implement the State strategy and action plan for the protection of wetlands, and integrate the strategy into local ecological protection plans;
• To compile and organize the implementation of plans for wetland preservation;
• To implement a system of wetland preservation zones, establish a province-wide network of wetland protection, conduct scientific research on wetland resources;
• Ensure the conduct of an Environment Impact Assessment and Ecology Impact Assessment System to rationally utilize the wetland resources;
• To undertake publicity and education, and raise the public awareness of importance and necessity for protection of wetland.

3.4 Implementation Plan of World Summit on Environment and Development

The World Summit on Environment and Development of 2002 (WSSD) highlighted sustainable development as a central element of the international agenda and gave new impetus to global action to fight poverty and protect the environment (WSSD 2002). One of the WSSDs major outcomes was recognition of the need to increase protection of the land as a major strategy against poverty eradication, reduce the loss of fertile soil and increase the effectiveness of use of water. The Legal Component contributes to the main objective of the WSSD Plan of Implementation by applying the IEM approach in the review of laws and regulations and outlining its role in the development of integrated land management plans, improving the productivity of land and adoption of policies and laws that commit to enforceable land and water use rights, and address security of tenure, particularly for disadvantaged people.

3.5 IUCN (The World Conservation Union) Environmental Law Program

The IUCN Environmental Law Program (ELP) advances environmental law by developing new legal concepts and instruments, and by building the capacity of societies to employ environmental law for conservation and sustainable development. Its primary areas of activity include environmental aspects of biodiversity, climate change and energy, ecosystem services, environmental governance, forests, protected areas, sustainable use of soils and water resources.[19] The Legal Component makes a significant contribution to the ELP by furthering the development of environmental law for LD control and addressing gaps and weaknesses in environmental law through its response to new environmental law challenges in China.[20] In particular, the Legal Component contributes to the specific aspects of the IUCN ELP by developing and testing a range of environmental law and policy techniques that are now available to other countries through the ELP imitative, including:

• Testing the methodology for determining the capacity of environmental law to manage widespread LD and desertification problems;

• Showing that a large mass of environmental law can be managed effectively and meaningful results can be readily obtained and interpreted;

• Developing a format to report environmental law, policy and institutional issues in comprehensive reform program;

• Successfully showing how the concept of IEM can be integrated within a regime of environmental laws, policies and institutional arrangements for LD control;

In particular, the Legal Component makes a significant contribution to the Montevideo Program III Area 5 -Strengthening and development of international environmental law through its following specific activities: [21]

(a) Undertaking assessments of existing and emerging challenges to the environment to identify gaps and weaknesses, including inter-linkages and cross-cutting issues in domestic environmental law and specifying the role in responding to those challenges;

(b) Developing criteria for determining the need for and feasibility of new domestic environmental instruments, taking into account existing instruments and practice;

(c) Reviewing the application of the principles contained in the 1972 Stockholm Declaration of the United Nations Conference on the Human Environment and the 1992 Rio Declaration on Environment and Development, and identifying the extent to which they applied nationally and disseminated the resulting information to the provinces;[22]

(d) Examining other fields of law for the purpose of identifying emerging concepts, principles and practices relevant to the implementation of environmental law for LD

---

[19] See http://www.iucn.org/themes/law/
[20] Includes verifying many aspects of the Draft Protocol for Protection and Sustainable Use of Soil, see Revised Draft 30 June 2007; the Draft was prepared by the IUCN Commission on Environmental Law Specialist Group for Sustainable Use of Soil and Desertification.
[21] UNEP 2001, Program area 1-5, 'Strengthening and development of international environmental law', see Actions (a)-(g), p6
[22] Principles and actions from these instruments were taken into account in the development of the assessment method used to analyze the provincial and central laws and regulations

control;

(f) Strengthening collaboration within the UN system as well as with other intergovernmental bodies in the development of instruments relevant to the environment and encouraging the integration of sustainable development in those instruments; [23]

(g) Encouraging efforts by academics and researchers towards better organization of international environmental law, as a step towards codification.

3.6 UNEP Strategy on Land Use Management and Soil Conservation

The UNEP Strategy outlines the critical issues in environmental assessment, policy guidance and implementation to improve the integration of environmental, land and soil aspects across other environmental focal areas and relevant international, regional and national development processes, in particular to meet the UN Millennium Development Goals (UNEP 2004). The Legal Component makes a direct contribution to many issues identified by the UNEP Strategy, including the eradication of extreme poverty and hunger, promotion of gender equality and empowerment of women, ensuring environmental sustainability and developing a global partnership for development. In particular, the Legal Component has contributed to the Objectives and Goals of the UNEP Strategy in the following specific ways:

It applies the ecosystem approach and develops inter-linkages and synergies within and across the different sectors of the administrative system. The provincial legal and policy frameworks reflect land management aspects in a functional and integrative manner and underline the ecological and sociological functioning of land resources. The Legal Component identifies many synergies, especially in environmental assessment, policy development and implementation and contributes to implementation of multilateral environmental agreements which provide opportunities to address impacts of LD (UNEP 2004, Goal and Strategy A, pp 19-38).

It recognizes that environment focused and development oriented policies on sustainable land use should be developed and implemented, and to be achieved through capacity building, information management and public participation. The Legal Component emphasizes policy development and guidance to prevent and mitigate the environmental and social impacts of LD, identifying: (a) constraints in policy, administration and culture, (b) capacity building and institutional arrangements for participatory partnerships, (c) ways access public information systems, (d) provide technical support to government and society for decision-making, and (e) mainstreaming of land issues into policies (UNEP 2004, Goal and Strategy C, p41).

It supports the use of legal processes for the integration of the environmental aspects of land management and soil conservation as a component of policy development. The Legal Component recommends a focus for policy on capacity building, environmental emergencies, developing guidelines, raising awareness, education and training. More effective legislation will play a critical role in accommodating aspects of, or implementing the UNEP Strategy at a national level, as part of an integrated approach to land

---

[23] Including Asia Development Bank, the World Bank, UNEP, AusAid

management (UNEP 2004, Goal and Strategy C, p41-42).

It recommends improvements to science-policy interaction and to strengthen knowledge systems. Dissemination of information on best practices in land management, including the development of databases, is an important component in policy implementation (UNEP 2004, Goal and Strategy D, p45).

It recommends mobilization of additional financial, institutional and human resources for LD management and more cost efficient development and implementation of policies; increasing the involvement of the private sector in the early stages of program and project development (UNEP 2004, Goal and Strategy F, p49).

4. CONCLUSIONS

As one of the most severely affected areas by LD in the world, the worsening condition of western dryland region of China is seriously impacting on its national economy and the sustainable social development of the country. These factors raised the awareness of international society to the situation in western China which led to the development of the Capacity Building Project under the framework of PRC-GEF Partnership on Land Degradation in Dryland Ecosystems. The Legal Component of this Project has shown that the legal system for control of LD comprises a complex of laws and regulations including the Chinese Constitution and many individual laws, administrative regulations, local regulations, governmental rules, and ordinances, and operates at both the national and provincial/regions level. In particular, the outcomes of the Legal Component, which assessed the capacity of existing laws and regulations to control LD, produced comprehensive and timely information not only to improve the legal and policy framework for control of LD in western China, but in doing so, makes a significant contribution to the development of international environmental law and policy for control of LD and other countries and regions of the world. The approach developed by the Legal Component has already been followed by 3 other projects/countries and is a useful model for more countries to follow. It has also been effectively shown that the Legal Component makes a significant contribution to the interpretation and implementation of many important international environmental law strategies.

5. ACKNOWLEDGEMENT

The role and contribution of Mr. Bruce Carrad (formerly Asian Development Bank, Manila) and Mr. Frank Radstake (Asian Development Bank, Manila) and various colleagues and professional acquaintances of the People's Republic of China, particularly Ms Wang Hong and Professor Du Qun (and others from the Central Project Management Office, Beijing, Research Institute Environmental Law Wuhan University, and from the six provinces and regions) to my role with the Legal Component of the PRC-GEF Partnership on Land Degradation in Dryland Ecosystems during 2004-2008, is gratefully acknowledged.

References

ADB (Asian Development Bank). 2002. Technical Assistance to the People's Republic of China for Preparing National Strategy for Soil and Water Conservation. Manila.

ADB (Asian Development Bank). 2004. Financial Arrangement for a Proposed Global Environment Facility Grant and Asian Development Bank Technical Assistance Grant to the People's Republic of China for the Capacity Building to Combat Land Degradation Project. TAR: PRC 36445. Manila.

Boer, B.W and I.D Hannam. 2003. 'Legal Aspects of Sustainable Soils: International and National', Review of European Community and International Environmental Law, Vol 12:2:149-163.

Chalifour, N.J, P, Kameri-Mbote, Lin Heng Lye and J. R. Nolon. 2007. Land Use Law for Sustainable Development, IUCN Academy of Environmental Law Research Studies, Cambridge University Press.

Grossman, M.R, and W, Brussaard. 1992. Agrarian Land Law in the Western World, C.A.B International, Wallingford, UK.

Hannam, I.D. 2003. A method to identify and evaluate the legal and institutional framework of water and land in Asia: the outcome of a study in Southeast Asia and the People's Republic of China. Research Report 73, Colombo, Sri Lanka: International Water Management Institute.

Hannam, I.D. 2004. A Method to Determine the Capacity of Laws and Regulations to Implement IEM, PRC-GEF Partnership on Land Degradation in Dryland Ecosystems, Capacity Building to Combat Land Degradation Project Component 1 - Improving Policies, Laws and Regulations for Land Degradation Control.

Hannam, I.D. 2005. Synthesis Report: the Legal, Policy and Institutional Aspects of Sustainable Land Management in the Pamir-Alai Mountain Environment. Sustainable Land Management in the High Pamir and Pamir-Alai Mountains GEF PDF-B Project United Nations University, Tokyo.

Hannam, I.D. 2008. Assessment of Environmental Laws, Strengthening Environmental Governance in Mongolia, UNDP-Netherlands Government Project Mongolia.

Hannam, I.D. 2006. Working Paper No 1: International Laws and Regulations for Soil and Water Conservation, Iimplementation of the National Strategy for Soil and Water Conservation TA 4404, Report and Recommendations on Revising the 1991 Water and Soil Conservation Law of the People's Republic of China.

Hannam, I.D., with B.W. Boer. 2002. Legal and Institutional Frameworks for Sustainable Soils. A Preliminary Report. IUCN Gland, Switzerland and Cambridge, UK, 88p.

Hurni, H, and K. Meyer (Eds). 2002. A World Soils Agenda, Discussing International Actions for the Sustainable Use of Soils, Prepared with the support of an international group of specialists of the IASUS Working Group of the International Union of Soil Sciences (IUSS), Centre for Development and Environment, Berne 63pp.

Hurni, H, Giger, M, and Meyer, K (Eds). 2006. Soils on the Global Agenda, Developing international mechanisms for sustainable land management, Prepared with the support of an international group of specialists of the IASUS Working Group of the International Union of Soil Sciences (IUSS). Centre for Development and Environment, Bern 64pp.

IUCN (The World Conservation Union). 2003, Subsidiary Body on Scientific, Technical and Technological Advice.

Jiang Zehui. 2006. Integrated Ecosystem Management, Proceedings of the International Workshop on Integrated Ecosystem Management, Beijing 1-2 November 2004, China Forestry Publishing House.

Millennium Ecosystem Assessment. 2005. Ecosystems and Human Well-being: Desertification Synthesis, World Resources Institute, Washington, DC.

United Nations. 1971. Convention on Wetlands of International Importance Especially as Waterfowl Habitat.

United Nations. 1992. Convention to Combat Desertification in those Countries Experiencing Serious Drought and/or Desertification, Particularly in Africa, Nairobi.

UNEP. 1995. Convention on Biological Diversity, Nairobi.

UNEP. 2001. Montevideo Program III - the Program for the Development and Periodic Review of Environmental Law for the First Decade of the Twenty-First Century, adopted by Governing Council of United Nations Environment Program (Decision 21/23 of the 2001).

UNEP. 2004. UNEP's Strategy on Land Use Management and Soil Conservation, a Strengthened Functional Approach, UNEP Policy Series, Nairobi.

WSSD (World Summit on Environment and Development). 2002. Plan of Implementation, United Nations.

15. Executive Summary and Overview of Strategy and Action Plan for Combating Land Degradation in Northwest Six Provinces of China

Zhang Kebin
Professor, College of Soil and Water Conservation, Beijing Forestry University
ADB consultant

1. Introduction

Facing large area of land degradation and huge economic losses and such social consequences of land degradation, Government of China has launched Western Development Strategy (WDS) in late 1990's. Under the direction provided by the WDS, the Government of the People's Republic of China (PRC) requested the Asian Development Bank (ADB) to take a lead role in facilitating the preparation of a long term PRC-Global Environment Facility (GEF) Partnership on Land Degradation in Dryland Ecosystems (the Partnership), under Operational Program (OP) 12 on Integrated Ecosystem Management (IEM). It was agreed that the Partnership would 1) tackle land degradation issues through an integrated, participatory, and cross-sectoral approach aimed at the root causes, and resolve inherently conflicting policies; 2) evolve effective mechanisms to coordinate policies, programs, and actions by various sectoral agencies operating in the areas of agricultural and rural development, land, forestry and water management, environmental protection, finance, and planning; 3) mainstream stakeholder participation and introduce effective and transparent monitoring and evaluation systems to assess the outcomes and impact of efforts to combat land degradation and reduce poverty.

Of which, *Integrated Ecosystem Management Strategies and Action plans to Combat Land Degradation* is one of the key output of project components. It asked the project province/region (which includes Shaanxi, Gansu, Ningxia, Qinghai, Xinjiang and Inner Mongolia) to formulate its own strategy and action plan for an IEM approach for land degradation control. After about two year hardworking by a group of experts from different fields and on behalf of different departments of government, the six provincial IEM strategy and action plans have been finished and it was approved by local authority or government.

The Integrated Ecosystem Management Strategies and Action Plans to Combat Land Degradation aims to (i) combat land degradation; (ii) promote productive, profitable and sustainable ecosystem management; and (iii) alleviate rural poverty. Assessment of the economic costs and benefits of land degradation control will support the strategy formulation process.

This report is the executive summary and overview of these six provincial IEM strategies and action plans and it can be thought as regional IEM strategy and action plans.

2. Background

PRC-GEF-Partnership on Land Degradation in Dryland Ecosystem (OP12) program and ADB 4357(G) Project of Capacity Building to Combat Land Degradation involve six provinces/autonomous regions which include Shaanxi, Gansu, Qinghai, Ningxia, Xinjiang and Inner Mongolia in northwest provinces in China with a total area of 4.282 million km^2, which take up 44.6% of the total land area of the whole county. The total population of six project provinces provinces/autonomous regions is about 220.8 million, which takes up about 17.0% of the country's total population.

Based on the national statistic data, project six provinces and regions are the most un-developed provinces and regions in China. In year of 2003, the total GDP of project six provinces and regions is only 844.19 billion Yuan RMB, which takes up 7.23% of the nation's GDP (compared with about 17.0% national population in six project provinces and regions, its GDP per capita is only 42.5% of national average). In 2006, China's GDP per capita was 16084 but in the project six provinces and regions, it was only 11960 Yuan RMB, which was only 74% of national average, of which, Inner Mongolia was 20053 Yuan RMB, Shaanxi 11762 Yuan RMB, Gansu 7232 Yuan RMB, Ningxia 8933 Yuan RMB, Qinghai 10085 Yuan RMB, Xijiang 13652 Yuan RMB..

Land degradation and poverty has closely relations worldwide as well as in China. The Following Fig. 1 shows the linkages between land degradation and poverty. Based on national poverty statistic data in year 2004, the total poverty population in six project areas is about 8.28 million, which takes up 31.7% of whole nation (26.10 million). Based on the state statistic data, half of poverty population in China is located in west China.

China is the socialism centralized government that most of laws were drafted by central government. At present, the laws related to combat desertification include Law on Combating Desertification, Forest Law, Land Contract Law, Land Management Law, Soil and Water Conservation Law, Water Law, Environment Protection Law, Grassland Law. These laws are related to land degradation control, forest protection and management, desertification combating, soil and water conservation, farmland protection, water resources protection and allocation, etc.. Following legislation system building and perfection, a lot of bylaws, regulations and outlines had been established in past years.

In central government of China, after signature of UNCCD in Oct. 1994, China National Committee for Implementing the United Nations Convention to Combat Desertification (CCICCD) had been established. 17 CCICCD member institutions had been

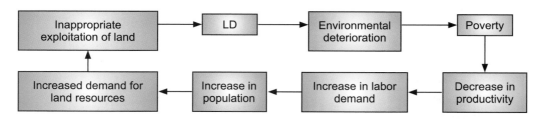

Fig. 1 Linkages between land degradation and poverty

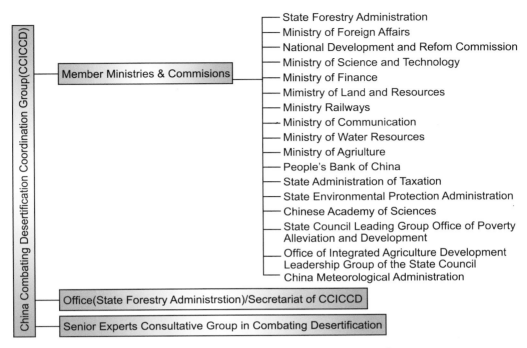

Fig. 2 Framework of institutions for combating Desertification in China

reorganized as Fig. 2. Related institutions for combating desertification had been setup in related provinces and autonomous regions.

3. Land degradation in Six Project Provinces

In this project, national data of Land degradation in project Six Provinces is from the results of SFA's national desertification monitoring. Based on national desertification monitoring (Fig.3) issued by China National Desertification Monitoring Centre, SFA, in 2005, the total area of desertification which includes wind erosion, water erosion,

Table 1 Land degradation in Northwest Project Six Provinces of China   Unit: in km²

Province	Desertification area	Wind erosion area
Shaanxi	29878	14344
Gansu	193476	120346
Ningxia	29745	11826
Qinghai	150700	52800
Xinjiang	1071583	746283
Inner Mongolia	622383	425936
Subtotal	2097765	1371535
Whole China	2636168	1739663
Ratio to whole China (%)	79	78

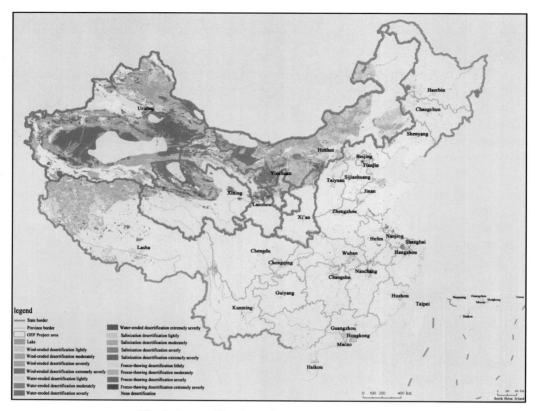

Fig. 3 Desertification of China (by SFA, 2005)

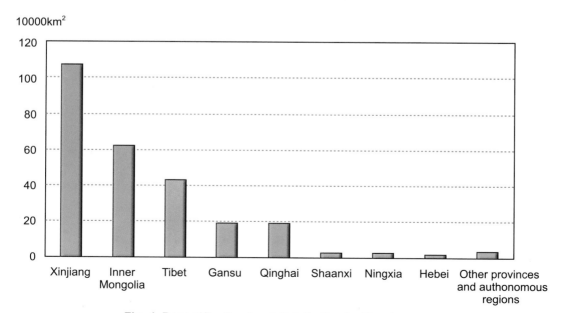

Fig. 4 Desertification land distribution by Provinces

salinazation and freeze-thawing (f-t) erosion in arid regions of six provinces is about 2.0977 million km², which takes up 79% of total desertification area of China. Of which wind erosion takes up 78% of the whole China (table 1: Land degradation in Northwest Project Six Provinces of China). Fig. 4 is the national desertification land distribution by Provinces in China.

Based on the results of the third term national desertification monitoring, desertification in China in 2004 decreased by 37924 km² compared with 1999, i.e. annually decreased 7585 km². Except Qinghai provinces, desertifications have decreased in 5 project provinces (autonomous region) both in area and degree compared with 1999, among which, Inner Mongolia has decreased by 16059 km², Xinjiang decreased by 14226 km², Ningxia decreased by 2329 km², Gansu decreased by 1900 km², Shaaxi decreased by 1257 km² (Fig. 5).

Compared with 1999, sandy desertification (desertification caused by wind erosion) has decreased by 6416 km², i.e. annually decreased by 1283 km² (Fig. 6).

Land degradation is caused by natural condition (such as climate variation and

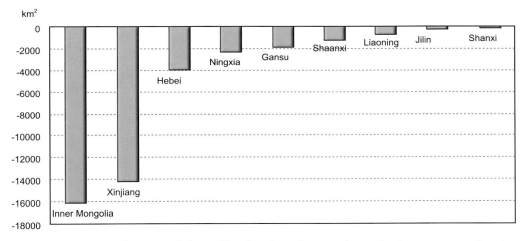

Fig. 5 Dynamic changes of desertification in major provinces (autonomous regions)

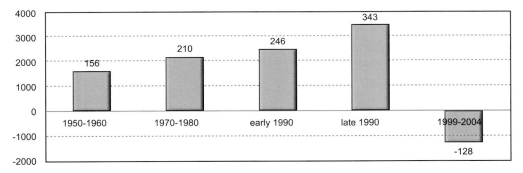

Fig. 6 Trend and dynamic of sandy desertification in China in past 50 years

easily eroded soil and sand) and human activities as well as global changes. For natural factors, dry climate, less precipitation with strong wind (gale) are the main factors in desertification. Un-rational uses of land resources, deforestation and fuel-wood collection, over-grazing, un-rational uses of water resources in dryland regions and mining are the main human factors in desertification worldwide. Emissions of greenhouse gas have led to global temperature increased in the past century and have threatened desertification worldwide.

Land degradation has caused seriously impacts in northwest of China in ecological, economic and social sectors. Research showed that the economic loss by desertification in China was 54.1 billion RMB each year in early 1990. It also showed that based on the national desertification data of 1999 and by means of hazard economic assessment methods, desertification economic loss in China in late 20 century is about 128.141 billion RMB, of which land resources loss take up 74.58 % of the total, desertification loss to productivities of agriculture and animal husbandry takes up 20.84% of the total. Studies show that the national total desertification loss takes up 1.41% of GDP in 1999 and it also takes up 23.64% of GDP of these key desertification regions of Inner Mongolia, Gansu, Xinjiang, Qinghai, Ningxia and Shaanxi Province.

In social sector, a large poor population and undeveloped economic are the result of LD in project region. The LD and poverty caused by LD have become one of the key factors effacing economic and social development in dryland of west China. As stated above that in China 50% of its poverty counties are located in west China.

4. Efforts in Land Degradation Control in Past Years

In past years, especially since the opening up policies were implemented, central government has made great efforts to combat LD. After the UN's Convention on Environment and Development in 1992, China took the lead in formulating China Agenda in 21st Century—White Paper of China's Population, Environment and Development in the 21st Century as a programmatic document to guide national economic and social development in 1994. The *National Soil and Water Conservation Outline* was proofed by state council in 1993. From the period of National Ninth Five-Year Plan to Tenth Five-Year Plan and Eleventh Five-Year Plan, the problem of LD control received great attention. *State Council's Decision on Further Strengthening the Work on Combating Desertification was issued in 2005*". In order to promote the development of China's western provinces the Chinese government put forward the West China Development Strategy in March 2000. It aims i) To narrow the economic gap between the western region and east China; ii) To ensure sustainable use and management of natural resources. Since then, many national projects related to LD control have been implemented. It involves the program of Natural Forest Protection Program, Conversion of Cropland to Forest Program, Key Protective Forest Program in Three North, Combating Desertification Program in Blown-sand Source Area Around Beijing & Tianjin, Wildlife Conservation & Natural Reserve Development Program, Soil & Water Conservation Program, Integrated Management in Yellow River and Inland Watersheds, Grassland Protection and Restoration Program, Eco-Agriculture and Countryside Energy, Integrated Agricultural Development Program, Combating Desertification and Poverty Alleviation, etc..

For local governments in six project provinces and regions, many thematic plans in combating LD have also been outlined and many project have been implemented or being implemented in combating land degradation in China and in six project provinces and regions. It involves fields or sectors of agriculture, forestry, soil and water conservation, poverty alleviation. Nation wide projects include: Natural Forest Protection Program, Conversion of Cropland to Forest Program, Key Protective Forest Program in Three North, Combating Desertification Program in Blown-sand Source Area Around Beijing & Tianjin, Wildlife Conservation & Natural Reserve Development Program, Soil & Water Conservation Program, Integrated Management in Yellow River and Inland Watersheds, Grassland Protection and Restoration Program, Eco-Agriculture and Countryside Energy, Integrated Agricultural Development Program, Combating Desertification and Poverty Alleviation, etc.. Provincial government also implemented related project in combating land degradation. Take Shaanxi province as an example, total 29 projects have been implemented since 1998 with a total investment of 75.69 billion Yuan RMB, of which 9 belongs to sector of agriculture with investment of 19.72 billion Yuan RMB, 7 in field of forestry with investment of 31.42 billion Yuan RMB, 13 in field of soil and water conservation with investment of 24.55 billion Yuan RMB. The successful experiences in LD control projects implementation include:

Great economic achievements have also been gained in control LD. Recent national desertification shows results and achievements in LDC in six project provinces and regions. Except Qinghai, LD enlargement trend in the rest 5 provinces and regions have been reversed.

5. Provincial Strategic Planning Framework

The objectives of the strategy and action plan for combating LD in project regions include (i) Formulate a policy and law framework related to LD control; (ii) Improve the capacity for utilizing the IEM approach; (iii) Enhance on-site LD control in the province and counties; (iv) Increase the provincial capacity to design investment projects for LD control; (v) Standardize the monitoring and assessment system for LD; and (vi) Establish foundation for integrated implementation of national planning framework.

Guided by use of IEM approach to the multi-dimensional problem of LD, coordination and collaboration among multiple sectors and agencies, application of the participatory approach, enabling policy and legislative environment and harmonization of the overlapping sectoral action plans, every project province/region has established its strategic goals and short term, medium term and long term objectives. Its aims are to prevent further increases in LD, reduce poverty, improve the living environment, and enhance social welfare; and achieve sustainable use of natural resources, land security, reduce the risk of natural hazards, and harmonize development with nature.

While in formulating strategy and action plan in combating LD, following key elements are carefully considered, which include: 1) Measures and proposals to improve legislative environment; 2) Options and recommendations for change within the policy environment; 3) Options and recommendations for change within the institutional environment; 4) Adoption of community-based participatory planning approaches; 5) Make the best technical demonstration through the demonstration sites; 6) The replaceable approaches for high yield, high effectiveness and sustainable living in rural areas; 7) Technical and policy options for geo-hazard/natural disaster mitigation; 8) Options and recommendations for bio-diversity preservation; 9) Measures and demand or coordination and cooperation between provinces; 10) Assessment on investment demand and source of short, middle and long terms; 11) Means of encouraging private enterprise to increase investment; 12) Cost and benefit of investment plan; 13) Monitoring and assessment index of project impact; 14) Standard of choosing preferential problems of land degradation and region; 15) Items need be considered for project design.

6. Action Plan
6.1 Priority target areas
While establishing standards for selection of priority problems and areas in the action plan in each province and region, the following factors are considered by most provinces and regions. It includes three indicators systems: a) Sensitivity indicator for the ecological environment; b) Poverty indicator, including the proportion of a population in poverty and the proportion of poverty villages and towns; c) Benefits indicator, including benefit-cost ratio, environmental benefit, social benefit, cultural benefit, and other benefits.

Shaanxi:
According to the status of degradation, and the feasibility of rehabilitation and /or actions needed to prevent degradation from occurring, the priority ranking of four eco-functional regions in Shaanxi Province (in terms of the prevention and control of LD) are: First priority - the windy and sandy grassland ecologically functional area along the Great Wall; Second priority - the key soil erosion and water loss control ecologically functional area of the Loess Plateau hills and gullies in the north part of Shaanxi Province; Third priority - the water source conservation and biodiversity protection ecologically functional area in the Qinling Mountains; and Fourth priority the Weihe River Valley Agriculture Ecologically functional area.

Counties and cities needed to take priority actions in Shaanxi: According to the flow chart which shows the priority area selection for LD, the ranking of counties that need to take priority actions on LD in Shaanxi Province is: Yuyang, Shenmu, Fugu, Hengshan, Jingbian, Dingbian, Suide, Mizhi, Jia Counties; Wuba, Qingjian, Zizhou, Zichang, Ansai, Zhidan, Wuqi, Baota Districts; and Yanchang, Yanchuan, Ganquan, Fu-xian, Luochuan, Yichuan, Huanglong, Huanglin, Yang-xian, and Foping.

Gansu:
(1) The north and the middle-east of the loess hilly area, in the middle of Gansu;
(2) The eastern Loess Plateau in Gansu;
(3) Oases and deserts in the Hexi Corridor;
(4) Alpine and wet zone on the Aerjinshan-Qilian Mountains;
(5) Gannan Plateau.
Totaled 12 counties have been selected as priority counties in combating LD. They are:
(1) Jingtai county in Baiying city;
(2) Jingyuan county in Baiying city.
(3) Huining county in Baiying city.
(4) Anding district in Dingxi Municipality.

(5) Dunhuang city in Jiuquan Municipality.
(6) Guazhou county in Jiuquan Municipality.
(7) Jinta county in Jiuquan Municipality.
(8) Minqin county in Wuwei Municipality.
(9) Gulang county in Wuwei Municipality.
(10) Dongxiang Hui Autonomous County in Linxia Municipality.
(11) Maqu County in Gannan prefecture.
(12) Danchang County in Longnan Municipality.

Ningxia: During the 11th 5-year period:

Priority Area for Soil Conservation: Supervision of soil conservation in the northern part of Ningxia. Rehabilitation of eroded land in the central part of Ningxia; Management activities for soil conservation in the southern part; The top priority will be given to soil conservation activities in the Anjiachuan watershed (Pengyang), Shuangjingzi and Yangdazi valleys (Guyuan), Majian River Basin (Haiyuan), Zhesi Valley (Tongxin) and Lanni River Basin (Xiji). The second priority for soil conservation activities will focus on parts of the central part of Mt. Helan as well as Mt. Liupan, Mt. Luoshan and Mt. Yunwu as well as Baijitan and Shapotou nature reserves.

Priority Area for Combating Desertification: The priority areas for land desertification control are the northern part of Yanchi County, the sandy lands in Lingwu County, Yueyahu (Xingqing District, Yinchuan), Shapotou and the places along the Beigan Canal (Zhongwei), Ningxia Ningdong Energy and Chemical Industry Base, and the eastern piedmont of Mt. Helan.

Priority Area for Soil Salinization Control: The priority areas for tackling soil salinization are: (i) Pingluo and Huinong Counties in the northern part of the Yinchuan Plain, (ii) Huanhe and Yongkang Townships in Zhongwei County, (iii) Nanliangtaizi (Helan), Yueyahu (Xingqing District, Yinchuan), uplands at the margin of the Huangyangtan Farm, (iv) Ganchengzi (Qingtongxia), Langpiziliang (Lingwu), and the Zhongning section of the Qingtongxia Reservoir.

Qinghai:

Priority target areas in Qinghai Province are 1) Qaidam basin sub-region; 2) Huangshui valley sub-region; 3) Lake Qinghai drainage sub-region; 4) Gonghe basin sub-region; 5) Qaidam basin sub-region; 6) Yellow River valley sub-region (Table 2).

Table 2 Priority target areas in Qinghai Province

Eco-region		Ranking
Loess hills in the east	Huangshui valley sub-region	2
	Yellow River valley sub-region	7
Qilian Mountain forest-alpine meadow	Lake Qinghai drainage sub-region	3
	Upper Heihe Catchment sub-region	6
Desert in Qaidam basin and Gonghe basin	Gonghe basin sub-region	4
	Qaidam basin sub-region	5
Sanjiang Yuan cold-alpine meadow	Sanjiang Yuan sub-region	1
	Kekexili sub-region	8

Xinjiang:
Priority target areas in Xinjiang are:
(1) Aibi Lake Wetland Ecosystem
(2) Desert Shrubland, Grassland, and Farmland and Oases Ecosystems in Southern Edge of Junggar Basin
(3) Desert Grassland, Shrubland, and Dry Farming Ecosystems in Eastern Edge of Junggar Basin
(4) Riparian Woodland, Desert Shrubland, Desert Meadow, and Desert Grassland Ecosystems on Alluvial Plains in the Tarim River Basin
(5) Oasis and Desert Grassland Ecosystems on Alluvial Plain in Southern and Northern Edges of Tarim Basin
(6) River Valley and Desert Steppe Ecosystems along the Ertis River and the Wulungu River
(7) The Cold Temperate Coniferous Forests and Alpine Steppe Ecosystems of the Altai Mountains
(8) The Alpine Cold Desert Meadow Ecosystems on the Pamirs Plateau, the Kunlun Mountain, and the Algun Mountain
(9) Mountainous Forests and Steppe Ecosystems on the Tianshan Mountain
(10) Mountainous Forest and Steppe Ecosystems on the Poluokenu and Alato Mountains

Inner Mongolia:
Priority target areas are:
1) The Integrated Controlling Area in the East of Inner Mongolia;
2) The Middle Sandstorm Source Areas in Inner Mongolia;
3) The Water and Soil Loss and the Desertification Area in the Middle and the Upper Stream of the Yellow River in Inner Mongolia;
4) Alxa Sand-dust Source Area in Inner Mongolia.

6.2 Priority component activities

Priority activities required to improve legislative environment: There is a need to improve the national, provincial and local laws and legal regulations to effectively implement the national LD control framework. These improvements include a series of activities. It involves 1) Reform and Enhancement of Relative Laws and Legal Regulations on LD Prevention by use of IEM ideas; 2) Enhancement of Functions of Provincial and Local Governments in Enforcement of Laws on LD Control; 3) Enhancement of the Reform of the provision of Long-Term Land User Rights;

Priority activities required to improve the policy environment: To establish land management policy for sustainable development, to make LDC strategy, establish poor reduction policy, widen employment opportunity and reduce population pressure of the land.

Priority activities for capacity building: The project aims to introduce IEM in natural resources management through capacity building. The expected outcomes are: (i) an improved legal and regulatory framework for LD combating; (ii) enhanced national and provincial coordination mechanisms and strengthened capacity for strategic IEM planning; (iii) improved provincial and in situ LD control; (iv) improved provincial capacity for formulating LD control investment projects; (v) improved LD monitoring and evaluation systems; and (vi) effective implementation of the National Program Framework projects. Therefore, in capacity building: 1) Cooperation of Multi-departments and Institutions; 2) Personnel Training; 3) Adequate Incentive Mechanism for Institutional Participation; 4) Encouragement of Cooperative Research amongst Institutions; 5) Technology Sharing Development for Practical Research.

Preferential activities for selecting and popularizing the best technology: After carefully selection, each provinceand region selected the best practices suitable for their provinces and regions. These practices can be divided into engineering measures, biological measures, recycle uses of natural resources and management measures. Shaanxi province has selected 31 methods for LDC. In Gansu Province 25 best practices have selected. Of the 25 best practices (techniques) those most effective for combating LD include: clay sand barrier, check dam, terrace, fuel-saving stove, artificial legume pasture, fenced pasture, solar energy stove, ponds, grassland pest insect control, water storage tanks, and household bio-gas energy.

Priority activities required to improve the provincial eco-system management and popularize the development-featured poverty alleviation: Development-oriented poverty alleviation, Pay attention to increasing the overall skills and knowledge of the masses, and build up their capability for sustainable development, to help poor households to increase their incomes in market economics, integrated and packaged inputs methods should be taken, helping poor households to build self-development capacity and facilitating the harmonious development of humans and nature, Continue with development-oriented poverty alleviation, increase financial inputs and provide social security for the poverty-stricken populations which still do not have basic food and clothing.

Priority activities required to protect and enhance biodiversity: 1) Improvement of Laws and Legislation for Natural Reserves and Enforcement Supervision; 2) Establishment and Improvement of Effective Management Mechanisms; 3) Establishing biodiversity M&E system; 4) Formulate plans for reserves, wetland protection and rehabilitation programs, and forest parks; 5) Enhancement of Publicity, Education and Training, and Establishment of Public Participation Mechanisms;

Priority activities for reducing and preventing disasters: The following activities need to be completed: (i) Strengthen publicity on anti-natural disaster activities and disseminate knowledge on disaster control; (ii) Tighten land use approval procedures in geo-hazard prone areas to ensure ongoing land utilization, and subject the land for development in such areas to geo-hazard risk assessments; (iii) Establish a natural disaster monitoring network relating to meteorology, earthquakes, and geo-hazards in the whole province, and set up a database to facilitate information-sharing; (iv) Set up a monitoring, prediction and early warning system for natural disasters; (v) Strengthen studies on geo-hazard risk assessment and risk classification, and conduct research into disaster prevention and combating systems for different regions including demonstrations on typical mud flows, desertification, and water and soil erosion control; and (vi) Raise funds to quickly resettle people from disaster-prone areas, and to implement mechanical disaster control projects in the highly threatened areas.

Strive for the Priority activities of investment: The investment framework should be as follows: (i) central government will play a leading role for some time. (ii) The provincial and municipal governments should invest a certain proportion of the funds, which could increase as the province's economy grows. (iii) Encourage private enterprises and entrepreneurs to undertake investments in LD control. (iv) Try to obtain investments from the middle and eastern provinces.

6.3 Priority project proposals

After carefully ranking, 26 project proposals have been selected for combating LD in near future, of which, 4 is in Shaanxi, 4 in Gansu, 2 in Ningxia, 4 in Xinjing, 5 in Inner Mongolia, 2 in Shanxi, 1 in Jilin, and 2 is inter-regional and should be coordinated by SFA. At present, Aibi Lake IEM project is implemented by World Bank and Ningxia IEM project and Silk Road Ecosystem Restoration Project are implemented by ADB.

16. Land Degradation Status and Control Measures in Xinjiang Autonomous Region

Cui Peiyi, Gao Yaqi and Liu Xiaofang
Xinjiang Academy of Forestry

Abstract: On account of problems in combating land degradation in Xinjiang Autonomous Region, the project adopted effective measures in land management through introducing and applying the Integrated Ecosystem Management Concept and Participatory Approach that is widely used abroad, in combination with the advanced local experiences in combating land degradation and sand prevention and control. Thereby, the project proposed the sustainable development strategies for integrated control and prevention of land degradation.
Keywords: Land Degradation Situation Analysis   Combating Strategies

Xinjiang Autonomous Region, located in the northwest border area of China, covers a vast expanse of land with an area of approximately 1.66 million km^2, about 1/6 of China's total land area. The mountainous terrain takes up about 637,100 km^2, making up 38.4% of the total land in Xinjiang, while plain area is 746,800 km^2, taking up 45.0% of the total land in Xinjiang. Specifically, the Altai Mountain Range in the northeast of Xinjiang Autonomous Region is an area distributed with Eurasian forest vegetation quite rare in China, amounting to 53,000 km^2. The Tianshan Mountain Range, spreading across the middle of the region, is one of the largest mountain systems in Asia with an area of 291,400 km^2. The Kunlun Mountain Range in the southwest is the driest mountain range in Asia, amounting to a total area of 262,400 km^2. The three mountains enclosing the Tarim Basin and the Junggar Basin form a geographic feature of "enclosure of two basins by three mountains". The Tarim Basin in the southwest of Xinjiang has a total land area of 524,400 km^2, and the Junggar Basin in the north has a total land area of 222,400 km^2, ranking second among basins in China.

Along with the economic development and the population increase, harsh natural conditions and human activities have stepped up pressures on the ecological environment, and land degradation problems (e.g. soil erosion, soil salinization, biodiversity loss, water quality deterioration and water shortage, etc.) continue to seriously impair the ecosystem, economic and social development of Xinjiang Autonomous Region. To date, 2/3 of the total land in Xinjiang is desert. About 600,000 hectares of fertile land and over 12 million people are affected by wind and sand annually, and more than 8 million ha of pasture is severely degraded. It is estimated that the direct economic loss caused by land degradation in Xinjiang amounts to RMB 9.24 billion yuan annually. Thus, there is a pressing need for a harmonious development between human and nature by exploring new measures in combating land degradation in accordance with a scientific outlook of sustainable development.

1. Main Issues and Characteristics of Land Degradation in Xinjiang Autonomous Region

Land degradation is the process in which land quality and productivity decreases under the influence of adverse natural factors and unsustainable land use by local peoples.

In terms of land characteristics, land degradation in Xinjiang can be categorized into four main types: physical degradation, chemical degradation, biological degradation, and productivity degradation. More specifically, ten subtypes of land degradation can be further classified: wind erosion, water erosion, ice erosion, gravity erosion, anthropogenic erosion, soil salinization, soil pollution, vegetation degradation, biodiversity loss, and transformation of land use types. Within the land use categroy, it has farmland degradation, pastureland degradation, forestland degradation, and wetland degradation. Generally, a certain degree of overlap exists among different categories, and land degradation in Xinjiang mainly shows in the following categories: wind erosion, soil salinization, soil pollution, vegetation degradation, and biodiversity loss.

In line with ecosystem types, land degradation can be divided on the basis of farmland ecosystem, forestland ecosystem, grassland ecosystem and wetland ecosystem. Land degradation proceeds at different rates in the four types of ecosystems.

1.1 Serious degradation of the farmland ecosystem: Xinjiang depends on irrigation-based agriculture, and its farming area mainly distributes in oasis plain areas. The oasis plain areas bear over 90% of the total population and produce 95% of the total local income, although they merely make up 8.85% of the whole dryland area in Xinjiang. The farmland ecosystem is the hub and control center of the dryland ecosystem. Since the setting-up of the 10th Five-year Plan, Xinjiang Autonomous Region has actively adjusted its agricultural industry structures, and focused to develop four bases of agriculture: grain, cotton, orchards, and animal husbandry. However, serious land degradation remains the main problem in rural areas, and the improvement of farming efficiency and increasing the income of farmers remains a large problem. Such serious land degradation typically shows in the following ways:

(1) Water resource shortage and imbalance in water distribution. There are 570 rivers of various sizes in Xijiang Autonomous Region with an amount of 87.9 billion m^3 of total surface runoff, among which an average amount of 22.62 billion m^3 runs out of the region annually. The production volume of the underground water is 15.3 billion m^3. Even though the total water resources rank the 12th in China, the average water production module is only 50,600 m^3/hm^2, ranking the third of the worst. Meanwhile, there is an extreme imbalance in water resource distribution in space and time. By 2030, the volume of usable water is predicted to be only 43.6 m^3/hm^2 although irrigated areas will reach 5,666,700 hm^2. From the perspectives of economic development, land resource development and eco-environment protection, Xinjiang Autonomous Region generally belongs to a category of resource-related water shortage.

(2) Severe natural soil erosion along with a large area eroded by wind, sand and water. Owing to influences from regional-topography and climate, deserts occur in 53 of the totally 88 cities and/or counties in Xinjiang, which consequently are designated as

key wind-sand-affected counties by the Three-North Shelterbelt Bureau. In Xinjiang, there is an erosive area of about 1.03 million km^2 accounting for 28.1% of China's total erosive areas. According to the desertification monitoring result by the State Forestry Administration from 1999 to 2004, desert land in Xinjiang Autonomous Region amounted to 1,071,600 hm^2, occupying 64.36% of the total land in Xinjiang, and areas suffering from desertification took up 21.26% of the total land in Xinjiang. Currently, 2.46% of the farmland in Xinjiang is desertified, 934,000 hm^2 is eroded by wind, and 129,000 hm^2 is eroded by water.

(3) Soil salinization being exacerbated. Serious soil salinization comes from a flucturation of underground water level and long-term irrational irrigation, which affects an area of 1,228,800 hm^2 with 148,000 hm^2 of farmland being severely damaged. In the 1960s and 1970s, soil salinization turned for worse for because of rising underground water level caused by the surface flow irrigation; later in the 1980s and 1990s, technologies were adopted to reduce damages from salinization, such as digging alkali-discharging channels, leveling the ground and carrying out furrow irrigation. In the past decade, secondary soil salinization has been primarily kept under control[1] by adopting modern water saving technologies, such as irrigation and drainage by pumped well, beneath-film drip irrigation and sprinkler irrigation. However, no fundamental breakthrough has been made for the technology of saline and alkaline elimination.

(4) Land degradation becoming worse owing to an inadequate cropping system and soil pollution. The mono-cropping agricultural practice (Fig. 1) is accompanied by a low ratio of high-efficiency cash crops, and farmers are slow to adopt the crop rotation techniques. Mono-cropping results in a decrease of land fertility and an increase of plant disease and pests. According to the statistics, the organic matter in the 0-25 cm soil layer in chernozem in Shaosu County was reduced from 16.69% to 14.14% after five years of reclamation, and then to 7.24% after 25 years of reclamation. Similarly, the organic matter in Qitai County was reduced from 18.2% to 7.24% after six years of reclamation.

Nearly 1/3 of the newly reclaimed farmlands are abandoned or wasted because of water shortage, secondary salinization, wind erosion and desertification. Vegetation on these farmlands is difficult to restore and most of the land then changes into barren

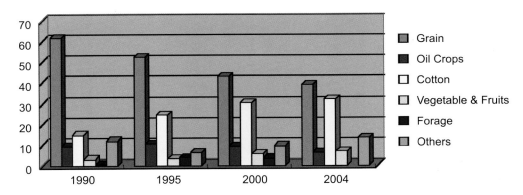

Fig. 1 Change of agricultural plantation structures

land, with some of those lands becoming new sources of deserts due to loosening soils by overgrazing. Consequently, all these phenomena directly threaten the survival and development of agriculture in the oasis areas.

The over-use of fertilizer, pesticides and plastic films gradually intensifies pollution of the farmlands (Fig. 2). According to the report issued by the Soil and Fertilizer Station of Department of Agriculture of Xinjiang Autonomous Region in 2000, the amount of plastic film residue in the region was 37.8 kg/hm^2 on average, with 52.8 kg/hm^2 in cotton fields and 22.95 kg/hm^2 in melon fields. In accordance with a correlation study between the amount of film residue and crop yield losses, the crop yield would reduce by 10% to 23% if the amount of residue films is over 58.5 kg per hectare. It can be seen that the economic loss from soil pollution is tremendous.

1.2 Decreased function of forest ecosystem. Forests in Xinjiang consisting of natural forests in mountainous areas and man-made forests in plain deserts (valleys) and oasis plain areas are important eco-shelterbelts of the region. However, spruce forests decreased by 25,000 hm^2 and larch forests by 26,000 hm^2 in Tianshan Mountains and Altai Mountains from 1950s to 1980s[1]. The natural loss of the natural forests reaches 770,000 m^3 and 959,000 m^3 respectively in Tianshan Mountains and Altai Mountains annually. The lower treeline in some parts of Tianshan Mountain has moved from 1,200

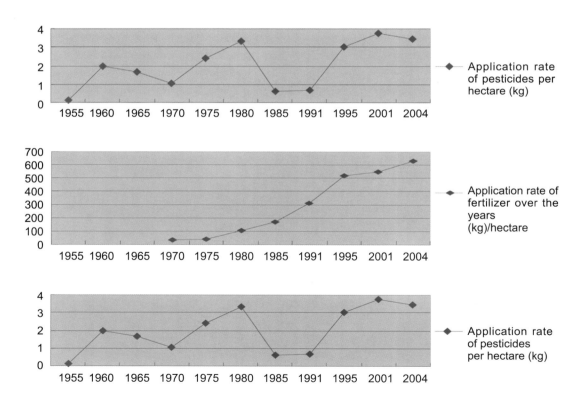

Fig. 2 Application of plastic films, pesticides and fertilizer in Xinjiang Autonomous Region over years

- 1,400 m to above 1,700 m due to decrease of the forest stand quality, canopy density and standing volume.

Along with a sharp reduction in the mountain forest areas, hilly wild fruit forest areas have also been drastically reduced. For example, Ili wild fruit (*Malus siever*) forests have shrunk from 9,300 hm^2 in 1950s to 4,000 hm^2 at present, and wild fruit forests in some places like Gongliu County and Xinyuan County have been replaced by the mountain-steppe landscape. In the meantime, the number of species is declining. For example, there are only four isolated populations for endangered species Prunus domestic, and *Juglans Regia* remains little in Daxigou Area and Xiaoxigou Area[2].

The species *Populus euphratica* on both sides of the Tarim River decreased from 460,000 hm^2 in 1950s to 175,000 hm^2 at the end of 1970s. Although a certain degree of restoration did take place, there are only 298,000 hm^2 in total currently, far less than that in 1950s. Furthermore, the species *Haloxylon ammodendron* and *H. persicum* at the Junggar Basin decreased by 80,000 hm^2 and the valley forests in Ili River and Irtysh River are only 53,000 hm$^{2[1]}$. The area of shrubbery in the desert in Xinjiang has decreased by 68.4%, and secondary valley forests are also seriously damaged due to grass collection and grazing[3].

1.3 Grassland ecosystem undergoing serious degradation. In Xinjiang, there are various types of lands, such as desert grassland, mountain grassland, alpine meadow, and marsh grassland, etc. Of these land types, the natural grassland occupies nearly 57 million hm^2, with more than 80% experiencing various degrees of degradation caused by climate change related to wind, sand and drought and/or by reclamation, overgrazing and excessive medicinal-herb digging. In particular, the severely degraded grassland accounts for over one third of the total pastureland with grass yield having dropped by 30 - 60%. According to statistics, usable grassland has been reduced by 2.4 million hm^2 in the last 40 years, and the grass yield per unit area has decreased by 30 - 50% compared with the 1960s' level. Furthermore, in some severely degraded areas the grass yield has been even reduced by 60 - 80%. Grassland degradation reduces the coverage of the dominant primary plant communities and facilitates an increase in harmful and invasive alien species, narrows the living space of pasture grasses and dwarf under-bush in a negative succession tendency, changes the physical and chemical property of grassland soils and incurs disease, pests, rats and locusts that severely damage grasslands[4].

1.4 Natural wetland ecosystems have degraded. Xinjiang used to be a region with numerous lakes and various types of wetlands. The total area of wetlands in Xinjiang was once 14.835 million hm^2, accounting for 0.8% of the entire area in Xinjiang; there were 139 lakes of the size over 100 hm^2, with total area of 970,000 hm^2, which take up 7.3% of China's total lake area and ranking the fourth nationwide as a province. However, at the end of 1970s, the lake area shrunk to 475,000 hm^2, losing almost 500,000 hm^2. Some lakes dried up in as early as 1970s and 1980s, such as the famous Lop Nur Lake (with a water surface area of 190,000 hm^2), Manas Lake (with a water surface area of 55,000 hm^2), Aydingkol Lake (with a water surface area of 12,400 hm^2, though being restored a little bit in 1998) and Taitema Lake (with a water surface area of 15,000 hm^2). Further, Ebinur Lake shrunk from 120,000 hm^2 to 53,000 hm^2.

Wetland resources have been damaged, mainly by water pollution, land reclamation, over-fishing and hunting. First, wetland degradation can be distinctly found in the succession of the wetlands, in overall succession trends from stable to unstable, from year-round to seasonal, from natural to artificial, from deep water to shallow water, from fresh water to salty water. Second, wetland degradation is related to the over-use of water resources. Since irrigation by surface water cannot meet the increasing demand of agriculture cultivation, local farmers have to turn to an alternative water resource (i.e., underground water) to satisfy water demands for expanding crop plantations. Such a transformation from the traditional surface water irrigation to pumped-well irrigation results in a sharp increase in the area of pumped-irrigated farmlands and a gradual annual increase in the use volume of underground water, therefore leading to a shrinkage of the wetlands and vegetation cover as well as a degradation of the natural ecosystems.

2. Analysis and Discussion of Causes of Land Degradation

2.1 Land Degradation Caused by Natural Factors

Xinjiang, far from the sea and having low precipitation, is typical of arid continental climate regions. The wind erosion causes loss of soil and land desertification. Sparece vegetation, rainfall and snow-melt result in the losses of the soils and their nutrients. High evaporation and a high underground water table lead to soil salinization and salt accumulation in inland rivers and lakes. Along with climate change, land degradation problems led by the adverse factors of the arid continental climate are found everywhere in Xinjiang,.

2.2 Land Degradation Caused by Human Activities

Harsh natural conditions and vulnerable eco-environments are the external factors of land degradation, while unsustainable land use and improper land management are the direct causes of land degradation.

(1) Direct causes of the farmland ecosystem degradation. In the agricultural production process, soil fertility decrease and soil nutrient imbalance are caused by inappropriate land use, such as a predatory method of "only cropping regardless of maintaining the soils" in some areas, an irrational cropping structure, an invariable crop plantation, and poor management of organic matters. Discretionary land reclamation and water resource use results in damanges in vegetation and an imbalance between water and soil. The overuse of fertilizer, pesticides and plastic films intensifies soil pollution. The lack of water efficiency in the irrigation system worsens the secondary salinization, which reduces the yield of cotton and grain by 20 - 40% or results in no harvest at all in some extreme cases.

(2) Direct causes of grassland degradation. Over the years, a sharp increase in livestock, practice of traditional grazing, and overgrazing have all contributed to land degradation. It is calculated that the theoretical carrying capacity of the grasslands in Xinjiang is 3,224.86 sheep-units (1 sheep-unit = 10,000 sheep), but the number of livestock such as cows and sheep in the region in 2004 reached 7,481.5 sheep-units, an overcapacity of 132%. As a result, high quality grasses are excessively grazed and little chance is left for grasses to restore. Normal mechanisms of vegetation and seed production are prohibited. The livestock also tramples on pastures and destroys vegetation. Soil seed

banks have proven impossible to be activated, which consequently causes a sharp decline in the resource storage and damage to the pastures.

(3) Intensive harvests of forests. The annual average lumber consumption exceeded 3 million m^3 before implementing the Natural Forest Protection Program in 1998 due to intensive over-felling, vegetation damages, inefficient prevention of deforestation for farmland, historically large area of commercial felling, road construction and urban development, etc. The timber-felling plan in mountain forest areas was between 80,000 and 250,000 m^3, but the actual consumption was 500,000 m^3. Consequently, the overfelling not only led to a decrease in forest area and the intensification in wind erosion, water erosion and salinization, but also made wild animals lose their habitats; microclimate changed and renewable resources including timber and non-timber lost their production potential.

2.3 Land Degradation Caused by Other Factors

In addition to natural and human factors, land degradation also relates to regional protection policies, social and economic status of land users, population increase, poverty conditions, the degree of dependence on eco-resources, skills of land managers, and inputs to the control and prevention of land degradation, etc. Presently, there are still 1,879,600 impoverished people in Xinjiang, and their awareness of a sustainable use of resources remains at a relatively low level due to the influence of a poor economic condition.

3. Land Degradation Control Strategies in Xinjiang

Combating land degradation is not only related to ecological conservation and restoration, but also closely related to social issues. Therefore, we must take into account both the ecology and economy and adopt an integrated treatment model that coordinates the harmonious development of the eco-environment and social economy as well as facilitates the mutual assistance of development and conservation. While focusing on an eco-treatment of the degraded lands, concerns should also be put on the follow-up industrial development and alternative livelihoods for farmers.

3.1 Land Degradation Control Projects and Financial Issues

Ecological projects under implementation have played a tremendous role in the protection and restoration of the ecological environments in Xinjiang, and accumulated rich experiences. Moreover, all departments should do their best, actively seek support from the Central Government Project, and attract investments from enterprises, which extend the scope of ecological treatment and have already achieved significant results.

(1) Over the years, industries and departments in agriculture, forestry, animal husbandry and water conservancy implemented a number of key projects related to the overall situation. Forestry departments had projects like the Natural Forest Protection Program, Conversion of Farmland to Forestry Use, the Three-North Shelterbelt Program, Wild Life and Nature Reserve Establishment Program, Public Welfare Forest Conservation through Compensation, and Fast-growth Forest Construction and Management Project, etc. Agricultural departments carred out Project on Water Saving and Fertilizing in High-quality Cotton Production Bases, Water-saving Demonstration Project for Dryland Farming, "Fertile Soil" Project, Project on Soil Fertilizer and Water Service for Cotton, etc. Water conservancy departments focus on the Integrated Eco-environ-

ment Treatment Based on the Tarim River Watershed, Water and Soil Conservation Project in Small Watershed, etc. Departments in animal husbandry had Natural Grassland Restoration and Rehabilitation Project, Natural Pasture Enclosure Project, Forage Grass-seeds Base Construction Project, and Conversion of Pasture to Grassland. Although those projects related closely with each other, there was no effective coordination and collaboration among them.

(2) On the basis of the analysis of funding sources for combating land degradation during the 10th Five-year Plan period, the Central Government invested RMB 18.5883 billion yuan in Xinjiang, the Autonomous Region invested RMB 769.01 million yuan, and governments at county level funded RMB 540.96 million yuan. It can be seen that the funds for combating land degradation in Xinjiang mainly came from the Central Government, accounting for over 90% of the total, while local governments in Xinjiang funded less than 10% of the costs, thus relying strongly on the state's investment. According to a preliminary assessment, the direct losses caused by land degradation in Xinjiang amounted to RMB 9.24 billon yuan per year, among which the direct losses by wind erosion, water erosion and soil salinization are RMB 5.83, 1.16 and 2.25 billion yuan, respectively. The funds used for combating land degradation were much less than the direct losses caused by the land degradation.

3.2 Land Degradation Control Strategies in Xinjiang

The main objectives in combating land degradation in Xinjiang should be the improvement of the local ecological environment, and increase of farmers' income and the regional economic development.

(1) Establishing an integrated ecosystem management concept. Efforts should be put into correctly dealing with ecological and service functions, and gradually transferring from a single-unit management to a multi-agency management pattern with the participation of all related departments and stakeholders. Issues on the predatory resource exploitation should be solved at the fundamental level.

(2) Guiding future eco-construction planning through the Integrated Strategies and Action Plan in Combating Land Degradation in Xinjiang. The future planning should take integrated water resource management, vegetation protection and rehabilitation, farmland ecosystem security in oasis areas, and harmonious development between human and nature as core objectives. It also needs to strengthen legal systems and straighten relationships among them so as to reflect the integrity, systematic nature and sustainability of the ecosystems. Meanwhile, great importance should be attached to multi-sector and multi-level coordination mechanisms that can facilitate inter-departmental coordination and participation. Thus, the following issues should be adequately dealt with:

• Conserving water resources, and promoting a recycling economy. Active efforts should be made in promoting modern and new technologies in water conservation, dramatically reducing the irrigation quota, and meeting water consumption in agriculture and ecological requirement with agricultural water conservation as a key.

• Prohibiting illegal forest and grass-collection and reclamation. It is necessary to strictly control the non-agricultural land occupation, implement the "Fertilizing Soil" Project, strengthen land arrangement and re-reclamation, and encourage construction projects in non-farmland areas like the Gobi desert and wasteland to ensure a gradual increase in the quality and quantity of the farmlands.

• Strengthening the integrated development and sustainable resource management, firmly prohibiting excessive and unauthorized mining, and preventing improper mining. Great attention should be paid to stick to the principle of compromising conservation and development, rational use and effective protection of mining resources, and preventing new human-induced destruction of resources and soil and water erosion.

• Strengthening eco-construction and environment protection to promote a harmonious relationship between human and nature. Enforcement should be put on the control of water and soil losses, and the integrated treatment of small watersheds. Meanwhile, eco-restoration in seriously degraded eco-environment areas should start from poverty alleviation and reducing people's dependence on, and damage to natural resources in order to ensure a smooth operation of their eco-functions.

(3) Adopting Integrated Ecosystem Management Concept to expedite the establishment of an eco-compensation mechanism, comprehensively implement the Forest Ecological Benefits Compensation System, and speed up the pace in restoring eco-functions in accordance with the regionalization (eco-sensitive region, prioritized treatment region, and privileged program region) by the IEM Strategies and Action Plan for Land Degradation Control in Xinjiang. It is necessary to continuously implement the Three-North Shelterbelt Program, Conversion of Farmland to Forest (Grassland) Use, Conversion of Pasture to Grassland, and Natural Forest and Grassland Protection Program, to complete the integrated ecosystem treatment of Tarim River watershed, and to carry out the eco-treatment around Tarim Basin, sand control and prevention in southern areas of Junggar Basin, and integrated watershed treatment in Ebinur Lake. It is also necessary to enforce priority projects such as the enclosure and protection of the vegetation in deserts, take an emphasis on nature reserve construction and wild animal conservation, and build up the green eco-shelterbelts.

References

[1] Fan Zili, Hu Wenkang et al. The ecological environment problems and protection management [J]. Journal of Arid Geography, 2003, 23 (4): 298-303.

[2] Bai Ling, Yan Guorong, Xu Zheng. Yili wild fruit tree species diversity and protection [J]. Arid Zone Research, 1998, 15 (3): 10-13.

[3] Qian Yi. Xinjiang biodiversity and its protection strategies. Journal of Xinjiang Agricultural University, 2001, 24 (1): 49-54.

[4] Cui Guoyin. Analysis on resources and degradation of marshy grassland and meadow in the north slope of Tianshan Mountains in Xinjiang. Agricultural Sciences, 2008, 45 (2): 347-351.

17. Participatory IEM Approach Used in Land Degradation Control in the Community: a Review on Mandulahu Gacha (Village) Pilot Site in Naiman Banner Inner Mongolia

Gao Guiying
Senior Engineer General, Forestry Bureau of Inner Mongolia Autonomous Region

Abstract: Since the pilot site in the village of Mandulahu, Naiman County of Inner Mongolia was initiated several years ago, evident accomplishments worthy of extension have been achieved in fulfilling its goals, though the project had some difficulties at the beginning as it was hard for the rural households to change their attitude toward accepting the IEM concept and to actively participate in the project. The IEM concept includes innovative ideas, identification of the starting point, explicit background information of the project, assurance of the participating people, fair opportunity of participation, flexible arrangement, provision of extensive training, and catering to the need of the local people, etc. The accomplishments of the project include conservation farming, testing of soil properties prior to fertilization, water-saving irrigation, utility of solar hot water tank and solar stove, extension on use of compact fluorescent lamps (CFLs), sorting of domestic wastes, and utilization of biogas in winter, in Naiman. One of the noteworthy accomplishements and highlights of the project is that the IEM concept has been integrated with former Ecology-Economy mode established for combating land desertification, so that not only Recycle Economy and sustainable theory were sufficiently taken into account in the project but also IEM concept was innovated and improved in theory as well as improved and extended in practice. Apart from the above mentioned accomplishments, the advantage of participatory management has been sufficiently exerted so that the IEM concept has been applied in rural community activitities and deeply rooted in their minds in less than two years. Anyway, the accomplishments of the project would not be made without supporting from the national and the local governments and hard works from the coordinators at county level. The concepts of IEM and land degradation control have been widely accepted by the rural community and will be widely applied to guide the production activity in the near future.

Key words: land degradation, Integrated Ecological Management, participatory, communities

The PRC-GEF Partnership on Land Degradation in Dryland Ecosystems selected 18 pilot sites in 6 Project Provinces/Regions for its community capacity building, aiming to develop an operational model for land degradation control at the village level. Mandulahu Gacha (Village), Baiyintala Sumu (Township) in Naman Banner (County) Inner mongolia is one of these pilot villages

Covering a total area of 6167.4 hectares, among which 466.7 hectares is cultivated land, Mandulahu Gacha has 214 households with a population of 840. The annual income per capita was 1800 Yuan, leaving the Gacha in one of the national poor counties. Located in the center of Horqin Desert, Mandulahu Gacha suffered from severe wind and sandy hazards due to serious drought caused by few rainfalls all year round. Land degradation in this area is severe and has become a threat to sustainable development as it is inhabited by Mongolian neighborhoods who used to live a settled pastoral life style with backward production approaches and strong dependency on the vegetation. In 1985, the local people set up a "grain, forestry, livestock" ternary structure, a development model in which each household reclaimed a piece of cultivated land on its pastoral farm, built its house close by the cultivated land, grew trees and pasture and set up stalls for livestock around the land. This management model became famous worldwide and was recognized by UN officials, and has been promoted to a wider area and positive results were obtained.

A range of activities were undertaken in the GEF pilot sites. Their processes were complex and are worth studies.

I. Great efforts made for promotion of the IEM approach

Farmers were suspicious upon the Project at the beginning, just like what happened to previous projects. However, this was normal as they lacked Project knowledge. It was the most difficult time for the Project Coordinator who was Han Chinese speaking no Monglian. At this time, farmers gave no cooperation, local staff lacked enthusiasm and no progress was made. For example, 1000 copies of project material were delivered but no one showed any interest. It was decided that the Project Coordinator and Township staff worked together to largely strengthen communication and promotion of the Project approaches.

1. Increased farmers' awareness by a large number of promotion campaigns

Promotion campaigns included increasing awareness of the Project including the role the GEF Project would play, the difference between the Project and other projects, pilot site activities and who (farmers/herders) would be beneficiaries of the Project. At the same time, farmers were informed that the Project would produce technical material on plantation, livestock, environmental protection and renewable energy and deliver for free. Farmers were also informed that they were encouraged to go on study tours (20 people for each tour) and the Project would be funding.

2. A GEF Project web page was successfully set up within the Naiman Banner Government website as a result of great efforts made by the Project. The web page has been updated regularly, through which information of important events/activities was published in a timely manner. It strongly enhanced promotion of the Naiman Pilot Site and largely increased villagers' awareness of the Project and their confidence.

3. Attention and Active Coordination from Leaders

Leaders of the Forestry Bureau of the City of Tongliao and Naiman Banner Government provided strong support to coordinate relevant departments to work together on

Sandy Area IEM and Sustainable Ecological and Economic Energy Flow Model in Mandulahu Gacha GEF Pilot Site

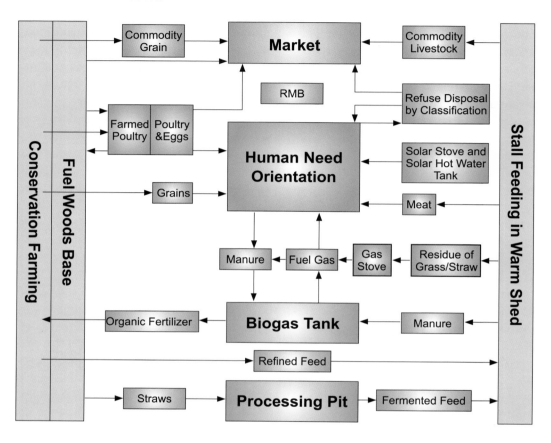

project difficulties and problems. The Banner Forestry Bureau provided nearly 40000 Yuan for the Project to purchase equipment including computers, GPSs and overhead projectors which were urgently needed by the Project for planning, training and filing.

The Agricultural and Animal Husbandry Bureau of the Banner provided new varieties and subsidy for cropping farmers. The Animal Husbandry Station and the Agricultural Extension Center seconded professionals to work for the Project. In particular, the three leaders of the Agricultural Machinery Management Station gave their best efforts in design, setting up signs/bulletin boards, commissioning machines, sowing seeds, spraying weed killers, delivering of monitoring tables and were liked by farmers.

II. Participation, Equality and Interaction

Large attention was paid to participation of farmers. They were involved in planning and implementation processes. They were empowered to take the leading role and were the main body and center of the Project and make their own decisions. Respect and equality were reflected throughout all project processes.

1. Establishment of Project Community Management Committee and Other Associations to Enhance Communication

1.1 Setting up Project Management Committee and Regulations. A Project Management Committee was set up, comprising 21 members including the Party Secretary, Village Head, Accountant, Director of All China Women's Federation, male and female heads of 5 demonstration households and male and female heads of one non-selected demonstration household. Those who had higher education level and new ideas, who showed interest in the Project and could drive other farmers to be wealthier, were accepted in the Project Management Committee. 'Equality, participation and interaction, transparency and voting must be used for decision making' were stipulated in management regulations, and several boxes including one for decision making, one for election of members of the Committee and one for collecting comments and recommendations were used. Work tasks and responsibilities were broken down into details and assigned to each demonstration household and responsible person. At the same time, indicators were identified to check performance and were open for villagers to observe, comment and record in tables. Such participatory methods used in management were new and surprising for farmers who really liked and supported it.

1.2 Establishment of Various Associations. The Project supported farmers to establish Guiwa Shawo Chicken Association, a Cattle Association and a Kashmir Goat Association, through which communication between the Project and farmers and farmers' enthusiasm were both increased.

2. Farmer Field School (FFS), an Approach for Participatory Learning

A Farmer Field School (FFS) was set up to enhance technology transferred onto farmers and farming activities, to increase exchange of experience and increase farmers' technical skills, marketing skills, legal awareness, teamwork and environmental management skills.

2.1 Applicable timing and contents, meeting farmers' needs. Participatory voting was used to identify timing, contents and methods for training to be undertaken in the FFS. It was decided that each training session should be between 2-3 hours to address a few urgent issues. Trainers should be experts and Project Coordinators, and farmers must be voluntary. Training venues should be in the field or on the farm.

2.2 New training methods, attracting participants. At the beginning of each training sessions, small presents such as towels, soups, gloves, drinking cups, pens and notebooks were prepared for participants. Sometimes, special arrangements were made such as giving participants energy efficient bulbs and red caps as presents at environmental education training sessions to increase project and environmental awareness. Such arrangements were not only complement of time, farming, gas for motorcycles lost by farmers but also ensured that they were happy to take time off work for training. Although most farmers live over 10 km from the training venue, the village committee office, over 150 participants attended each training session. At the EE training workshop held in May 2008, 10 out of the 18 participants (the required number) were women and none of them was late or quit earlier. Each participant actively spoke at the workshop, showing their teamwork and other awareness had increased obviously.

2.3 Requirements set for leaders and expert trainers. To create a relaxing and equal environment for farmers and to increase interaction, it was required that the training venue was set as a round table. Trainers, project staff and farmers became friends by sharing experience, ideas and even the same meals.

2.4 Participatory learning, a 'bottom to top' interactive training model. Interested farmers were encouraged to undertake practical operation together with experts. Participants were encouraged to raise problems and answers were given right away. Difficult problems without right-away answers were addressed later on until participants were satisfied. Such a participatory training model provided opportunities for farmers and trainers to dialogue, share experience and to benefit each other. Farmers were also encouraged to present their own stories and experience, which was good demonstration to inspire others.

III. Systems and Regulations were Developed to Improve Project Management and Promote Project Approaches

1. Project Management at Mandulahu Pilot Site

1.1 Each level had clear responsibilities and objectives, including (i) The Banner PMO was responsible for overall project management; (ii) Village leaders were responsible for implementation; (iii) Activities assigned by CPMO, PPMO and City PMO must be completed at required quantity and quality, and in a timely manner, with no argument; (iv) All villagers were beneficiaries; (v) Farmers involved in the Project must be voluntary and attention must paid to each participant; (vi) Project funds were used as incentive to encourage and promote innovative activities.

1.2 Participatory Approaches

1.2.1 A Project Management Committee was set up. A 'Voting System' was used for ranking project activities. Activities with more votes were prioritized by the Village PMO. The voting system decided project activities and their investments. Only the activities winning a greater half of the votes were accepted for investment in the project system.

1.2.2 Project activities and funds must be approved by Banner PMO.

1.2.3 Responsible persons pay for activities in advance;

1.2.4 Village PMO reported to the Project Management Committee on project activities regularly. Every payment must be explained in details and then participatory methods were used to review, question and secretly vote.

1.2.5 Payments with greater half votes were approved with signatures from Director of Village PMO and Director of Township PMO, and reimbursed.

1.3 Equality and transparency. Each one was treated equally and decision making was transparent. All information was open and every one had rights and opportunities to speak. Information was published at meetings, on posters, in materials and on notice boards to ensure transparency of the Project.

2. Regulations for the FFS

Details include: objectives, timing, contents, participants, training venue, training methods and financial source of the FFS. Training specialists were required to put themselves at the same level as farmers, treat every one equally, open information, respect different personalities and pay attention to every detail. It was also required that there should be as little instructions and limit as possible but as much suggestion and patience as possible. Special attention must be given to women, poor households and the disabled. Means must be tried to reduce their inferiority, build their courage and confidence and appreciate their advantages and potentials. It was prohibited that trainers gave orders or complaints. Special attention must be paid to 'participation'. Dialogues in an equal manner and interaction must be used to create benefit for all parties and languages used must be farmers' which was simple, direct and easy to understand. Problems to be solved must be practical. Confidence with each other was increased so every one trusted others.

3. Development of Village Regulations for Environmental Protection and LD Control

Details included: (1) Grazing was banned all over the Gacha. Stall feeding must be adopted for all livestock to reduce damages over the environment; (2) Special protection must be undertaken upon natural willows and elm trees, and pruning and felling were prohibited; (3) Application for business of forest land and grassland must be submitted for review and approval. Business operation must follow the approved time, location, area and technical process (forest and pasture cultivation). (4) All wildlife was under protection and no hunting was allowed. No damages should be made to wildlife habitats; (5) No fire was allowed in the wild. Burning of unutilized land, straws and wastes was prohibited; (6) Digging and extracting of sand and soil, which damaged the vegetation were prohibited; (7) Water-efficient irrigation and saving water must be promoted; (8) Conservation farming, little tillage or non tillage must be promoted. Autumn tillage was prohibited; (9) Using birds, insects and fungus to control pests must be promoted; (10) Recycling of wastes must be promoted.

IV. Comprehensive Coordination and Innovations Resulting in Evident Achievements

1. The high-level interest from all project stakeholders

As a result of the project promotion activities such as GEF project handouts and the visits to the pilot sites from all sides of society, there was an increasingly number of people and agencies starting to show interest in the project. The Youth League Committee of Naiman Banner used the project pilot site as the "environmental education base for teenagers in Naiman Banner" and All China Women's Federation of Naiman Banner named the project pilots as "Capacity building base for women in Naiman Banner". A large number of promotion activities and social activities were carried out through the two platforms as mentioned above. The GEF approaches were delivered through various of meetings and campaigns several times in all relevant departments and at all levels including the Party School, Forestry Bureau meetings, training courses of the Personnel Bureau, Township Government meetings and training sessions, Township middle schools and primary schools, involved government officials at all levels, farmers and school children.

The project received delegations from various units, such as Agriculture and Animal

Husbandry Department of Inner Mongolia, Inner Mongolia Agricultural University, Lanzhou Desert Research Institute, Shenyang Ecology Research Institute, Beijing Institute of geography and Remote Sensing. The experts and scholars from home and abroad showed particular interest in the application of IEM, the cycling ecology-economic system with sustainable features and the participatory project management model in which the interactive bottom-up community based participatory approaches were applied. Their perspectives towards the project model changed from being suspicious at the beginning, to surprised and convinced in the end.

2. Number of line agencies to cooperate with each other increased

The efforts made for project promotion and coordination, and the application of innovative approaches have extended cooperation among more and more line agencies. Line agencies actively participated in implementation of the project jointly. The project has incorporated more than 20 line agencies including sectors of forestry, agriculture and animal husbandry, land resources, communication department, poverty alleviation and development, water resources, education, civil affairs administration, sciences and technology.

The project endured a large number of difficulties from the very beginning, including difficulties in finance, coordination (to coordinate higher level and line agencies), inherent culture and understandings among the villagers. However, with strong sense of responsibility, the Project had successfully overcome these difficulties through flexible innovative approaches and ongoing hard work, which eventually created significant outcomes.

V. Project Resulted in Good Achievements: Improved Infrastructure and Increased Capabilities

1. Good impact of demonstration households on others and infrastructure over fulfilled

The demonstration activities included: forage and pasture development, soil improvement, soil testing before fertilizing, water-efficient irrigation, introduction of new species, conservation farming, forage pit construction, breed improvement, solar energy development and biogas. All project demonstration activities were over fulfilled.

2. The best platform for technical training: FFS, resulting in significant outcomes

Good achievements were received at FFS. Training courses were organized for various groups including engineers, technicians and farmers. 78 engineers, 288 technicians and 2430 farmers were involved. A total of 2520 copies of project materials covering the GEF project concept, conservation farming, animal raising and environmental education were delivered. Currently, the project is drafting *A Training Manual on Animal Husbandry in Warm Pens* in Mongolian language.

Free-of-charge training courses based on farming timetable were delivered by the FFS and well welcomed by the local people. Positive responses regarding the training were received from the villagers. In particular, training on conservation farming and livestock management in the warm pens were well commented. On average, each training session had more than 120 attendances. The community financed on their own a set of computer in order to provide better access to information for villagers and to keep the related record electronically. Farmers also raised fund on their own and purchased 165 sets of satellite TVs, which enhanced information flow, broadened their minds and improved their competencies.

3. Good Results from Environmental Education (EE): Increased Awareness in Energy Saving and Environmental Protection

The project design document enlisted EE as an important component and EE was ran throughout all project activities. Energy saving and environmental protection activities were promoted through all parts of the life of the community. Ideologies were penetrated into practices, following the IEM approach.

3.1 Strengthening EE to change thinking

Firstly, the Project designed and distributed "Environmental Protection Proposal". The proposal was also made into display boards and showed in the community, township government, near the major road, in middle schools, primary schools, the forestry station and Banner-level government agencies. The project concepts are presented through various meetings and training sessions. In addition, the project objectives (care the land, save energy and protect the environment, economic development and independent management), project targets (focus on farmers and herders, care the stakeholders and respect the nature), project approaches (participate equally, bottom to up, communication and interaction), and the project processes (integrated management, longer-term view, planning for future and working from the roots) were also made into signs put up in the public to scale up project impacts.

Secondly, village regulations on environmental protection and LD control were produced and distributed to every household, and was made obvious in the public by display boards.

Thirdly, Comparison analysis of local TM images since 1980 was produced, providing an opportunity for villagers to visualize the changes of ecological systems and their relations to the human induced activities. It was also good EE to develop villagers' understanding of the causes for changes in ecological systems and their impact on farming and herding activities.

3.2 Introduction and scale-up of energy saving and environmentally friendly actions

The Project promoted refuse deposal in classifications to develop such awareness among farmers: garbage is resource put into a wrong place. The differential treatment of garbage is a small activity in terms of scale. However, it well represents a great concept and is worthwhile for more attention and efforts. Many experts and scholars within and outside of the Region were amazed by such behavior as many people in the urban area fail to do this.

3.3 Saving energy and reducing emission

Solar stoves, energy-efficient bulbs, solar hot water tanks, new fuel wood/gas stoves were promoted or introduced to enhance energy saving and reduction of emission. Manuals on how to use power, fuel and water efficiently were produced and distributed. The manuals also covered disasters fireworks would bring along and the harm of wasted batteries.

4. Conservation farming

Villagers financed nil-tillage seeders on their own. A study tour to observe conservation farming at the GEF Project Site was organized by the Banner. Conservation farming was presented and explained to participants and villagers to increase their understanding of conservation farming: wind blows and sand storms were often seen in the past. Conservation farming fixed soil, secured soil fertility and is drought resistance, resulting in better output but less input is required). Participants were inspired by these demonstrations and were willing to follow and practice in their land. Soil and water loss control in the mountainous areas is equally important.

VI. An Additional Bonus, the GEF Project Facilitator Became an "Unofficial Village Leader" Who Villagers Trusted and Relied on

The Project Facilitator experienced a large number of difficulties at the beginning of the project, including suspiciousness from the villagers, lack of impress fund and lack of coordination among line agencies. However, he overcame all these difficulties due to his enthusiasm towards and good understanding of the Project and continuous hard work. The Project Facilitator provided direct assistance to all project activities including planning/implementation of the pilot activities, making display board, drafting materials, undertaking training and promotion activities, coordinating inputs from line agencies, managing the archives and relaying messages from bottom to top or vice verse. He eventually gained respect and trust from village leaders and villagers, in particular the villagers in Mandulahu Gacha. The Project Facilitator was regarded by villagers as an "unregistered villager leader".

The Project Facilitator was also directly involved in compiling training materials and managing the archives, which built his good relationship with villagers. Trusting him and appreciating the benefits from the project, villagers called the Project Facilitator an "unregistered villager leader". It is expected that the Project would settle in this place for ever to enable scale up of the GEF approaches to wider areas.

18. Challenges Facing Legislation for Land Degradation Control and Their Countermeasures in China

Wang Canfa, Feng Jia
Institute for Environmental Resources Law
China University of Political Science and Law
100088 Beijing

Abstract: undergoing a rapid development period for over 30 years, China has set up a relatively integrate legal system for land degradation (LD) control and many administrative systems that are relatively effective. However, China is still facing challenges: legislation concepts do not completely meet integrated management concepts; lacking a sound legal system and some of the necessary legislation is still a blank field; the administrative system constrains integrated management of land degradation; some necessary administrative systems have not been set up and some have not been improved; legal punishment intensity has not become deterrence to law breakers; laws are not observed or enforced strictly; the public has not played a role as it should do in land degradation control. Facing such challenges and following the need of land degradation control in China, the legal system for land degradation control needs to be improved and strengthened, a necessary administrative system and measures need to be established and strengthened, using integrated management approaches; the punishment intensity needs to be identified to make sure there is no benefit from illegal behaviors.

Key words: land degradation, integrated management, legislation and administrative system.

Land degradation (LD) is a process in which land environment deteriorates and land loses its original productivity due to internal structure and physiochemical property of land disturbed, damaged and changed by human induced or natural factors. Legislation for LD control is a general term for legal norms developed by the country for combating and managing LD. China is one of the countries suffering from severe LD due to geographical, climate and human induced causes, at the same time it is one of the countries with rapid legislative development. Facing a large number of problems and challenges which need to be addressed, China's legislation for LD is undeveloped, although it has been preliminarily established as a system, and as well a series of management systems have been set up and to some level, integrated ecosystem management (IEM) has been used.

I. Challenges and Problems Faced by China's LD Legislation

The main challenges and problems China's LD legislation is facing include the following:

1. Challenges and problems in legislative concept

Legislative concept is the soul of a law and the most fundamental value all contents of a law must follow. For a system, different legislative concepts result in different design and different implementation with different social effects. A law taking enhancement of economic development as its legislative concept will pay large attention to establishment of mechanisms increasing economic benefits; a law aimed to promote social equality will emphasize equal use of social resources and equal allocation of social incomes. Therefore, legislative concept plays a fundamental role in directing the value orientation of a law.

Integrated ecosystem management (IEM) is an integrated management strategy and approach for natural resource and environmental management. It requires a holistic approach to treat all parts of an ecosystem, take social, economic and natural (including environment, resource and biology) needs and values into consideration, and use multi disciplines and methods and an administrative coordinating mechanism as well as the market and the society when addressing problems in resource development, environmental protection and ecological deterioration. It recognizes the linkages amongst ecosystem functions and services and human social, economic, and production systems. It also recognizes that people and the natural resources are inextricably linked. Rather than treat each resource in isolation, IEM offers the option of treating all elements of ecosystems together to produce multiple benefits.

However, legislative concept of China's LD control legislation has not yet met the requirements of IEM, which can be summarized as follow:

First, components of an ecosystem and elements of the environment are not treated as a whole for integrated management. Laws are developed for individual departments in isolation in accordance with classifications of natural resources. For example, a Forest Act is developed as it is needed to strengthen forest development and protection; to strengthen land resource management, then a Land Resource Management Act is developed; to stifle desertification, water and soil loss, a Desertification Control Act and a Water and Soil Conservation Act then are established. The result of this approach is segmental management of different components of an ecosystem. For example, protection and management of water and land resources should be integrated into one system. However, legislation for development and management of water resource hardly involves land resource development and vice versa.

Second, some legislation has not integrated social, economic and natural needs. In legislative value orientation, the existing laws involving LD control consider only one of the values from economic, social and natural values. They either stress economic gains and social benefit from development of water and land resource and biodiversity, or emphasize pollution control. This management approach covers a single issue, i.e only development or only protection. Although some laws stipulate integrated management of the ecosystem, society and economy, it is only in-principle expression with no details refined for operating rules, let alone establishment of an appropriate management system. Such in-principle expression is usually hard to implement resulting in good effects.

Third, no regulation mechanisms involving administration, market, society and laws have been adopted. Most of the existing LD control legislations adopt administrative orders, a 'ban-penalty' model, ignoring market and social mechanisms. Although some legislation considers the important role public involvement and the market mechanism can play, and include stipulations for market and public involvement in land degradation, such stipulations are macro and absorb, lacking applicability. Legislation for public involvement is very careful, which makes it face problems in implementing these mechanisms.

2. Challenges and Problems in Legislative System

Although, a relatively complete legal system for LD control has been set up in China, different sectors followed their own needs and situation in drafting laws and regulations, with little consideration given to other sectors, leading to several deficits in the legislative system for LD control. These deficits include: lack of specific comprehensive act for integrated LD control, lack of necessary special act for LD control, conflicts among existing laws and regulations and lack of enforcement details for some important special laws.

3. Challenges and Problems in the Administrative System

In the administrative system for LD control, challenges and problems are that necessary administrative systems have not been established and some of the existing systems have not been improved.

(1) Some necessary management systems have not been established. First, it is necessary to strengthen implementation of ecological compensation measures in LD control efforts as implementation of LD control measures will cause right main bodies' loss of economic benefit. However, most of the existing laws for LD control give very little consideration to ecological compensation. Second, mountain closure for revegetation which has been proved to be an effective approach should become an important part of law. However, this approach has not been upgraded into the law system. In legislation for water and soil conservation, there have not been stipulations for systems for conservation programs.

(2) Deficits in administrative systems lead to ineffectiveness of legislation. Many of the existing laws for LD control have many deficits including:

Implementation of the land use regulation system is not linked with the overall land use planning. Therefore, land use planning lacks legal effect. Regions, sectors and different fields have their own plans which are not consistent with the overall land use plan. The public is not involved in the land use planning process.

In the natural reserve system, the ownership relation between land and farmer collective land ownership and use right inside the reserve is not clear. Therefore, it is hard to implement effectively the management measures for the natural reserves and funding is not guaranteed. The lack of a sound administrative system for natural reserves, the combination of administration with tourist business have resulted in obscure organization property, functions and authority, and weak legal liabilities, which cannot effectively punish illegal activities.

In the environmental impact assessment system, environmental impact assessment is only applicable in planning and construction projects. Legal liability is much too light and narrow. The maximum penalty for a construction company who breaks the law is only 200,000 yuan, which is not strict punishment to meet the requirements from strict environmental protection and management. In addition, it has been stipulated that construction projects under implementation without going through all environmental assessment processes should be ordered to cease production or construction and go through the procedures again. If this is not done before deadline, a penalty for 50,000-200,000 yuan must be imposed. Actually, it makes no sense to set up an environmental assessment system if construction projects are required to go through the procedures again after implementation has started. It will not prevent environmental damages, which goes just against the purpose of setting up such systems.

(3) Challenges and Problems in Administrative Systems

A good administrative system is the key to support establishment of a legislative system to accommodate the IEM approach, the development and improvement of a refined legal system. However, LD control administration has not been well coordinated in China as it can be seen that some institutions are overlapped, some of the administrative establishments do not follow scientific rules so there is not sufficient stipulation for public supervision of law enforcement by legal operation departments.

(4) Legal punishment intensity has not become deterrence to law breakers

The legal liabilities stipulated in the existing LD control legislation in China do not have the weight to meet the severity of the behavior of law breakers. For example, it is stipulated in the 31st clause in the Environmental Impact Assessment Act that construction projects start without environmental impact assessment are required to go through the procedures again; those who refuse to go through the procedures again will be punished by a maximum penalty of 200,000 yuan. Such legal stipulation leads to legal consequence at two aspects: construction companies who are found by the environmental protection department to have started implementation without environmental impact assessment are only required to go through the procedures again, without being asked to take any extra legal responsibilities, unless they refuse to go through the procedures again. If such legal behavior is not found by the environmental protection sector, a project can be implemented without undertaking environmental impact assessment. This actually means it costs far less for construction companies to break laws than to observe laws. Objectively, it encourages construction companies to skip over environmental impact assessment and enhances the ineffectiveness of legal stipulation for environmental impact assessment. On the other hand, a penalty for 200,000 yuan is not heavy punishment for large and medium-sized companies and is hard to become deterrence.

Such situation as described above is common in laws and regulations for pollution control and natural resource management. In legal liabilities, the stipulation in China's environmental resource laws is far too weak to support effective implementation.

II. Countermeasures for Improving China's Legislation for LD Control

An overview through the issues in China's legislation for LD control with IEM taken into consideration indicates there is a need to improve the legal system:

1. Improving legislative concept

To achieve sustainability, social, economic and natural needs and values need to be taken into consideration. Change must be made at three aspects in legislative concept for LD control: i. change from environmental protection balanced by economic development to economic development coordinated by environmental protection, and in some cases, priority must be given to environmental protection to stifle worsening of environmental degradation; ii. change from scattered control administrative systems into a coordinated administrative system to enable establishment of a comprehensive administrative system for LD control; iii. increase the weight of both administration and civil actions instead of paying more attention to administration than to civil actions, to provide opportunities for public involvement in LD control.

2. Improving the legal system for LD control

Precondition for improving and strengthening the legal system for LD control is to identify what an integrated legal system for LD control is like. In accordance with the IEM concept, we consider a sound legal system for LD control should include at least the following contents: legislation for integrated LD control; specific law for water and soil conservation; specific law for desertification control; specific law for land pollution control; specific law for wetland resource protection; laws and regulations for LD control management system; legislation for compensation for land pollution and damages; legislation for processing procedures for land disputes and other legal stipulations for LD control such as stipulation in the Environmental Impact Assessment Act. The following shows the prospect structure of such a system:

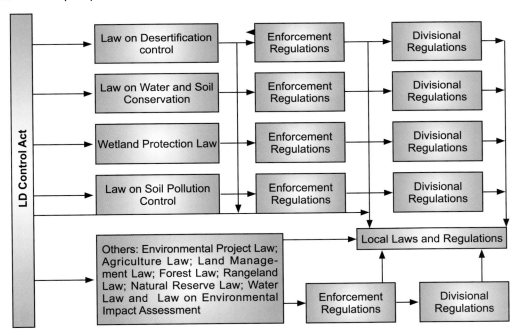

If such an integrated legislative system is established for LD control in China, it shall be firm legal foundation for LD control.

3. Establishing or improving a/the necessary system and measures

Although, a large number of administrative systems and measures have been set up for LD control in China, some necessary systems and measures are missed out and need to be established, and some need to be improved.

In land use regulation, legislation for land use planning is needed to clarify legal effect on the overall land use planning and coordinate the linkages among urban, rural, transportation, water resource, energy, tourism and environmental development plans made by various regions, sectors and industries.

In the nature reserve system, the property right relation between land and farmers' collective land ownership and use right within the nature reserve must be clarified. If the farmers collectively have the ownership and the use right of the land within the nature reserve, local government should take them over, at the same time provide timely and sufficient compensation, or local government and the collective economic organizations come to agreements on development. Local villagers should be allowed to undertake appropriate development activities, e.g. tourist services. The local government should provide compensation accordingly.

In establishing ecological compensation systems, there is a need to make the principles for ecological compensation clear. These principles should include identification of 'the beneficiaries and the compensators', diversified compensation approaches and public involvement. The main compensation approach should be based on government financial transfers and complemented with ecological compensation tax and ecological compensation foundations.

In the mountain closure system, it is needed to clarify the scope, the financial support mechanism, the methods, laws, measures and the local land ownership relation for mountain closure. It is also needed to identify the local government's role and responsibilities and legal liabilities.

In the environmental impact assessment system, first, the applicable scope should be expanded and extended into government' decision making, which will make it the real strategic environmental impact assessment, and second, the stipulation of the legal liabilities of the Environmental Impact Assessment Act should be improved to increase the weight of legal responsibility for those who break the environmental assessment system and to ensure construction companies breaking the law get no benefit from their illegal behaviors.

In improving the water and soil conservation program system, its applicable scope should be clarified. The categories, levels and the contents of reports on water and soil conservation programs must be made clear. The main body formulating water and soil conservation programs and their qualification need to be stipulated. The last one which is also the most important is the legal force for water and soil conservation programs must be clear.

4. Improving the LD control regulative and administrative system

To improve China's LD control regulative and administrative system, efforts need to be given to two aspects: i. specific legislation should be developed for the LD control regulative and administrative system which is essential in improving the LD control legislative system. In this specific legislation, it is needed to make clear whether a comprehensive administrative department should be set up, and the role of such a department, its institutional structure, functions and coordination approaches should be identified.

At a higher level, it is specific legislation for the whole environmental administration system; ii. There is a need to reform the existing system in which different sectors are leading the legislative drafting process, or set up or strengthen a/the mechanism coordinating different sectors under such a sector-leading system. Legislative documents which are drafted by a single legal department following their own studies will not be operational due to the complexity and high specialization of the administration system. To change the existing system in which departments are leading in legislative drafting, in theory, it is conducive for addressing the existing problems in the legislative system. However, it is not practical in implementation, which will lead to ineffectiveness. In another word, it will become obstacle to national legislation. Therefore, the second proposal would be more practical: establish and/or strengthen a/the mechanism coordinating all sectors within the existing sector-leading legislative system.

5. Identify the weight of punishment for law breakers to ensure they make no benefit from their illegal behavior

Enforcement of China's LD control laws has not been effective, which is considered to be related with weak stipulation of legal liabilities. It is needed to make sure law breakers make no benefit from their illegal activities if we are to stifle illegal behaviors.

First, the weight of punishment of each single legal liability should be increased. A maximum penalty has been set in most stipulated legal liabilities in the existing LD control legislative system, which means no matter how severely one has broken the law the penalty has a limit. The existing maximum penalty is not much of deterrence to the large-scale companies/industries. The severity of an illegal activity, the damages and consequence must be considered when identifying legal liabilities. There should not be a maximum of penalty for the illegal behaviors which are of severe consequence and damages. If there is a need to set a maximum penalty, it must be increased to meet social economic development level and affordability of law breakers.

Second, continuous illegal behaviors should be 'punished by day'. An 'only-once' punishment instead of consistent punishment is not adopted in the existing LD control legislative system. Law breakers are imposed with one compensation or one penalty for their illegal activities. Another penalty is imposed only when the damages from their continuous illegal behaviors reach a certain level. Such punishment, although creates certain pressure to lawbreakers, it is easy for them to endure as there is an interval between two penalties. Law breakers can earn enough profit to pay off the penalty during an interval, which means illegal behaviors are not completely stopped. To avoid such situation, some countries take a 'punish-by-day' approach in affixing legal liabilities, i.e, if a law breaker continues his illegal behaviors he is imposed with a penalty or compensation for each day, apart from the penalty he has received. The longer the illegal behavior lasts, the sum of penalty or compensation gets larger. This approach has very strong deterrence to law breakers as there is no interval between two penalties. Law breakers will not attempt to defy such laws. In civil damage compensation, a punishing damage compensation approach should be adopted.

Finally, there is a need to increase the severity of administrative detention and criminal liabilities. Some of the illegal behaviors against LD control laws which damage the environment have serious social consequences. For example, illegal installation of pipes to discharge pollutant or poison into soil or water does not only damage the land (constrain LD control), it creates threat to people's lives and property and can cause social instability. Penalty is not enough punishment in such case. More severe punishment including punishment to personal freedom and life sentence should be adopted.

6. Increase public involvement in LD control

Public involvement and right guarantee include guarantee for public access to information, public involvement in decision making and access to relief. Legislation at this aspect should be improved to ensure the public has access to information. Clear stipulation should be made for information the government intend to open or the public ask to open. Careful stipulation must be made for exceptional cases apart from open information. The relation between protection of national and business confidential and information opening should be clear. More details are needed in identifying processes for public application for opening government information. In involvement in decision making, more efforts should be made to diversified approaches for public involvement. Procedural stipulation for public involvement including public hearing needs to be further improved. In designing procedures, consideration should be given to how to mobilize the public's enthusiasm, how to guarantee the public involvement in decision making and their access to relieves and how to ensure public opinions are fully considered by decision makers and implementers. Consideration also needs to be given to guarantee that the public has proper access to relief when public involvement or related rights and benefit are violated..

19. Practice and Application of IEM Concept in Legislation for Combating Land Degradation in Gansu Province

Wan Zongcheng
Deputy Director, Legislating Affairs Committee of the Gansu Province Peoples' Congress

Abstract: GEF/OP12 project is the first partnership project of Global Environment Facility (GEF) in the prevention and control of land degradation, which is aimed at preventing and restraining land degradation of the ecosystems in arid region by means of IEM, so as to eliminate poverty, preserve the biological diversity, promote sustainable development of western China, as well as to provide technical and administrative support for integrated management of natural resources across different sectors and regions.

The practice of IEM concept suggests that firm implementation of IEM concept in light of the real condition of Gansu Province is the very momentum for practicing prevention and control of land degradation; and it is the long term objective of persistent prevention and control of land degradation by legislative measures. Yet the "two focuses and two promotions" is the fundamental approach for prevention and control of land degradation. By the same token, it's the basic guideline to implement the IEM concept in the whole process of prevention and control of land degradation; and it's the ultimate objective to emphasize practice efficiency so as to improve our legislation capacity.

We are better off in an all-out manner from practicing IEM concept. The innovative IEM concept is engraved by the public, and officials at various levels of government are more conscious of implementing the scientific development concept, together with increasing strength of environmental control and protection. A set of regional laws and governmental regulations regarding environmental protection and construction helps promoting law guarantee capacity. One group of the IEM professionals has been established through practice and becomes the solid basis for prevention and control of land degradation. The brand new IEM concept and methods of assessment render conducive reference for establishment and assessment of laws and regulations by the legislature. To sum up, the IEM has become the driver of ecological civilization in Gansu Province. Government at all levels are paying more attention to environmental rehabilitation than ever, and the policies and measures established are closer to the practical demand.

Proposal: Application the IEM approach in should be in compliance with local needs. The laws for prevention and control of land degradation should highlight integrated management. The coordination and cooperation mechanism for prevention and control of land degradation should be perfected further. Relevant laws and administrative regulations should be subject to prompt revision on the basis of GEF accomplishments.

Gansu Province is featured by vulnerable ecosystem, frequent occurrence of meteorological disasters and serious land degradation. In Gansu Province prevention and control of land degradation have never stopped with the rapid social and economic development. GEF/OP12 project introduces the IEM concept, making a threshold for the mission of prevention and control of land degradation in Gansu Province. New practice is launched for legislation for prevention and control of land degradation. For the five years' efforts, new achievements have been made in the field of legislation on prevention and control of land degradation in Gansu Province, which in return pushes forward sound development of prevention and control of land degradation. Our understanding of IEM concept and method is gradually deepened during the practice.

Firstly, it is not only necessary to have instructions by proper training, guideline and policy, but also scientific concept and methods, so as to prevent, control and reduce land degradation in arid region and eliminate poverty. GEF introduces the brand new concept of IEM, which is scientific, systematic, integrated, sustainable, innovative and flexible, and application of the IEM concept in prevention and control of land degradation is helpful for us in terms of more consciousness for implementation of scientific development concept.

Secondly, it is a systematic project to prevent and control land degradation in arid region and eliminate poverty. Agriculture, forest, water, water and soil conservation, grassland, prevention and control of desertification, biological diversity, environmental impact assessment and environmental protection are included in IEM as research subjects, with which a new perspective on research of land degradation is rendered, widening the vision of sustainable management of natural resources for various sectors, industries and regions. It is constructive to further implementation of the ideological line of emancipating the mind, being practical and realistic and keeping pace with the times in the process of construction of ecological civilization.

Thirdly, it is critical to improve the legal system, so as to prevent and control land degradation in arid region and eliminate poverty. Of various measures involving prevention and control of land degradation, the legal system is more substantial, sustainable, stable and authoritative. The 12 principles, 20 basic categories of legal elements as well as assessment methods and steps of the IEM provide a new mode for post-assessment of local rules, regulations, and laws regarding prevention and control of land degradation. Apploication of the IEM concept is helpful for improving the local legislation quality. The legal system regarding prevention and control of land degradation develops more efficiently in terms of guidance, standardization, administration, promotion and guarantee through revision and amendment of local legislation. The legal system framework for prevention and control of land degradation has been improved by local legislation.

Fourthly, GEF makes organized and overall assessment of laws, policies and institutions for improving prevention and control of land degradation capacity by the IEM concept and methods, with which the capacities of the laws, policies and institutions are coordinated harmoniously, so that the evaluation is more inclusive, objective and realistic. Assessment with this perspective is conducive for us to avoid partial implementation, and is practically significant in terms of improved administrative capacity and evaluating quality as well as scientific legislation quality and decision-making by administrative organization.

I. Basic Practices

The purpose for learning the IEM concept is for better application, and the law survives by proper enforcement. Based upon this understanding, we regard the legislation on prevention and control of land degradation in terms of the IEM concept and methods as the core of project implementation. To this end, we persist in the "two focuses and two promotions": namely, strict assessment of existing laws and regulations on the one hand, and persistence in establishing new laws on the other hand, pushing forward establishment of new laws by making assessment of prior ones, so as to demonstrate the capacity building in legislation by newly established laws as well as to improve on successful implementation of the project. The basic practices include (1) extension of the IEM concept by means of a great varieties of disclosure, so as to apply the IEM concept in the whole process of project; (2) paying more attention to planning in terms of coordination, for the purpose of establishing new law as well as timely revising legislature's legislation planning and annual planning draft relevant to prevention and control of land degradation by the IEM concept and method;s (3) trying to pass the decision-making; the legislative planning is subject to discussion by standing committee of the provincial People's Congress and approval by the CCP committee of Gansu Province before implementation, so that the project implementation is coordinated with related decision-making; (4) organization for implementation; cooperation is required among governmental sectors in terms of strengthened aspects such as proposal and confirmation of draft law, research and investigation, drafting, amendment, review, revision, and discussion, etc., to ensure successful enactment of new laws, so as to continuously make improvement in legislation regarding prevention and control of land degradation.

II. Accomplishments

Total 13 local regulations and rules have been established regarding prevention and control of land degradation using the IEM concept in Gansu Province as of July 2008, with efforts made since of the GEF/OP12 project,was commenced, including *Regulations of Prevention, Control & Quarantine of Forest Pests & Diseases of Gansu Province, Regulations of Gansu Province on Management of Crop Seeds, Measures for Implementation of the Water Law of PRC of Gansu Province (Revision), Regulations of Environmental Protection & Construction of Liujiaxia Reservoir Area of Linxia Hui Autonomous Prefecture of Gansu Province (Approved), Regulations of Gansu Province on Environmental Protection during Oilfield Exploitation and Development, Regulations of Gansu Province on Voluntary Tree Planting by the Public, Regulation of Gansu Province on Management of National Lianhua Mountain Natural Reserve, Regulations of Gansu Province on Grassland Management, Regulation of Gansu Province on Integrated Utilization of Resources, Regulations of Gansu Province on Management of Water Resources of Shiyang River Valley, Regulations of Gansu Province on Protection of Agricultural Environment, Regulation of Gansu Province on National Natural Reserve of Extremely Arid Anxi Desert and Regulations of Gansu Province on Management for Prevention of Meteorological Disaster.* In addition, 14 local legislations are listed for annual planning, investigation and drafting, including prevention and control of water pollution of the Yellow River in Gansu Province, woodland protection, forest park, fertilizer, Taolai River valley, crop seed base, quarantine of agricultural vegetables, collection and distribution sewage emission fees, aquatic inspection & quarantine, preservation of scenic spots, and urban greening. These regulations are important legal basis for the formation of legal framework regarding prevention and control of land degradation by the IEM concept and methods as well as speeding up of prevention and control of land degradation in Gansu Province.

III. Capacity Improvement

The new laws are characterized as follows: (1) wider coverage of prevention and control of land degradation; (2) thorough incorporation of the IEM concept in terms of different measures of IEM principles in legislative purpose, scope of regulations, basic contents, system disposal and detailed measures; (3) improved efficiency of law enforcement; (4) better harmony between law and policy, so that the capacity of law and policy regarding prevention and control of land degradation is improved remarkably.

Presently, 14 projects listed in legislation plan are under orderly implementation. The IEM principles are widely used in new law for improved law quality. We are committed to integrating IEM concept and practices in Gansu Province for improving the project quality. During legislation investigation and amendment, we have paid much attention to guide for the best application of the IEM concept in all respects of legislation. Below, we take the Regulations of Gansu Province on Grassland Management as an example.

For sound protection, construction and utilization of the grassland, improvement of the environment as well as preservation of biological diversity, the Standing Committee of Gansu Provincial People's Congress re-constituted the *Regulations of Gansu Province on Grassland Management* (Hereinafter abbreviated as the Regulation) and promulgated it on Dec 1, 2006 on the basis of 12 IEM principles. The revised version of the regulation has the following characteristics: (1) purpose of the law and scope of application are explicit, embodying the equal significance on protection and utilization stipulated in IEM (Principles 1 and 2); (2) it stipulates that various levels of the people's government should incorporate the protection, development and utilization of grassland into the national economy and social development planning, incorporate the grasslands that are subject to degradation, sandization, salinization and desertification into the land control and development planning, classify control area, organize related sectors for special control, with the governmental support of fund, resources and technology, representing the sustainable development conception for integrated protection and utilization (Principles 3, 9 and 25); (3) the new law stipulates that various levels of the people's government should organize scientific research institutes and professionals to conduct fundamental research such as grassland degradation mechanism and biological evolution law, etc., intensify research and development of advanced technologies such as restoration of grassland environment, selection and breeding of high quality and stress tolerant pasture varieties, methods for breeder optimization and breeding and so on, as well as active extension of research accomplishments of the grassland, representing the principles of the IEM oriented to the grass-root and wide public participation (Principles 4,

5 and 11); (4) it defines the principle of "determine livestock carrying capacity according to grass, balanced grass and livestock," with stipulation that the grasslands subjecting to degradation, soil erosion, and desertification and that of environmentally vulnerable grasslands should be protected from grazing, and seasonal grazing ban should be implemented on the sites with minor degradation, and integrated improvement measures should be adopted according to the extent of grassland degradation, so as to restore the grassland vegetation; the grassland for grazing prohibition shall be marked clearly, representing the principles of environmental restraining management (Principles 22, 23, 24 and 32); (5) it prescribes proper grassland use and protection of grassland vegetation; prevention of excessive grazing, prevention and control of grassland pests and rat outbreak; prohibition of grassland reclamation as well as digging of greensward and turf for prevention of damage of new turf, sandification of grassland and loss of water and soil; expropriation or use of grasslands for mining and project construction shall be subject to procedures of environmental impact assessment and other relevant approval for construction project by the law; the environmental impact assessment of construction project shall include plan on grassland environmental protection. These provisions represent the principles of priority of environmental protection and win-win of economy and environment (Principles 21, 27, 28, 29, 30, 31, 32, 33, 34, 35, 36, 37, 38, 39 and 40); (6) it states that the contracted grassland shall be concentrated, with preserved common-land for grazing path, drinking spot, and breeding spot, etc, so as to make the production and living of the farmers and herders as well as integrated construction of grassland more convenient; various levels of the people's government should mend the grasslands by means of the mode for protecting original vegetation of the grassland by popularizing and adopting make-up sowing without plowing, broadcast sowing or air seeding according to local condition and on the basis of the plan on grassland protection, development and utilization, and the farmers and herders are expected to lead to change their mode of production and living by development of artificial grassland, base of forage grass and feed, infrastructure for prairie water conservancy and domestic water supply projects. Measures for protecting grassland vegetation shall be made for activities such as geological surveying, road construction, prospecting, pipeline erection (laying), construction of tourist spots, military drills, film shooting and vehicle running on the grassland, and vegetation recovery fees shall be paid to grassland supervising and administrative institution, representing the principles of activity with consideration to environmental protection (Principles 10, 20 and 37).

For two years since the *Regulation of Gansu Province on Grassland Management* was enacted, various levels of grassland administrative departments and grassland supervising and administrative institutions have performed the legal responsibility earnestly in terms of increasingly strengthening law enforcement, standardizing enforcement activity and intensifing enforcement supervision, with remarkable effects, which have played an important role in protection, development and sound utilization of grassland, environmental improvement, reservation of biological diversity, preservation of national environmental security, development of modern stockbreeding, promotion of sustainable development of provincial economy and society, advancement of national solidarity of the minorities Gansu Province, stability of the border area and society, speed-up of economic development in pasturing area, as well as improvement of standard of living of the vast herding farmers, with the positive effects reflected in the following four aspects:

(1) The administrative departments have achieved better understanding of the significance of strengthening grassland supervision and management by law.

Various local governments have profound understanding of the significance of strengthening grassland supervision and management by law in the new trend from the strategic perspective of implementing the fundamental national management policy and overall practice of legal administration, set up the steady and scientific development conception, persist in the principle of "attach equal importance to the objective of social economic development and environmental protection, with priority given to environmental protection," so as to realize equal consideration to grassland environmental security as that of the crop production, protection of the basic grassland as that of the prime farmland as well as construction of grassland environment as that of forestry environment, so as to facilitate overall, coordinated and sustainable development of the grassland.

(2) The public has promoted their awareness on legal protection and construction of the grassland.

In 2008, the "publicity month" activity for the dissemination of the regulations was launched in Gansu Province. During that month, more than 138,000 pamphlets and 36,000 handbills were distributed to the public, and 13,000 brochures printed in Mongolian, Kazakhstan and Tibetan languages were distributed, in addition to 420 banners, billboards and publicity plates, 260 times (Vehicles) of publicity cars, more than 3,000 hyper, 220 issues of column, showcase and blackboard posters, 24,000 slogans, 30 TV reports, tape broadcasting for more than 500 times, more than 130 videos, 139 newspaper articles, 51 law lectures, 175 training courses, more than 50,000 trainees, involving more than 360,000 people. By wide publicity, the public has better understanding of the significance, urgency, complexity and enormity of grassland environmental protection and management, and the people from all circles and the herding farmers have more consciousness to pay more attention, recognition and protection of the grassland. Still, it's helpful to improving the image of grassland supervision industry of Gansu Province, encouraging and driving the grassland administrative officials and law enforcement personnel for learning, understanding, following laws as well as strict law enforcement and administrative capacity by law.

(3) Grassland protection and development have been remarkably strengthened.

The following major measures are adopted in various regions for grassland protection and construction, including: (1) more investment in grassland protection and construction, establishment and improvement of stable financial fund mechanism for grassland development and a large number of farmers and herders are mobilized to invest in grassland development; (2) more publicity regarding grassland contract and proprietorship transfer is carried out according to the good will of farmers and herders for legal contract of grassland and long term steady production; the farmers, herders and contracted grassland operators are encouraged and supported for legal management and use of the grassland; also, the farmers and herders are guided for legal transfer of contracted grassland management, so as to gradually strengthen their skill for protecting against their legal rights and interests; (3) activities involving grassland use such as mining, project construction and grassland tourism are effectively standardized, and strict procedures of auditing and approval are implemented; the measures for restoration of grassland vegetation are executed, so that the grassland users can consciously perform respective obligation for grassland protection; (4) strengthened publicity and protection for collection, selling and purchasing management of key grassland wildings for protection such as licorice, ephedra, Boschniakia, snow lotus, aweto, radix gentianae Macrophyllae, parsnip, radix scutellaria, radix bupleuri, songaria cynomorium herb, saffron, and red-spotted stonecrop, etc.; any activity involving illegal picking, purchasing and sale of key grassland wildings for protection shall be prohibited, so as to guarantee sound wilding resources management of the grassland without failure; (5) grassland vegetation restoration fee is collected by law; the standard for collection of grassland vegetation restoration fee (Notice of Gansu Provincial Pricing Administration and Financial Office on the Standard for Collecting Grassland Vegetation Restoration Fee, Gansu Provincial Price & Charge No. [2007]320) that is basically suitable to the provincial situation is established and enacted, in which 14 types of pasture distributed all over Gansu Province are included for collection of grassland vegetation restoration fee. Gansu becomes the first province in China where it practices vegetation restoration and environmental compensation for expropriation and use of grassland, so as to make a solid basis for active research and construction of grassland environment by adopting compensation mechanism suitable to Gansu Province.

Besides, the Gansu provincial people's government is coordinated to summon related departments for establishing auxiliary standard document Minutes of the Meeting on Processing the Problems regarding Approval Power for Grassland Expropriation (Gan Zheng Ban Ji No.[2008]36), which defines in more detail the auditing and approval power of grassland administration and land and resource administration, so as to effectively stop the activity of approving expropriation and occupation of grassland by exceeding the authority or violating the legal procedure, which is significant for strictly controlling the quantity of grassland use for project construction. To standardize auditing and approval procedures for grassland expropriation and occupation, for the purpose of processing the auditing and approval matters according to legal power, the Notice of Gansu Agriculture & Stockbreeding Office on Printing & Distribution of the Guide for Administrative Permission of Integrated Handling & Processing (The 2nd lot) (Gan Nong Mu Circular No. [2008]124) was issued by Gansu Agriculture & Stockbreeding Office, in which matters and processing guide are published, including "auditing of grassland expropriation (Collection) and use for mineral mining and project construction," "approval of temporary occupation of grassland" and "approval of construction of grassland protection and stockbreeding service project facility," and the power of auditing and approval, processing condition and procedures by various levels of grassland administration are defined clearly. The measures create favorable conditions for effective implementation of graded auditing and approval of grassland expropriation and occupation as well as standardization of auditing and approval activity of administrative approval subjects.

The leading officials from the United Land Expropriation Office for Construction Project of Gansu Provincial Department of Land & Resources are consulted with for coordination for problems relevant to grassland use.

The grassland expropriation and use for construction projects of mineral mining and project construction are checked in key cities and counties of Gansu Province. The overall check is carried out in 16 cities (Counties), including Suzhou District of Jiuquan City, Guazhou, Subei, Akesai, Ganzhou District of Zhangye City, Shandan, Liangzhou District of Wuwei City, Minqin, Jinchuan District of Jinchang City, Yongchang County, Huachi of Qingyang City, Huan County, Xiahe of Gannan State, Luqu, Maqu and Zhuoni, so as to understand the basic situation of construction project involving grassland expropriation and occupation in the whole province. At present, large-scale check is under organization and implementation, and the activities of expropriation, use and temporary occupation of grassland for mineral mining and project construction are treated as per different periods of time. Preliminary effect is preferable according to treatment of impeached cases at the present time. Again, the management measure of balanced livestock carrying rate is implemented in Gansu Province, and excessive grazing is prohibited. The farmers and herders are trained with knowledge of balanced livestock carrying rate and applied technology, and they are encouraged to take measures such as artificial grassland construction, mending natural pasture and breeding in pen, improvement of the forage grass supplying capacity, relief of the grazing pressure of natural pasture, so as to achieve appropriate grassland utilization.

(4) Grassland law enforcement and supervision have been greatly reinforced.

The Regulation of Gansu Province on Grassland Management and related rules and policies are followed strictly, and sectors such as the public security, environmental protection and industrial and commercial administration are coordinated and organized for special control and rigorous punishment of various activities violating the grassland law. Many serious and major cases involving severe grassland damage and bad influence are investigated and treated by mobilized forces, which enhances the deterrence power of grassland law enforcement.

The proprietorship and access of grassland shall be defined by law, so as to implement the family contract operation responsibility system of grassland. The contract supervision and management shall be strengthened, and the contractors shall be guided and standardized for legal and payable transfer of grassland contract operation. Any activities of willful adjustment and forced transfer of grassland contract shall be corrected strictly; unapproved alteration of proprietorship and purpose of contracted grassland are prohibited. Any dispute caused by grassland contract shall be treated appropriately; the legal rights and interests of the farmers and herders shall be protected by law. The grassland protection and construction project shall be organized and implemented seriously. The pressure of over-grazing shall be relieved by implementation of construction project (such as turn grazing land back to grassland), speeding up of enclosing grazing and rotational grazing. Also, maintenance of grassland construction achievements shall be strengthened, and various activities of damage to grassland construction achievements shall be punished severely. Prevention and monitoring of grassland disasters shall be reinforced. According to the policy of "prevention foremost, integrated prevention and control," the prevention, elimination and control of grassland disasters shall be practiced during routine work processes, The fast response capacity shall be enhanced; and advance forecast and preparation, timely discovery, report and rescue are required for unexpected disasters, so as to minimize the casualty and property loss. A solid basis for grassland supervision and management shall be made by extensive education on disaster prevention, establishment of emergency mechanism, mobilization of social mechanism and development or improvement of preplan on disaster prevention.

Positive measures are adopted for promoting development of a grassland supervision system. The principle of "adjusting measures to local conditions, classified instructions, gradual standardization, and improvement of quality" shall be followed to speed up construction of grassland supervision system. According to the provisions of the regulation, grassland supervision organization shall be established in the large grassland area without grassland supervision institution. In the region of small grassland area, organizational establishment and personnel quota shall be suitable to local conditions. The region with grassland supervision organization will be standardized and improved gradually. Besides, the group quality and system establishment will be enhanced. As required for "standardization of activities, enhancement of responsibilities, improvement of efficiency," the grassland law enforcement system and internal management system of grassland supervision organization shall be established and improved, in addition to further standardization of grassland supervision procedures. It shall strengthen construction of working style and improve the moral within the sector through "civilized law enforcement, quality service, uncorrupted and with high efficiency" according to the basic requirement of "internally strengthening quality, externally creating image." The capacity building is reinforced by organizing training and performance assessment for grassland supervision personnel, overall implementing employment for the grassland supervision personnel with certificate, as well as continuously improving the political thinking quality and management capacity of grassland supervision personnel.

The organization and leadership of grassland supervision and management are strengthened by clearly defining responsibility and strict implementation. The plan on construction of scientific and sound grassland supervision and management are established and perfected, and the working condition for grassland supervision and management are improved gradually; the facility and equipment for grassland supervision and management are supplemented and perfected for improving the efficiency of grassland supervision and management.

IV. Wide Influence

The project has extremely wide influence: the innovative IEM concept has been engraved in the public mind; the awareness of officials from various national organizations for implementing scientific development concept is intensified, and the environmental control and protection are strengthened; a set of local rules and governmental regulations regarding environmental protection and rehabilitation are created; a group of professionals engaged in the IEM acitivities are fostered. As an innovative mode, the IEM concept and assessment methods are conducive reference for the legislature's establishment and assessment of rules and regulations. The IEM concept has become the driver of environmental rehabilitation in Gansu Province, and the provincial government attaches unparalleled importance to environmental restoration, and the policies and measures are more practical and realistic.

V. Proposals on Legislation

(1) Necessary improvement of the IEM evaluation methods is advisable in light of existing legislative system in China. The principles shall be featured by keeping with the pace of modern times, particularly adaptive application of local rules in terms of different scopes and subjects. In the aspect of legal elements, complete adoption of the IEM concept is untenable, and it is impractical for the local rules pursuing complete elements and the full system. The legislation shall satisfy the necessity and completeness for the sake of local features and applicability.

(2) It is needed to develop the legal system regarding prevention and control of land degradation on the basis of the IEM concept and methods, and make innovation in the system and mechanism, so as to highlight the integration.

(3) It is needed to draft planning and carry out revisions on the Environmental Protection Law of the People's Republic of China and relevant laws in terms of the imperfection and shortfall of elements found in law assessment and in accordance with the IEM concept and principles.

(4) It is needed to establish a sound coordination and cooperation mechanism among various sectors for prevention and control of land degradation, and clarify the responsibilities among functional departments, particularly settling problems such as overlapping responsibilities.

20. Achievements of Hundan Watershed Pilot Site in Huangyuan County Qinghai Province and Lessons Learned

Cai Chengyong
Governor, Huangyuan County, Qinghai Province

Abstract: It has been 4 years since Hudan Basin Pilot Site was established, resulting in positive achievements. The main activities undertaken in this pilot site include institutional capacity building, development of Project approaches, participatory community development plan for IEM land degradation (LD) control and annual plans, through which the limited project funds were optimized and became 'seed fund' in mobilizing involvements from all relevant departments and sectors in the development of the Project. Over 17 million yuan was raised for the pilot site by integration of resources, combination of funds, intensive investment and management, which has obviously improved the environment, stifled water and soil losses, improved the livelihood and environmental awareness of the local community.

Key words: Hundan Basin, pilot site development, achievements, lessons learned

1. General information of the pilot site
General information of Huangyuan County

Huangyuan County is located in the origin of Huangshui River, which is one of the major tributaries of the Yellow River. It is in the transition zone of Loess Plateau and Qinghai-Tibet Plateau and the natural boundary between agriculture and animal husbandry in Qinghai, and serves as an important gateway linking Qinghai and the inland and animal husbandry of western area and Tibet as well as the combination point of Chinese-Tibetan culture.

The county has a plateau continental climate, with altitude varying in a range of 2470-4898 m above sea level and large differences in vertical height. The climate is characterized by long sunshine hours, strong solar irradiance, dry and windy springs, short and cool summers, rainy autumns, and long and dry winters; the daily temperature fluctuation is large, while annual temperature fluctuation is small; freezing period is long while frost-free season is short. Annual mean temperature is 3°C, annual rainfall is about 400mm. Huangyuan County is under the jurisdiction of Xining City, with a total land area of 2,263,500 mu. Its forested land is 1,047,000 mu, accounting for 44.97% of the total land area of the county. Its forest cover is 26.8% and cultivated land area 256,000 mu, which accounts for 10.96% of the total land area. The county consists of seven townships and two towns, 146 administrative villages, and seven communities. The total population is 136,400, of which 30,300 are town residents. There are 16 nationalities including Han, Tibetan, Hui, and Mongolian, etc. The average net income per capita was 2694 yuan in 2007.

Situation of the pilot site

Hudan watershed is located in the Bayan Township of Huangyuan County, covering Shanghudan and Xiahudan administrative villages. There are 608 households and 2501 people. The altitude varies between 2760 m and 4150 m. The total area is 3456 hm^2, including 2667 hm^2 grassland, 2033 hm^2 utilizable grassland, 633 hm^2 severely degraded grassland, and about 200 hm^2 alpine shrubs. Due to the natural constraints and unsustainable use, natural vegetation of the watershed has been seriously destroyed by various factors. The first problem is serious water and soil loss. Sparse vegetation and continuous drought, little rainfall and great evapotranspiration have caused decline in the ground water level and soil moisture. Statistics by the relevant departments show 60% of the cultivated land in this watershed has been damaged by various degree of wind erosion. The second problem is severe grassland degradation. Statistics shows 85% of the natural grasslands are degrading, and about 40% of the grasslands have been severely degraded. The third problem is severe soil nutrient loss in cultivated land. Soil hardening, increase of soil bulk density, large decrease of granular structure, soil nutrient loss, and low and crop yields due to decreased use of organic fertilizer by villagers.

2. Development approaches of the Pilot Site

Development of the Pilot Site was listed as one of the priorities of the County as soon as it was identified as a project pilot site. Following the development approaches identified by the GEF Project, including coordinated planning, integrated resources, combined funds, intensified investments and management, we mobilized all relevant departments and sectors to participate in the development efforts. A total investment of 17 million yuan was raised for the Pilot Site and good results have been achieved.

2.1 Extension of Project approaches and concepts

In order to fully accomplish project objectives, several advocation campaigns on increase of the project awareness were undertaken among villagers and government officials to extend project approaches. First, during the pilot site tender bidding process, project approaches were advocated through village meetings and interviews with farmers, and in surveys and questionnaires undertaken to identify villagers' willingness for the Project. Second, in order to develop stakeholders' better understanding of the project approaches and obtain their support to the project, extensive advocation and education campaigns were undertaken. Australian Youth Ambassador for Development (AYAD) Helen, the Strategy and Action Plan Team and the Legal and Policy Expert Team of the Project were invited to deliver training sessions/workshops on participatory methods, integrated ecosystem management (IEM), project approaches and environmental education for government officials at all levels and villagers, which was firm foundation and guarantee for successful implementation of the Project.

2.2 Development of participatory community plan for land degradation control using IEM approach

Following the objectives of the Project and with support from the Provincial Project Management Office (PPMO), the Project Leading Group undertook detailed field surveys and analysis on LD types, causes, ecosystem forms, land use and social and economic situation in Hudan Basin, and developed environmental and community development plans with the local community, using the 'bottom to top' method in which villagers were empowered to propose their comments and recommendations. Environmental rehabilitation, courtyard economy, new energy, agriculture and animal husbandry demonstrations, infrastructure, environmental education and vocational training which affect the local environment and community development were listed in the management plan and annual plans. Relevant departments were mobilized and villagers' enthusiasm for LD control was increased. These efforts assured villagers' strong support to the development of the Pilot Site.

2.3 Demonstration of the IEM approach and application

2.3.1 Grassland management

Taking the severe degradation of the grasslands within the watershed into consideration, the Project identified grassland management as a breakthrough, followed by reaching agreements with households who have a large number of livestock and 8 administrative villages (AVs) to work together on grassland management. The grazing period within the Watershed was then shortened to 2 months and full time personnel was assigned to guard the grasslands. At the same time, Major households with a large number of livestock were encouraged and funded to adopt stall feeding. 54 livestock pens with each covering 60 m^2 were set up and 24 hm^2 oats were established. 15000 Chinese zokors were killed by villagers to reduce damages to the grassland, which also reduced reverse impact on the environment due to reduced application of pesticide.

2.3.2 Revegetation

To improve domestic water security within the Watershed, the Project invested 5480 yuan to fence the domestic water source area which covers 3.33 hm^2 and banned grazing. At the same time, wetland protection demonstration and vegetation monitoring activities were undertaken: 126,000 Yuan was invested where water and soil losses were severe in establishment of 2800 *Populus cathayana* seedlings and 37,000 Picea crassifolia Kom seedlings; 195,000 Yuan was invested in fencing of 266,000 mu, afforestation for 324 hm^2, mountain closure and replantation of *Caragana Korshinskii* Kom for 40 hm^2.

2.3.3 Improvement of infrastructure

The Project invested 52,500 Yuan to improve infrastructure including road upgrading for 21.237 km and 2 bridges, which was integrated into the Government's Integrated Village Development Program. Meanwhile, villagers themselves raised funds for field roads for 12 km to improve road access to their farmland.

2.3.4 Renewable energy

To reduce dependency on the limited natural resource, the Project invested 47,000 Yuan to assist pilot villages to introduce 31 biogas tanks which are more efficient, safer to use and are able to generate more biogas, and provide a electricity-efficient stove for each demonstration househould (total 608 stoves) and set up a solar stove for each household within the Watershed.

2.3.5 Establishment of Farmer Field School (FFS) and demonstration activities

In early April of 2007, a Farmer Field School (FFS) was respectively set up in Shanghudan and Xiahudan Villages. It was the first farmer school in Huangyuan County in its history. At the same time, a Livestock Farmers' Association (FA) was established in Bayan Township in Hudan Watershed, with FA systems set up and staff members elected. The FFS assists farmers to address problems raised from income generation activities and restructuring of plantation. Several technical training courses on warm sheds and greenhouses establishment, mushroom development and management, pest and disease control and other communication activities were organized for farmers by the FFS. The Project also invested 43,800 Yuan to assist 159 households in Xiahudan Village to set up greenhouses for development of vegetables by providing part of the necessary materials and upgrading the irrigation facilities. In addition, other training courses including mushrooms, wooden-board-bed pig raising and demonstration of breeding pigs were organized for farmers.

2.3.6 Improvement of infrastructure for entertainment and cultural activities

The Project invested 24,700 Yuan to set up basketball courts and Ping Pong tables respectively in Shanghudan and Xiahudan Villages to improve facilities for sports and cultural activities. This meets cultural needs of the villagers.

2.3.7 Project monitoring

Monitoring activities were undertaken to monitor the development of the pilot sites. Monitoring was strengthened by undertaking soil testing before fertilizing, observing the change in rainfalls, gullies and the number of livestock.

2.3.8 Environmental Education (EE) campaigns

EE campaigns were organized in the pilot villages to increase environmental awareness with assistance from the Australian Youth Ambassador for Development (AYAD). Training workshops on LD, environmental protection and management were organized using participatory methods which were easier for farmers to accept and villagers' environmental awareness and their willingness for environmental protection were developed and increased.

3. Achievements of the pilot sites

The GEF pilot sites in Huangyuan County have provided experience in participatory approaches and developed the IEM concept among the Project areas. The objectives of the pilot site were successfully achieved: the environment of the Project areas has been improved; water and soil loss has been stifled; wildlife resource has been protected; villagers' knowledge on LD control, environmental protection and management has been strengthened; villagers' thinking has been changed and their willingness for environmental protection has been strengthened. It can be concluded that prominent achievements have been made at the Project pilot sites.

3.1 Project approaches fit well in the New Rural Development Program and environmental management efforts in the Watershed

The approach advocated by the Project was the multi-sector and multi-disciplinary participatory approach. To meet this requirement, the Project Leading Group of Huangyuan County undertook advocation and education campaigns on participatory approaches, the IEM approach, the New Rural Development Program, ecological development and relevant legislation at all levels, which largely increased the leaders, staff and villagers' awareness and understanding of the Project approach and environmental issues, and through which the IEM approach was integrated into other programs of the County, including the New Socialist Rural Development Program, forest development programs, agricultural and animal husbandry development programs and infrastructure improvement activities.

3.2 Multi-sector and multi-disciplinary participation in the Project

The Pilot Site provided an opportunity for Huangyuan County Government to undertake coordinated planning, in which all relevant departments were mobilized to get involved in and support development of in the Project by using their advantages, integrating resources, combining funds and intensifying investments for key management activities. To date, over 17 million Yuan has been raised for project activities, among which a combined fund for 1.758 million Yuan was provided by the food program of the County Government to support setting up 400 greenhouses and part of the irrigation facilities, upgrading electricity facilities in two villages, training courses on science, technology and environmental development were provided for the Project areas, with 1850 participants involved, increase of migration workers, with over 560 labor force in each of the two villages migrated into urban areas; a combined fund for 324,000 Yuan was provided by the Agricultural and Animal Husbandry Bureau and the Agricultural Extension Center to assist villagers to set up 64 60-square-meter livestock pens; the Water Resource Bureau provided combined investments for 8.456 million Yuan to support the Project to build one flood control pond, flood control walls for 50 m, a domestic water supply project in Shanghudan Village, upgrade of tap water pipes for 17.8 km in 7 sub-villages, irrigation ditches for 70 km to improve irrigation for 73.33 hm^2 dry land, provide domestic water for 7500 people in the area and improve irrigation for terraced fields for 252.33 hm^2; the Education Bureau provided 800,000 Yuan to set up 4 Hope Schools; the Transportation Bureau provided 2.8 million Yuan to upgrade roads in the Project areas; the Poverty Alleviation Office invested 1.371 million Yuan to set up two Tibetan Carpet Processing Centers and 308 biogas tanks, and offer 130 jobs; the County Forest Bureau invested 85,000 Yuan to support afforestation for 25.33 hm^2 in Shanghudan Village and 31.33 hm^2

In Xiahudan Village, assisted farmers to convert land use from cropping to afforestation for 9.33 hm^2; the Energy Station of Huangyuan County invested 47,000 Yuan and gave solar stoves to each household in the Watershed for free.

21. Building Information Platform Serving for Ecological Development

Wang Zepeng
Deputy Director, Forestry Planning and Design Institute of Ningxia

Abstract: Ningxia is located in the western part of China. It is one of the provinces in China that suffer from serious problems of land desertification. In order to accelerate the progress of land desertification control, a provincial platform on comprehensive information of land desertification is required. With the support of GEF-OP12 Project, Ningxia IEM Information Center has been set up to provide scientific and systematic data services in favor of formulating policies and plans for land desertification control in Ningxia. At the same time, the Center manages the data flows within the IEM information system, facilitates the data exchange between the local IEM institutions and the international IEM institutions, and helps analyze the benefits/impacts from the efforts of land desertification, so that the integrated capacity against land degradation is upgraded.

Since the establishment of the Center, lots of work has been done, including (i) that the required equipments have been procured and the Center has been professionally staffed; (ii) that training has been strengthened to upgrade technical capacity; (iii) that the data sharing agreement of Ningxia IEM Information Center was collectively signed by all the member institutions; (iv) that the technical personnel of the member institutions have been trained, data inventory on land degradation has been formulated, and a list of meta-data collection has been prepared; (v) that a system of meta-database has been set up in the Center; and (vi) that the Center has begun its due services for the ecological construction in Ningxia. Also, it has provided lots of information services for the Project of Ecological Construction around Liupan Mountain, Sino-German Afforestation Project in Ningxia, and Overall Planning of Ecological Construction over the Loess Plateau in Ningxia.

In addition to the above-mentioned achievements, there are still some problems. For example, there are some difficulties in data collection, the technical force is still weak and there is deficiency in budget for the operation. With powerful support of Ningxia GEF-OP12 Executing Office, the Center will actively communicate with the relevant institutions to collect more information, so that the Center will make better services for the ecological construction in Ningxia Province.

Key words: information center, data sharing

Enhancing ecological development and protecting ecological security are not only common global problems we face in the 21st century, but are also the important foundations for China's economic and social sustainable development. Ningxia is located in the western part of China. It is one of the provinces in China that suffer from serious problems of land desertification. Due to its geographical position and relatively poor development, Ningxia faces great challenges in land desertification, which is one of the most serious threats to Ningxia. Therefore, the immediate concern for Ningxia Hui Autonomous Region is to establish and put into effect the outlook of scientific development, and initiate a new overall prospect of combating desertification to accelerate combating land desertification within the scope of land degradation prevention and control. Due to division of sectors and the relative limited knowledge regarding land degradation, the data related to land degradation are usually acquired and processed through different methods, and are seldom shared among different departments. The investigation and monitoring conducted by each department can only reflect one aspect of the region's problems in land degradation. Up to now, the integrated information of the region's land degradation cannot be offered from the viewpoint of integrated ecosystem management, which greatly restricts the formulation and implementation of the strategic plan for combating land degradation. Therefore, Ningxia Hui Autonomous Region actively strives to support the GEF-OP12 Project on Combating Land Degradation of Dry-land Ecosystems, and established Ningxia IEM Information Center to provide scientific and systematic data services in support of formulating policies and plans for combating land desertification, and to further improve integrated capability in combating land desertification in Ningxia.

For the mankind, information is valuable, especially when it is viewed as a type of resource. Information represents the nature of knowledge, and tells people the motioin state of things and the change in the motion state. In this sense, information is the raw material for all knowledge, and people's knowledge comes from information processing. There could be no knowledge without information. That is why information becomes the basis for decision-making and the core support for management. With continuous social development, people will need more information resources apart from more and better materials and energy resources. Moreover, the inevitable trend is that the need for information resources will gradually surpass the need for materials and energy resources. Ningxia has set up the land degradation database and information system based on a GIS platform aiming at collecting the original data from the original data source information system, surface data of the classified original data of each discipline, integrated eco-factor data, and catalogue data of the ecosystems in Ningxia, and establishing monitoring and evaluation models related to integrated ecosystems, models of long-term population dynamics of the important species, models of system successions and gene evolutions, land degradation models, and the models of water resource dynamics by utilization of various databases in the system. The above-mentioned models and modeling environments together constitute the information-coordinated library of models, which not only provides information services for integrated ecosystem management projects, but also provides information and decision-making support in relation to ecosystem change and land degradation tendency for national and provincial decision-making sectors.

Since the establishment of the Information Center of Ningxia Integrated Ecosystem Management (IEM), great efforts have been actively made to development of the functions of the Center.

(i) Purchasing Equipments

The following facilities and software have been purchased, including desktop computers, laptop computers, workstations, servers, large-format inkjet printers, AO-scale scanners, ARCGIS9.2 and ERDAS9.1 software provided by the Central Government's GEF Project Office.

(ii) Choosing Staff for the Organization

Based on the Center of Geographical Information of our Institute, the organization has been established and the center has been assigned with technical personnel.

(iii) Strengthening Trainings

Training has been strengthened to upgrade abilities and technical capacities of the technical personnel of the IEM Information Center by self-learning and special training. Five technical officers participated the training held by the Central Government's GEF Project Office based on their systematic study in Remote Sensing and Information Center of State Forestry Administration's Northwest Institute of Forestry Inventory, Planning & Design, and Institute of Remote Sensing of Ningxia. By the short-term intensified technical trainings, the technical officers of the IEM Information Center have basically learned the software operation of geographical information system (GIS) and can apply the knowledge and technology to land degradation monitoring and evaluation.

(iv) Signing of the Data Sharing Agreement of Ningxia IEM Information Center

The information center has signed the "Data Sharing Agreement of Ningxia IEM Information Center". Upon coordination of the Executive Office of the GEF Project of Ningxia, this agreement was collectively signed by 13 member institutions including the Finance Department of Ningxia Hui Autonomous Region, the Commission of Development and Reform, Department of Land & Resources, Forestry Bureau, Environment Protection Agency, Agriculture and Animal Husbandry Department, Water Conservancy Department, Poverty Alleviation Office, Legislation Office, Bureau of Reclamation, Department of Science & Technology, Academy of Agriculture & Forestry, and Ningxia University.

(v) Training Technical Personnel of Member Institutions and Collecting Data Inventory on Land Degradation

After the establishment of the Ningxia IEM Information Center, data collection has been actively carried out. Since May 2007, two training activities have been held, including "Training Class on Meta-data of IEM Information Center of the Ningxia GEF Project", and "the Seminar of Meta-data Input of IEM Information Center of the Ningxia GEF Project". Officers of the Central Government's GEF Project Office, an international expert Mr. Douglas, Prof. Liu Rui, chairman of the 5th panel group, and technical personnel of the member institutions were invited to attend the training activities. The Information Center trained the technical officers of the member institutions on meta-data with concrete examples in practicing meta-data input. As for the seminar, topics discussed included the data inventory for combating land degradation in Ningxia that could be offered by every member institutions and arranging the data inventory collection and input of the data into meta-database. Following the seminar, meta-data has been collected according to the data inventory of each member institution and then arranged as the standard pattern of the meta-data. The Information Center has also collected and filed the data inventory including basic geographical information data, topographic map, satellite images, DEM, forestry resources of thematic data, desertification, sandification, wetland, and wildlife provided by some member institutions including Ningxia Forestry Investigation & Planning Institute, Ningxia Agricultural Reconnaissance Institute, Ningxia Development & Reform Commission, and Ningxia University.

(vi) Establishing a Meta-Database System in Ningxia IEM Information Center

The meta-database system in Ningxia IEM Information Center has been established and posted on the website of Ningxia Forestry Information according to the collected meta-data inventory of land degradation provided by each member institution as well as in accordance with the requirements of the GEF OP12 Project of China-GEF Partnership on Land Degradation in Dryland Ecosystems.

(vii) Ningxia IEM Information Center Provided Due Services for Ecological Construction in Ningxia

Invited by Ecological Construction Office of Ningxia Development & Reform Commission, Ningxia IEM Information Center has provided trainings on how to set up databases and how to draw project maps related to eco-economic development surrounding Liupan Mountain for the project management staff. There were eight base maps offered by Ningxia IEM Information Center for the training courses. During implementation of the Project of Integrated Treatment of Desertification in Ningxia under the framework of the Sino-German Financial Cooperation Project of Integrated Desertification Prevention and Control in Northern China, Ningxia IEM Information Center has done lots of work in participatory land-use planning, grassland restoration, erosion prevention and control, rural development, and farmer training. Further, the Information Center has also collaboratively established a publicly-shared database, geographical information system, drawing maps recording current situation and future planning, carrying out surveys on training demands, formulating training plans, and vegetation monitoring of Luo Mountain.

Although much has been achieved, difficulties in implementing this project still exist, such as data collection, weak technical forces, and deficiencies in the operating budget of the project. So, Ningxia IEM Information Center will actively collaborate with the Ningxia GEF Project Office, and communicate with relevant industries and departments to collect information related to the Ningxia ecological environment. The Ningxia IEM Information Center will do its best to accomplish the goals through a full preparation in office spaces, infrastructures, staffing and training, data collection and processing and by means of mutual support of work and publicity. Thus, the Center will offer better services for the ecological construction in Ningxia.

22. Community Capacity Building and Achievements in GEF Pilot Sites, Shaanxi Province

Wen Zhen
Deputy Director, International Forestry Cooperation Center of Shaanxi Provincial Forestry Department

Abstract: The pilot sites of the GEF-OP12 Project in Shaanxi was commenced in early 2007 and along with implementation activities great attention has been paid to community capacity building on land degradation control and extension of integrated ecosystem management concept. The community capacity building activities include participatory community development planning, a process in which community members are empowered to make their own decisions for their future prosperity; Farmer Field School (FFS) where farmers improve their ability by identifying, analyzing and addressing problems using their indigenous knowledge; and environmental education and participatory monitoring, aiming to increase farmers voluntary involvement in land degradation control and other environmental rehabilitation activities by increasing their environmental awareness. Implementation of capacity building activities has improved community's overall capacity and quality, which is of significant contribution to improvement of the local ecological environment, land degradation control and promotion of sustainable community development.

Key words: GEF, pilot sites, land degradation, capacity building

PRC-GEF Partnership on Land Degradation in Dryland Ecosystem (GEF-OP12) is an integrated ecosystem management project implemented by GEF in China aiming to control land degradation, alleviate regional poverty, and improve and restore ecosystem which human being relies on. Shaanxi Province is one of the provinces involved in the implementation of the project. With the hard efforts of both strategic team and law team, Shaanxi has achieved important results in legislation, policy and institutional capacity building for land degradation control. Among these achievements are the IEM Strategy and Action Plan for Land Degradation Control in Shaanxi drafted by the Project and application of the IEM approach at the pilot sites. Pilot site activities include afforestation to improve ecological environment, establishment of biogas pits to promote clean energy, water-saving irrigation works to save water resources, farmland improvement to control land degradation, and community capacity building to improve community capability.

The proverb "Giving man a fish and you have fed him for today. Teaching a man to fish and you have fed him for a lifetime" emphasizes the importance of capacity building. The community capacity building at pilot sites of the GEF-OP12 Project in Shaanxi is the important component in pilot site development, and it covers participatory development planning, farmer field school, environmental education, and participatory monitoring. During the implementation at the pilot sites, we paid great attention to fostering and improving community capability, which is welcomed by community members and evident effects have been achieved.

1. Community Capacity Building Activities

During the period between July 2006 and June 2008, many training sessions on participatory planning were carried out at the pilot sites of the GEF-OP12 Project in Shaanxi, and about 400 managerial and technical personnel and community members were trained. With the assistance of the technical personnel, the pilot site communities implemented participatory development planning and worked out their development plans; 2 farmer field schools were implemented with 60 farmer participants, which significantly improve the capability of the communities; three environmental education sessions were conducted at different levels with more than 100 perople directly involved; Shaanxi Provincial environment education plan was developed, and over 90 households at the pilot sites participated in the participatory monitoring activities.

To effectively improve the capability of the communities, many documents concerned were compiled, printed and distributed. They include development of training material on Participatory Development Planning which serves as guidance manual for training class and for the implementation of community's participatory planning at the pilot sites, development of Participatory Monitoring Manual to guide the participatory monitoring activities implemented by the communities, and development of Environmental Education Manual for elementary school pupils at the pilot sites to guide implementation of environmental education. The development and distribution of these materials played an important role in guiding the pilot site development and community capacity building.

2. Theory and Practice of Community Capacity Building

2.1 Participatory approach increases farmers' self-confidence

Participatory approach is a very scientific method for the village development planning currently. Its core is to respect communities' wishes, and implement the project activities according to what the communities most care for. In its process, communities played decisive role and they decided what activities should be done themselves. It allows the local communities and the stakeholders to fully participate in determination, design, planning, implementation, and monitoring of sustainable ecosystem management project. This is an important achievement of the pilot site development.

At the early stage of the pilot site development, we trained technical personnel on participatory development planning at the pilot sites. With organization of technical personnel, the communities to undertake in-depth discussions and analysis of the existing community problems. Further discussions were held and their problems were screened and ranked resulting in a list of prioritized community problems. According to their ranking, we discussed with village households about solution and then made scientific overall arrangements by means of integrated ecosystem management concept. Participatory development plans at the pilot sites were finally formulated.

The development planning for the pilot sites of Wuqi, Jingbian and Yuyang of the GEF-OP12 Project in Shaanxi was undertaken in participatory approaches. During the participatory process, communities had sufficient opportunity to confirm their capacity and potential, and mention their problems and express their wishes. They had deep understanding of their environment and the feeling of ownership, so their responsibility and self-confidence were both increased. Participatory approaches provide the land owners a platform to explore common development problems. Since community development issues are most cared for by the land owners and they reflect their wishes and need, the land owners are interested in them, and they are familiar with them and wiling to discuss them. During the get-together discussions, the farmers further realized their great potential for development. The targets determined by the participatory approach can truly reflect local situation, and express land users' wishes and needs. Communities take their developed community development planning as their achievement and organize its implementation with their great enthusiasm.

2.2 Farmer Field School improves the capacity of farmers' team work

Farmer field school is farmer-centered, where the field is taken as class and informal adult education was adopted. In the season when crops are growing, the training sessions were carried out in the field with heuristic, participatory and interactive characteristics. It enables farmer trainees to find problems and seek solutions by themselves. Through active and participatory learning and practice, farmers' self-confidence, ecological awareness, team work, production capacity and decision-making capacity were increased. In the meanwhile, it was also enlightening and fostering of trainee's awareness of learning initiatives to improve farmers' overall quality and wealth-acquiring capability. Farmer field school is a new model currently for agriculture technical extension, farmer training and farmer quality improvement; and it is an effective way to foster new farmers under the market economy. Using farmer field school to extend agricultural techniques plays an important role in improving farmer's management quality, degree of organization and overall capacity of agricultural practice.

According to the requirements of the Central Office of the GEF Project and spirit of multiple training, and based on our understanding on field schools, we held two farmer field schools at the pilot sites of Haizetan of Jingbina and Shayan of Yuyang in 2007. Field schools provide permanent schoolhouse for trainees, provide experimental field for trainee's study. The schools also have technical assistants, well-established learning system and institutional discipline, clear work division, and commonly interesting study topics. The schools also have learning and study equipments financially supported by the GEF and select interested and motivated community members to participate in the learning.

Field school at the Haizetan pilot site of Jingbian had 30 trainees. After full discussion between technical assistants and trainees, maize growing was selected as the main study topic , as it is the most important local agricultural activity. On June 13th, 2007, the field school held opening ceremony and leaders from the Project Central Office, the Provincial Project Office and the county government attended the opening ceremony. Afterwards, under guidance of technical assistants, the trainees had 6 times studies together. They learned and discussed about problems such as fertilizer application, non-fecundity and incomplete fecundity, and conducted comparative experiments. The field school provides the trainees with opportunity to observ, analyze and solve the problems together in the production practice.

Field school at the Shayan pilot site of Yuyang District had 30 trainees, and management of the local main cash crop "Kernel Apricot" was selected as study topic. In May of 2007, under the organization of District Forestry Bureau and technical assistants, the opening ceremony of the field school was held. Get-together learnings were organized five times to learn several key technical points concerning freezing damage control during flowering period, fertilizer application, and pest insect control, etc. Through the collective activities organized by the field school, the team work and cooperation of the trainees have been enhanced, their potential capacity was developed, and their participatory learning capacity, self decision-making capacity and self-management capacity were improved.

2.3 Environmental education raises the public's environmental protection awareness

To control anthropogenic destruction on environment, change human being's ill living habit and Change the worsening trend, we followed the overall requirements of the GEF Project and developed Shaanxi Provincial environmental education plan.. It covers environmental education goals, subjects, methods, etc. It is not enough to change environment by relying on only one department or an organization; it has to mobilize the whole society and take common action together then it could create impact on environment.

Since the apperance of mankind, environment has everlastingly provided human being with necessary resources and energy, and carried us from primative to civilization. At the same time, human being has also sought for the harmonious coexistence between man and environment to achieve sustainable development.

"The water that bears the boat is the same that swallows it up" and so does our environment. Environment has its self development pattern, and when the pattern is broken, its internal balance would be breached and its function of maintaining mankind would be lost, eventually causing harms to human being. With social and economic development and accelerating pace of industrialization, the human being's behavior would have tremendous impact on environment no matter it is intended or unintended, and it causes the environment that maintains our subsistence to deviate from its intrinsic development pattern, resulting in losses of its self-balance. Such environmental factors as global climate change, decrease of biodiversity, land degradation, shortage of water resources and frequent natural disasters are becoming the fatal hindrances affecting human social development.

Having environment to maintain its intrinsic development pattern and its balanced development, the society has launched many influential environmental protection campaigns. The United Nations has developed many international treaties on environmental protection, and each country has issued a series of policies on environmental protection one after another and launched regional environmental protection programs. These measures fully reflect the human being's unremitting efforts for their subsistence and for the improvement of the environment.

For environmental improvement, international treaties and national policies are no doubt very important. However, personal behavior should also not be ignored and joining together of everyone's action would form a tremendous power. How to have this power to play its role, how to have public to be aware of importance of environment to us and what impact it has on our life, and what impacts our actions would have on environment will need us to have broad environmental education on public. Only if the public are aware of the environment and its problems, their awareness has created impact on their thinking and action, and they would also self-consciously think about environment factors in their life and work, forming the whole society protects environment together, it will then create far impact on environment improvement. This is why we carry out environmental education.

The GEF integrated ecosystem management demonstration project takes environmental education as one of its important component to increase the public's awareness of environment problems, especially land degradation, when implementing land degradation control demonstration activities. It is to increase the public's awareness of environment and promote their active participation to support environmental protection and restoration activities. Protecting environment is not only the national or government's business, but is also everyone's. Issues from domestic waste management to ozonosphere protection in air are closely linked with every activity of ours. We hope everyone who has received environmental education could start taking environmental protection action from every piddling thing at his/her hand. We are the host on the earth, and every speech and activity of ours will have big impact on environment. The nature will react to everything we do to it.

In June of 2007, we held an environmental education activity in Xi'an which was targeted at government officials. Officials from the provincial departments concerned attended the activity. Through this activity, if leaders at each level of government departments could take environment factors into account in decision-making process, use integrated ecosystem management concept and the formulated policies could cover certain environmental protection part, it would make immense contribution to the environmental improvement. In April of 2008, we held an environmental education activity at the Shayan pilot site in Yuyang District which was targeted at both the county government officials and farmers in the hope that environmental protection measures, skills and integrated ecosystem management concept could be promptly applied in production practice and could effectively guide farmers' production activities to achieve the goal of environmental protection and land degradation control. In August of 2007, we held the environmental education at Haizetan of Jingbian which was targeted at elementary school and middle-school students, hoping that through their environmental protection activities, the students could build their awareness of environmental protection from their early age and then forward environmental protection knowledge, concept and activity to their parents or even broader communities. It would also have impact on their future growth and they could become the new-generation successors of environmental protection.

2.4 Participatory monitoring enables farmers to learn how to observe and analyze

The participatory land degradation monitoring and assessment implemented at the pilot sites enabled communities to learn how to observe, diagnose, understand and assess agricultural ecosystem which the communities live in, such as agricultural land, grassland and water resources, and other relevant indicators.

Participatory monitoring activities provided the communities with opportunity to analyze their risks, cost of input and results achieved so they are clear about the actions they will take or are taking are what they would expect. It is to timely provide reliable and understandable information so that both the communities and external project managers acquire more effective information. This could enable them to self-consciously apply integrated ecosystem management concept into natural resources management and land degradation control, and to form sustainable land management awareness and to help better and comprehensive understanding of land degradation issues so that effective land degradation control technology and methods could be selected and extended with consideration of the local situation.

In the pilot villages, we selected 30 farmer households for participatory project monitoring in each village, made clear of the tasks for each household and set their responsibility, purchased monitoring equipment, established monitoring sites and set up monitoring signs. According to the requirements of participatory monitoring manual, each one is responsible for monitoring 1-2 indicators, and conducts long-term observations and measurements and records data as basis for the analysis and assessment of the project.

3. Impact of Community Capacity Building

In the production practice of land degradation control, the pilot sites fully used the results of community land degradation control capacity building, emphasized on local conditions, and integrated air, vegetation, soil and human factors. The designed project activities are connected and rely on each other which reflect integrated ecosystem management thinking. Aiming at grassland degradation problem, the pilot sites designed the project "stable feeding livestock project based on available quantity of fodder grass". The concentrated excrement of cow and sheep then provides raw material for biogas tank. The biogas production is a clean and renewable energy, and it provides fuel for water heating and cooking for the local communities which reduces their demand for fuelwood. The solid and liquid residues from biogas production provides, in turn, organic fertilizer for the improvement of the arable land for growing maize and increases in grain yield. When the farmers' incomes are increased, they could implement more environmental protection activities and land degradation control measures. This is a scientific cycle under the integrated ecosystem management, and it is closely linked with land degradation control topic. It insists on environmental protection tenet, and it is the important reflection of significantly improved land degradation control capacity of the communities at the pilot sites.

Through capacity building of land degradation control, the GEF Project in Shaanxi at the pilot sites has already achieved preliminary results. There are well-established laws and regulations serving as the basis, the coordinating capacity of community institutions has been improved, the partnership of land degradation control is preliminarily established, the right of communities participating in public cause is respected, and participation interesting is fully exerted. Through implementation of the participatory community development planning, the local farmers had opportunity to express their wills, their participating consciousness, cooperation capacity and decision-making ability have been improved, which have been applied to various development planning in villages. Through farmer field school, the local farmers learnt how to observe and understand problems, learnt the team work and cooperation, learnt how to analyze and discuss problems together, and learnt how to solve the problems through comparative experiments. They are able to conduct field experiments themselves to improve their production skills. Through implementation of environmental education, it has improved communities' understanding on environmental problems, especially on land degradation. They are able to recognize the big impact of their production activities on environment, which promote them to actively participate in and also strongly support environmental protection activities. Through implementation of the participatory monitoring, it enabled the local public to directly recognize actual situation of surrounding environment, feel the advance of land degradation and threat of worsening environment, and increase their awareness of environmental protection so that they could self-consciously take actions for land degradation control and environmental construction. Integrated ecosystem management concept has been recognized and applied widely in the pilot project areas, and ecological, economic and social benefits are gradually emerging. The improvement of the community capacity will create far effects on sustainable development of communities, and the models established and experience acquired will play an important, demonstrating and leading role for global land degradation control and environmental improvement.

References
[1] Ren, Baozheng, & Sun, Zuow. 2006. Participatory Farmer School. China Agriculture Press, Beijing, PRC.

[2] Piao, Yongfan & Chen, Zhiqun. 2005. Theory and Practice of Farmer Field School. China Agriculture Press, Beijing, PRC.

23. Study on Evaluating Forest-Related S&T Innovation Capacity in Western Areas Stricken by Land Degradation of China

Liu Xilin, Hu Zhangcui
School of Management, Northwestern Polytechnical University, Xi'an, 710072, P.R.China

Abstract: Land degradation has been a crucial ecological concern increasingly threatening human subsistence and well-being. The pressures on ecological environment and the quality crisis of forest resources cause to the urgent need of S&T innovation to provide strong support. The degraded lands in the West China are the ecologically vulnerable areas of China, with rather weak forest-related S&T innovation capacity. In such context, this paper selected 7 greater regions and 28 provinces/regions and undertook evaluation on their capacity respectively by setting up an evaluation indicator system on such capacity. Through comparative analysis, it presented an objective evaluation on the forest-related S&T innovation capacity in degraded lands of West China and proposed corresponding countermeasures and suggestions to improve the capacity there.

Key words: Land degradation; S&T innovation; evaluation indicator system

1. The general situation of land degradation in West China and the analysis on demands of S&T innovation

1.1 The status quo of land degradation

Currently, land degradation, in which desertification is most severe, has been a crucial ecological concern threatening all human beings' subsistence and well-being. China is one of the countries suffering from the worse desertification in the world. Results of the Third National Desertification and Sandy Land Monitoring issued in 2005 showed that the total area of desertification lands was 2.6362 million km^2, taking up 27.46% of the total national land area. Among which the six provinces/autonomous regions of Xinjiang, Inner Mongolia, Qinghai, Gansu, Shaanxi and Ningxia in West China were the main areas for the distribution of desertification lands (Figure 1), taking up 81.13% of the country's total area of desertification lands. The total area of sandy lands was 1.7397 million km^2, taking up 18.12% of the total national land area and among which 82.53% were distributed in the above six provinces/regions in West China (Figure 2). It is evident that the west is the worst stricken area by land degradation in China. Here the poor population exceeds 30 million. Land degradation is not only directly threatening the local people' subsistence and well-being, but also affecting the living environment in the west.

1.2 Analysis on demands of S&T innovation for land degradation combating in west China

(1) A range of key scientific and technological topics are still waiting to be studied and addressed for and in the course of making the land degradation control efforts.

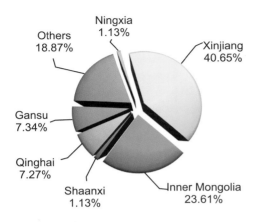 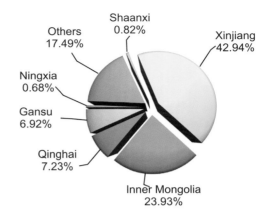

Figure 1 Proportions of desertification land area of the six provinces (autonomous regions) in the country's total desertification area

Figure 2 Proportions of sandy land area of the six provinces (autonomous regions) in the country's total sandy land area

The long cycle and heavy difficulty of land degradation control call for continuous reinforcement of scientific researches to tackle key issues while providing combating efforts. They include key technologies for land degradation control and revegetation in degraded areas; the rule to stabilize forest ecological communities; the cultivation of quality tree varieties with strong resilience and suitable for growing in degraded areas, the acceleration of quality forest tree variety regeneration and the improvement of the use level of quality varieties; efficiency of water resources in degraded areas and the guarantee of eco-system's hydrological balance; and etc.

(2) Scientific and rational planning of combating measures is required.

The composition of forest and grass species should be determined scientifically. The plantation of arbor trees, bushes and grass should be site-specific and in rational arrangements, together with the organic combination of biological, engineering and agricultural approaches. The achievements of S&T innovations should be assembled and fit in the characteristics of different areas, with the priorities given to extension and application of improved plant varieties with strong resilience and advanced and practical integrated control technology and modes to establish a range of high-level demonstration bases for scientific control step by step.

(3) Scientific and technological achievements and applied technology are needed to be largely extended and applied in land degradation control.

During the long-term experience and practices in preventing and controlling land degradation, China has developed a range of advanced and applied technologies and modes of control by researches, e.g. water saving and conserving technology, technology of afforestation in sand-blowing areas, aerial seeding technology for planting trees and grass in sandy areas, development of technology of shelterbelt forest system and structural model arrangement, technology for integrated small watersheds management, technology for ameliorating saline and alkali lands. Such advanced technologies need to be largely promoted and scientifically and effectively applied in land degradation control efforts.

In a word, in land degradation control in West China, both scientific planning and the extension and application of advanced and practical technology and the continuous and new resolutions to key scientific issues will need the forest-related S&T innovations to provide strong support.

2. The evaluation on the forest-related S&T innovation capacity in degraded lands of West China.

The forest-related S&T innovation is an organic part of the national scientific and technological innovation. It plays an important role in improving forest and forestry productivity, and is a dynamic and sequential innovating process. It includes not only innovations in forestry production and seed/seedling cultivation, but also in natural resource management and environment protection adaptive to socio-economic development. In particular, the forest-related S&T innovation capacity is more needed to be continuously improved in West China with severe land degradation, backward economic development and fragile ecology, so as to minimize the gap between the west and other areas in terms of forestry development and ecological improvement.

2.1 Evaluation indicator system

To improve the forest-related S&T innovation capacity, firstly, sound environments including market, knowledge and technology are needed for forestry development, so as to create a sound atmosphere for innovation. Secondly, the guarantee of formal or informal frameworks is needed to stabilize the anticipation of innovations. Thirdly, sufficient supplies of substances, human resources and knowledge capital are needed to ensure sufficient resources required for innovations. Finally, a robust extension system of science and technology is needed to be formed to benefit the west to the end. Therefore, an evaluation indicator system on the forest-related S&T innovation capacity in China is set up in 4 aspects including the basic environment, the capital supply, the innovative activities and the extending capacity, as Table 1.

2.2 Evaluation on regional forest-related S&T innovation capacity

Chinese forestry development is greatly characterized by regionality. There are factors causing regional differences. On one hand, the difference between natural conditions being zonal or non-zonal determines the adaptable extent and degree of forest biomass. On the other hand, the differences of socio-economic conditions determine the utilization direction, productivity outlay, structure, management modes and level of production development of forest resources. Also based on geographical and economic characteristics, China's administrative division has been specified as 7 major regions including North China, Northeast, East China, Central China, South China, Southwest China and Northwest China. This paper has used the principal component analysis to evaluate the forest-related S&T innovation capacity of the above 7 regions covering years of 2001 to 2007. Results are shown in Table 2.

Results of the evaluation on regional forest-related S&T innovation capacity during 2001-2007 showed, in term of the east-west distribution of the Chinese administrative division, the innovation capacity of the east is obviously better than that of the west. While there is no obvious change in the general ranking of the regions, north China en-

Table 1 The Evaluation Indicator System on Forest-related S&T Innovation Capacity of China

Objective Level	Grade -1 Indicators	Grade -2 Indicators
Forest-related S&T Innovation Capacity	Basic Environment (F1)	1. Number of key labs
		2. Amount of fixed assets of forest research institutions
		3. Amount of apparatus and equipment for forest researches
		4. Number of library collections and digital books and literatures of forest-related science and information institutions
	Capacity of Capital Supply (F2)	5. Constant income of forest research institutions
		6. Proportion of governmental capital investment in total incomes
		7. Proportion of capital for forest-related science and technology in governmental investment
		8. Proportion of funding for forest-related science research projects in total capital investment
		9. Proportion of personnel engaged in scientific and technological activities
		10. Proportion of personnel above master degree
		11. Proportion of R&D personnel
		12. Proportion of scientists and engineers
	Capacity of Innovation Activities (F3)	13. Proportion of funding for R&D in total funding for forest-related projects
		14. Amount of project funding invested by government
		15. Amount of project funding invested by non-governmental sources
		16. Funding for international scientific exchange and cooperation
		17. Amount of forest-related technical incomes
	Extension Capacity (F4)	18. Number of patent authorizations per 10,000 persons
		19. Number of published scientific papers of the concurrent year
		20. Number of published scientific books of the concurrent year
		21. Consultation activities on forest-related technology (person/year)
		22. Training activities on forest-related science and technology (person./year)
		23. Literature services and activities on forest-related scientific information (person/year)
		24. Demonstrating and extending activities for forest-related scientific and technological achievements (person/year)
		25. Amount of expenses of funding for applying forest-related technical research and trial development results
		26. Amount of expenses of forest-related scientific extension, demonstration and services

Table 2 Ranking of Regional Forest-related S&T Innovation Capacity during 2001-2007

Regions	2001	2002	2003	2004	2005	2006	2007
North China	1	1	1	1	1	1	1
Northeast China	3	3	3	3	2	4	3
East China	2	2	2	2	3	2	2
Central China	6	6	5	5	6	6	6
South China	4	5	6	6	5	3	5
Southwest China	5	4	4	4	4	5	4
Northwest China	7	7	7	7	7	7	7

joys remarkable advantages in its innovation capacity and its status is unshakable by other regions in a short run. The capacity of the northwest is extremely weak and lags far behind other regions.

The Northwest shares the typical development characteristics of West China. It is the most ecologically vulnerable region. As mentioned above, the five provinces/autonomous regions in this region include Shaanxi, Gansu, Qinghai, Ningxia and Xinjiang which are those suffering from the most serious land degradation in China, also with a rather weak forestry basis. For a long term, factors including changes of natural conditions, economy, society, and history have caused to the tense conflicts among resources, environment, economy and people of the west. Due to the inborn deficiency for development, there are more restricting factors for the forest-related science development, including relatively backward economic development level, slow steps in infrastructure and environment for S&T innovations and low attractions to scientific professionals, which lead to the failure to adapt to the requirements of S&T innovation development. Among these areas, some provinces also have rich forest resources, but most of which are located in poor mountainous regions, where there are conflicts between economic development and ecological improvement. These conflicts are shown in the fragile ecological environment and unsustainable production modes, the serious demolishment to the ecological environment in poor areas by long-term and frequent human activities and over-development and thus the conflict between man and nature. The real picture is that the industry in most of poor areas has small production scale, high energy consumption, low efficiency of resources utilization, large production of abandoned and wasted materials and poor product quality. The unsustainable production modes also deteriorate the conflict between man and nature and affect the advancing steps of S&T innovation to a certain degree. Meanwhile, the development of these areas is also restricted by backward economic development, insufficient investment in education and poor quality, slim number of professionals, the majority of labor force only adapting to the traditional production modes and the lack of quality professionals. As a result, the forest-related S&T innovation capacity of the northwest also lags behind other regions.

2.3 Evaluation on the forest-related S&T innovation capacity of the 5 western provinces/autonomous regions

Based on the above established evaluation indicator system, combining the applications of principal component analysis and analytic hierarchy process, the forest-related S&T innovation capacities of 28 provinces/autonomous regions have been evaluated. The five western provinces/regions suffering from serious land degradation including Inner Mongolia, Xinjiang, Gansu, Shaanxi and Ningxia (Qinghai Province was not included due to the loss of part of indicator data statistics) were graded in the 28 provinces/regions in term of their innovation capacity during 2001-2007, as in Table 3.

The evaluation results of the provincial/regional forest-related S&T innovation capacity shows that the five seriously stricken western provinces/regions still lag behind others in terms of innovation capacity. Although Inner Mongolia and Gansu comparatively were ranked better, they just fluctuated between No. 13-19 and are still in a lower status in the national overall forest-related S&T innovation. This is consistent with the above evaluation on the regional innovation capacity, which shows the innovation capacity of western areas stricken by land degradation is rather weak and there is still a large gap with other provinces/regions enjoying better scientific development.

Table 3 Ranking Results of Regional Forest-related S&T Innovation Capacity

Province/Region	2001	2002	2003	2004	2005	2006	2007
Xinjiang	21	17	18	22	18	18	19
Inner Mongolia	17	16	15	14	14	15	13
Gansu	16	15	16	19	17	14	15
Shaanxi	18	27	26	28	22	25	22
Ningxia	20	21	25	20	25	21	21
Southwest China	5	4	4	4	4	5	4
Northwest China	7	7	7	7	7	7	7

3. Countermeasures and suggestions to improve the forest-related S&T innovation capacity of western areas stricken by land degradation

To improve the innovation capacity of western areas stricken by land degradation and provide robust scientific support for the modern forestry development, the core is to reinforce the basic environment for forestry and provide institutional guarantee for forest-related S&T innovation; to continuously cultivate innovation talents and strengthen the innovation team building; to effectively increase the capital supply for S&T innovations and enhance the investment in and support for forest-related science and technology by public finance; to strengthen the regional S&T innovation and serve for the regional forest development; and to extend the innovation results, reinforce the linkage between forest-related science and technology and economy and further improve the extension system for forest-related science and technology, so as to comprehensively enhance land degradation control in West China.

(1) The environment improvement for regional forest-related S&T innovation

It should depend on the research and experiment resources such as key labs and site ecological research stations and the natural S&T resources like germplasm banks as priorities to continuously strengthen the resource base and capacity building for innovation and to improve the basic facilities and conditions in the western areas hit by land degradation and with relatively backward innovation environment. There are fewer national key labs distributed in the western areas severely hit by land degradation and the research apparatus and equipment there are rather poor. In such context, the government will need to enhance the investments and inputs in forest-related S&T infrastructure construction. Besides the increasing inclining of national and public finance investment towards the region's basic construction, the local governments in those areas also need to raise their attention to the basic environment rehabilitation for forest-related S&T, for example, by appropriating special funding for the regional S&T environment development.

Modern technologies such as IT and Internet should be fully used and the scientific data accumulated by national S&T program projects should be emphasized to set up the sharing and service network of scientific data and graded and classified sharing and service systems and develop a synergy and sharing network for regional S&T resources. S&T literature resources should be expanded and collected and digital libraries should be developed to form a guarantee and service system of forest-related S&T literature resources with complete varieties and reasonable structure.

(2) Building a regional forest-related S&T innovation team

The advantages of national and provincial forest research institutions and forest-related universities and colleges distributed in the western areas stricken by land degradation should be fully used to foster professionals for forest S&T innovations. As for the western areas with weak capacity of research institutions and universities and colleges, representative provincial research institutions or universities and colleges such as Northwest Agriculture and Forestry University and Inner Mongolia Agricultural University should be selected to build up a quality innovation team, so as to drive forward the S&T innovation team building in the stricken western areas. Meanwhile, capacity building of the grass-root professional team and the high-quality professional team of the S&T personnel of forestry departments at county level or below should be strengthened . Their overall quality and operation skills should be improved so as to fully function in the front line of land degradation control and forestry development.

Forestry universities and colleges and forest research institutions are the base to foster forest-related S&T professionals. According to the needs of modern forestry development, the higher forestry education should be coordinately developed and the forestry vocational education should be strongly promoted in the western stricken areas to deliver a large number of administrative, technical and skillful professionals for forest-related S&T development. The western stricken areas should be encouraged to open forestry universities and colleges and forest-related programs in multiple forms, to strengthen the development of related disciplines including desertification control and water and soil conservation and to enhance the delivery of forest-related S&T professionals in support to the western stricken areas. Meanwhile, the government should reinforce the implementation of various preferential policies for the forestry education development in the western stricken areas to enhance the development of local forest-related S&T.

(3) Optimization of capital supply structure for regional S&T innovation

Based on the current situation with the serious lack of inputs in forest-related S&T, the national public finance should dramatically increase the inputs in the forest-related S&T

innovation of the western stricken areas. The forestry administration departments at all levels should adopt powerful measures to ensure the amplitude investments in forest-related S&T, which should be arranged with priorities, is more than that of the investment in forestry development. Those local forestry departments in the western areas who have implemented the national, ministerial and provincial forest-related S&T projects should actively seek for the support to increase the level of co-finance As for the local governments, they should use the national and local public finance mainly for the innovation conditions including S&T infrastructure, experimental bases and professional teams.

The roles of market investing and financing mechanism and incentive mechanism should be fully used, and preferential policies such as subsidies and interest discounts as well as economic lever should be employed, so as to encourage and lead various regional social capitals to invest in forest S&T. Organizations, corporations and individuals with rich capital should be actively attracted to invest in the forest-related S&T development. Meanwhile, the national long-term and low-discount interest loan support capital should be sought to use for the forest-related S&T. The main bodies in forest-related S&T innovation such as the regional scientific researches, education and enterprises should set out from the strategic height at upgrading the core technological competitiveness and increase the R&D investment year by year. A regional diversified and multi-layer investing and financing system for forest-related S&T covering public finance investment, bank loan, self-collecting funds of enterprises and public institutions, risk investment, cooperative R&D and international funding should be finally formed and the regional forest-related S&T funds should be increased through various sources.

(4) The enhancement of regional forest-related S&T innovation

Aiming at the technological R&D and innovations in the western stricken areas, a regional forest S&T center should be established to develop the R&D, integration, experiment and demonstration of common ecological improvement technology and industrial development technology in the region as priorities, so as to provide strong S&T support for the regional forestry development. Typical provinces and regions in the western stricken areas should be selected to form the regional forest S&T center to play a role in demonstrating and driving forward the improvement of the innovation capacity of other western areas. During the establishment and improvement of the regional forest S&T center, based on the continuous reinforcement of the existing provincial forest research institutions and full exertion of their functions, the enthusiasms of all involved parties should be mobilized to innovate the management mechanisms, gather S&T resources, and strengthen S&T collaboration to gradually form the regional forest S&T center and the innovation team in typical eco-economic areas and improve the overall strengths and development level of the regional forest-related S&T. This should rely on the existing national and local forest research institutions and universities or colleges, abide by the national outlay of ecological improvement and industrial development as well as the principles of mutually supplementing with own advantages and conducting mage-merger, and make use of projects as linkage.

(5) The improvement of forest-related S&T extension system

The extension and application of applied forestry technologies especially the land degradation control techniques should be continuously enhanced. The forest S&T demonstration areas, counties and sites should be established and improved to form a demonstration network and extension system covering the western stricken areas and make full use of S&T in demonstrating technologies, transferring results and driving others forward in a radiation. This should rely on various S&T extension programs including the key forestry programs and the land degradation control programs and abide by the strategic outlay of forestry development in the western stricken areas. The Activities of Sending Science and Technology to Countryside should be actively promoted to organize a wide range of S&T personnel to go deep into the front line of forestry production to popularize the scientific knowledge and apply and extend the S&T achievements. Forest research institutions, enterprises, various universities, colleges and schools and societies or associations at various levels should be mobilized to form a pluralistic forest-related S&T extension system involving various powers at national, local and social levels and combining national support and market orientation, free and paid services, and professional teams and social service organizations, in order to create advantages for the forest-related S&T innovation capacity of the western areas stricken by land degradation.

References

[1] Zehui, Jiang et la (2008) *Modern Forestry of China (Second Edition)*. Beijing: China Forestry Publishing House.

[2] Yucai, Li (ed.) (2006) *A Blueprint for Rapid and Sound Development – Collection of Chinese Forestry Development Plans for 11th Five Year Period and Medium-and-Long Term*. Beijing: China Forestry Publishing House.

[3] Shu, Liu et la (2001) *Study on Integrated Evaluation of S&T Progress*. Hebei University Publishing House.

[4] Fuji, Xie (2004) *Technical Progress and Its Evaluation*. Shanghai Scientific and Technological Education Publishing House.

[5] Runlong, Huang (ed.) (2004) *Data Statistics and Analysis Technology – Applied SPSS Software*. China Higher Education Press.

24. Combating Land Degradation in Mediterranean Coastal Areas: Using Erosion Mapping for Assessment of Land Degradation in O. Rmel Watershed

R. Attia, S. Agrébaoui and H. Hamrouni

Abstract: The Mediterranean coastal areas are undergoing both natural and human activities which tend to affect its resources and the community's quality of living. In addition, the agriculture activities of population have led to noticeable environmental degradation including soil erosion. This document has been prepared with the technical and management assistance of the Priority Actions Programme Regional Activity Centre (PAP/RAC) of the Mediterranean Action Plan. Two pilot areas were chosen in O. Rmel watershed according natural and physical diversity: lithological, morphological, and hydrological behaviour. Obviously, the two areas have lots of differences in terms of their of problems which make the assessment of their degradation processes more meaningful and educational for the purpose of the project. The PAP/CAR approach integrates biophysical and socio-economic components of land degradation at different scales, recognising that socio-economic issues are also driving forces of pressures that impact on land conditions.

This study starts with the predictive soil map (result of diagnostic analysis) according to physical parameters: slope, soil cover, land use and agro practice, which allows later to produce maps showing erosion risk and land degradation processes in the two regions. The second component is to analysis of the human impact and economic activities that cause land degradation on stabilised and unstable areas, in order to identify prioritise areas for intervention. Finally the third component was to consolidate the outline of management planning activities based on remedial measures desired. These activities were supported by Geographic Information System (GIS) which includes all produced data.

Key words: Degradation; Soil erosion; Erosion risk ; Physical parameters; Land use; Socio-economic factors; Human pressure; Prioritise areas; Management planning.

Introduction

Land degradation in arid, semi-arid areas is resulting from various factors, including climatic variations and human activities (UNEP, 1995). The application of the CAR/ PAP mapping erosion methodology processes in pilot areas allows identification of erosion risk and actual erosion processes. The integration of socio-economic factors provides a detailed analysis of causes land degradation whith relevant preventive or corrective measures. Although it provides the opportunity to select areas for priority interventions and development of management plans. This study is conducted in O. Rmel watershed located at N-E of Tunisia (Fig.1).

The two pilot areas chosen are the Sbaihia watershed and Bouficha valley, the former

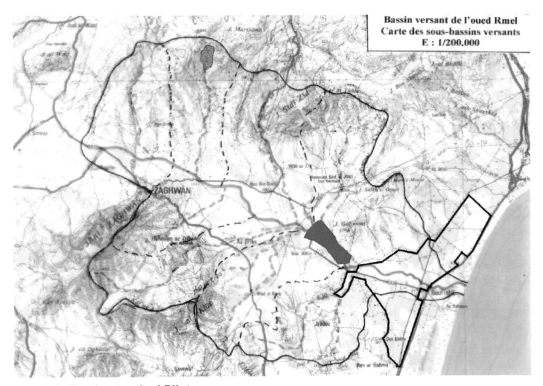

Fig. 1: O. Rmel watershed Pilot areas

is in upstream part of in O. Rmel watershed (6500ha), while the latter is in the south, it is a valley, with areas of 3000ha. They were chosen according to several criteria including location, size, natural characteristics like topography, vegetation cover, soil, erosion processes and human interference. Relevant features of these pilot areas are: great variety of ecosystems, great human pressure, and large areas affected by soil erosion processes, overgrazing, cropping on sloping lands (Toumia, L. 2000). Downstream, in valley, the agriculture activities of population, land use have led to noticeable environmental degradation caused by the lack of water irrigation reduced grazing space and water logging on clay soil.

Materials and methods

We start with the predictive mapping. For this step various thematic map were finalised topographic, land lithological, land cover and land use map. The final erosion risk map combines 3 maps together (PAP/CAR 1998). This map provides most important factor in assessing erosion risk and give useful tools for the knowledge and interpretation of the erosion processes that are presenting the pilot areas. The procedure used for this application is photo interpretation and field observation.

Descriptive map is elaborated too; it provides the main land units, the dominant types of erosion processes. Finally stabilised and unstable areas are delimited in order to identify prioritise areas for intervention. Stable areas need preventive measures, unstable areas need curative measures (Attia et al 2001). The second component is to analysis of the human impact and economic activities that cause land degradation.

The prioritisation procedure provides priority map, it integrates the results of the physical assessment and related descriptive mapping with the aggravating socio-economic conditions, and actual land use. For the different criteria is affected a rate from 1 (lowest possible score) to 3 (highest possible score).A: Physical instability risk related to the descriptive mapping code, B: Extent of area affected by a specific degradation process, C-D – E…. are criteria detailed in the study report. After giving a score for each criteria the final prioritisation is calculated in the following way:

For stable areas: *[(A * D + E) * F * G * H * 1] + [(J + K) * L * M* N]*

For unstable areas: *[(B * C * D + E) * F * G * H * 1] + [(J + K) * L * M *N]*

- High priority for intervention measures (class 3): 60 points as final score.
- Medium priority (priority class 2): 21 to 59 points.
- Low priority, class 1: 20 points and less as final score.

The development of remedial measures (preventive and curative) lead to promotion of future implementation of management plans for the pilot areas.

Results

The pilot areas show large variety of ecosystem, erosion process, and land degradation. Sbaihia watershed and Bouficha valley show unstable and stable intervention areas, which are described and evaluated for application of remedial measures. For the detailed analysis, relevant socio-economic factors are given in study areas two.

In Sbaihia watershed, the slope value constitue the most erosion risk value. According the erosion risk map about 90% of the area are seriously affected (Fig.2) .This is the result of slope gradient factor in fact wide slope areas are occupied by friable soil and covered with inappropriate agricultural practice: crops and cereal.

The watershed shows large variety of erosion process and land degradation given the fact that the central area is distinguished by dominance of unstable units. The rest of the western and northern area was classified in stable units. The dominant processes are gully erosion network that have a trend to widespread expansion. Bad lands affect sloping areas and free or clay soil.

The total population is about 1500

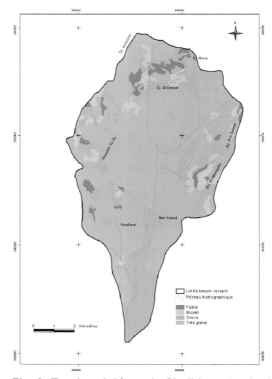

Fig. 2: Erosion risk map in Sbaihia watershed

person, 279 family distributed mainly upon hydrologycal network, human activities inducing most of degradation processes (Toumia, L 2000). The main human stresses are overgrazing and the abusive consumption of trees. The degradation of the vegetation cover leaves the soil unprotected. Brushwood for fuel leads progressively to deforestation. Cereal culture developed on sloping area (Fig.3) increases erosive power of water. The joint possession of lands does not allow the access to the credit. The lack of equipment of the farmers in tractors, the harvesters and the others is also noted.

Fig. 3: Cereal culture developed on sloping area

Priority map using prioritisation procedure prove that the watershed shows large variety of erosion process and land degradation given the fact that the central area is distinguished by dominance of unstable areas, the rest of the western and northern area were classified in stable units (Fig.4). Stabilised and unstable areas are delimited in order to evaluate the effects and to identify prioritise areas for intervention.

• Stable areas – priority 1: 114 ha. Stable areas priority 2: 2805 ha
• Unstable areas; priority 2: 986 ha. Unstable areas; priority 3: 2600 ha

Stable areas need preventive measures, unstable areas need curative measures.

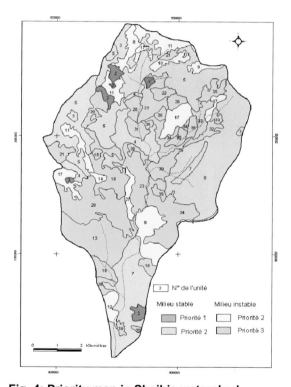

Fig. 4: Priority map in Sbaihia watershed

The second pilot area is Bouficha valley located downstream of O. Rmel watershed. It is plain fitted out with irrigated perimeters. Wide areas are occupied by clay soil (50%) that is affected by water logging process. The analysis of field observation shows flat to gently slopping agricultural land which is affected by seasonal flooding and silting in cultivated flood plain. Large amounts of soil deposit are accumulated in central part with an important local alluvial valley network. According to the erosion mapping methodology, various thematic maps were finalised: topographic, land lithological cover and land use map. Morphological

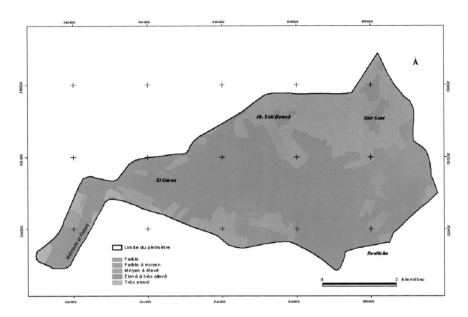

Fig. 5: Erosion risk map in Bouficha valley

units are classified as low, medium, highly, extremely highly susceptible to erosion.

The erosion risk map shows about 70% are seriously affected (Fig.5).This is the result of dominant erosion processes: sheet erosion, rill erosion combined with human activities inducing most of degradation processes.

The farmers are, for the greater part, old and limit themselves to the immediate and easy profit. Population density leads to parcel fragmentation, agricultural intensification, and generalized extension. Reduced grazing space induces consumption of land cover and leaves the soil unprotected. The lack of irrigation water leads to inappropriate agricultural practices. Trees plantation is very limited, all farmers practise cropping and cereal culture. Inadapted mechanisation and equipment have very negative impacts on the biological features of soils. Poverty and low budget for agricultural inputs lead to extensive farming this reduce soil fertility.

The stable or stabilised areas account for
• Stable areas: priority 1: 1105ha (32%); priority 2: 811ha (23%).
• Unstable areas priority 2: 1161ha (38%); unstable areas priority 3 : 251ha (7.3%)

The central region of Bouficha area is characterised by unstable area, it's the most affected area (Fig.6).

Recommendations for management of the study areas

The remedial measure: preventive, curative, and protective measures was made and adapted to the two pilot areas according the prioritisation procedure, this lead to the preparation of management plan adapted for area affected by erosion process and human interference and land use, major parameters in land degradation. In general, the first objective of these measures is to prevent, or stop the spread of degradation, re-

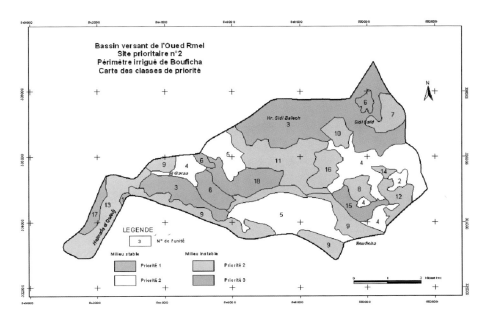

Fig. 6: Priority map in Bouficha valley

habilitate the degraded lands and recuperate their productivity wherever possible. The second objective is to restore and reserve soil.

The identification of preventive, protective and curative measures was made according to 4 specific themes dealing with land degradation: soil, agriculture, forest and human interventions and prove great variety of ecosystems, great human pressure.

Preventives measures
- Land use changes, such as cutting for agricultural purpose must be totally prohibited
- Uncontrolled fuel wood collection and excessive grazing need to be reduced.
- Control use of soil cover and grazing space, decrease of browsing (Fig.7).
- Choice agricultural practices according the characteristics of soil (Fig.8).
- Maintain of conservation practices: bancs, retaining walls

Fig. 7: Overgrazing

Fig. 8: Crops on slope area

- Prevent forest destruction
- Diversification of production
- Protection of stable areas

Curatives measures adapted to the local conditions
- Cover slope area with trees plantation and some kind of conservation practices
- Optimisation of forest
- Weak lithology must be protected with use of retaining walls
- Bad lands must be treated with forestation
- Improving soil structural stability by forestation of areas by specific species, this prone to mass movement

Protective measures
- Introducing modern living requirements notably water electricity, roads and construction -create job opportunities GCP/TUN/028/ITA
- Protection of infrastructure by avoiding sedimentation downstream
- Intensification of the meetings of formation of local population in order to improve their knowledge as regards conservation/ production

Specification of institution al and administrative arrangements is very important for the success of theses measures. The outcome is then influenced mainly by the following factors (Janicke and Weidner, 1997 in PAP/CAR2000): Actors, strategies, structural framework conditions, specific context and problems .

Conclusion

The project covered two pilot areas in O.Rmel watershed which have different characteristics, natural and human interference. A component of the project assessed the general land degradation in the study zones due to natural and human interference. Throughout the work, the project allowed assessment of interaction between biophysical and socioeconomic factors to improve dynamic and hydraulic behaviour of the study area.

Detailed analysis was done to compare and find relation ships between upstream (Sbaihia) and downstream (Bouficha) part of O.Rmel watershed.

The predictive erosion map was checked in the two pilot areas with focus on stability/soil erosion criteria reflecting the above characteristics. A detailed erosion map has been produced for the two areas. This allows prioritisation of intervention areas. Priority areas were determined among the unstable and stable areas, for which evaluation of remedial measures are given. Those measures were studied with respect to current and optimum use, and what best to apply, preventive, protective or curative approaches. They are assessed and rated as best applicable from field work in the two areas, and accordingly recommendations are given for that purpose.

The problems, priorities, remedial measures and institutional issues are linked to develop management plan. Thus, management planning activities are given, with focus on involvement by the community, as their contribution to the work in both areas. Furthermore, indicators on different factors affecting the output of the plan are given, and approval procedures are presented as well.

References

Alali, Y. (2003) , Paramètres hydriques des sols dans un aménagement en banquettes anti-érosives (El Gouazine, Tunisie). Mémoire de D.E.A. de S.E.E.C. à Montpellier, 63 p.

Attia, R; Agrebaoui, S . (2001), Contrôle de l'érosion et de la désertification dans les bassins versants pilotes en Algérie, au Maroc et en Tunisie. Cas du bassin versant de l'oued Lobna, 24 p.

Bennour, H., Bonvallot, J. (1980) carte de l'érosion de la Tunisie; échelle 1/200000. Bulletin des sols n° 11, 1980 ; Direction des Sols. 93 p.

Collinet, J., Testouri Jebbari,S. (2001), Etude expérimentale du ruissellement et de l'érosion sur les terres agricoles de Siliana (Tunisie). Institut National de Recherches en

Génie Rural et Eaux et Forêts (Tunis), Direction des Sols (Tunis), Institut de Recherches pour le Développement (mission de Tunis).48p.

Direction de la C.E.S . (2000) La planification participative dans le bassin versant de l'oued Sbaihia. 67 p.

Gilbert, S. (1995) Cartographie de l'érosion à l'aide d'un système d'information géographique. Application au bassin versant de l'Oued Joumine (Nord de la Tunisie). Mémoire de Fin d'Etudes. Faculté des sciences de Tunis. 87 p.

Mtimet, A., Agrebaoui, S. (1993) Cartographie de l'érosion potentielle des bassins versants de l'oued el Khirat et oued Rmel.

PAP/CAR. (1998) Directives pour la cartographie et la mesure des processus d'érosion hydrique dans les zones côtières méditerranéennes. PAP/8/PP/GL.1. Centre d'activités régionales pour le Programme d'actions prioritaires (PAM/PNUE), en collaboration avec la FAO, 115 p.

PAP/CAR, (2000) Directives pour la gestion de programmes de contrôle d'érosion et de désertification, plus particulièrement destinées aux zones côtière méditerranéennes. 115 p.

Rapport de projet GCP/TUN/028/ITA programme de la conservation des eaux et des sols dans les gouvernorats de Kairouan, Siliana et Zaghouan. – GCP/TUN/02.

Razzeg Dit Guiras, W. (2000) Cartographie de la sensibilité à l'érosion ravinante par unité litologique dans les sous-bassins versants des Oueds Ettiour et Hjar. Mémoire de DEA de Géologie. Faculté des sciences de Tunis; 75 p.

Toumia, L. Khelifa, A. (2000) Méthodologie de planification participative des aménagements C.E.S., cas d'étude bassin versant de l'oued Sbaihia. 67 p.

25. Legislation on Prevention and Combating Desertification Caused by Land Use in China

Yu Wenxuan[1], Zhou Chong[2]
1 Tsinghua University; 2 Chinese Politics and law University

Abstract: China is one of the countries facing most serious desertification problem in the world. Among other causes, excessive and inappropriate use of land is the most important factor. It is necessary to deal with the problems based on legal mechanisms. Although with some related laws, regulations and rules, China still has to take more efforts during the further legislation on such aspects as planning, ownership system, monitoring and evaluation, market mechanisms, management system, legal liability, etc. Most of these problems are the negative consequences of defects in management philosophy, administrative system and legal mechanisms. The solutions for these problems are proposed in this paper.

Key Words: Desertification; Prevention and Control; Land Use; Legislation; Solutions

1. Deficiencies of Existing Legal System

• *Plan of Land Use*

The Land Management Law does not pay enough attention to land protection. There exists some difference between the Land Management Law and the Law of Prevention and Control of Desertification. For example, the Land Management Law provides that the plan of the prevention and control of desertification should be approved by the State Council, while some departments designated by the State Council have the same power according to the Law of Prevention and Control of Desertification. Similar situation can be found in Environmental Impact Assessment Law and the Law of Prevention and Control of Desertification.

• *Monitoring and Evaluation*

Land Management Law and other laws and regulations provide that monitoring system should be established. However, the monitoring responsibilities of the government are not clearly defined. As the direct observer and mostly affected by the ecological changes, farmers have advantages over those high-tech monitoring equipment. Current legal system does not give them enough opportunities to play their role in land monitoring. The evaluation is only limited to the reclamation of the unused land, and adopts single evaluation approach, because the ecological value of land and the impact on environment are hardly considered.

• *Market Mechanisms*

Prevention and control of desertification has long been implemented mainly by the government. The market mechanism can not play its full role, which causes at least two problems:

(1) Investment channel. The funds for prevention and control of desertification are

mainly provided by the government in China. For other investment, neither promise nor investment channel is expressly stipulated.

(2) Incentive mechanism. Prevention and control of desertification is closely related to the sustainable development of agriculture, but agriculture is a low-profit industry. The government does not pay enough attention to encouraging preventing and controlling of desertification through incentive mechanisms.

- Land Consolidation

For China, land consolidation is necessary to achieve sustainable use of land resources. Land Management Law only provides "The State encourages land consolidation". Because of lack of sound planning and neglect of implementation, the effect is unsatisfactory.

2. Causes of the Problems

- *Management Philosophy*

China is now taking effective efforts to build up a resource-conserving and environment-friendly society. Although the State has recognized the importance of ecological value and paid a considerable effort, land use management still focuses more on the production value of the land than ecological value.

- *Administrative System*

The most significant problem is the unclear function of relevant administrative departments. Recurrent "other relevant departments" provided in the Law of Prevention and Control of Desertification reflects this problem. Another problem is lack of sound and effective coordination mechanism.

- *Legal Mechanism*

(1) Market mechanism. The prevention and control of desertification is considered as an undertaking more related to social benefit, and thus the governments should play a leading role. So a great amount of legal provisions focus on the powers and responsibilities of governments instead of market mechanism crucial for the prevention and control of desertification.

(2) Public participation. It is difficult for public to participate in the land planning and to learn the relevant information, especially the statistical data of land use. Besides, the feedback mechanism needs to be improved in some existing laws and regulations.

3. Solutions

- *Plan of Land Use*

(1) The ecological and environmental conservation should be well considered in the plans of land use to protect the ecological benefits and provide adequate policy basis for the prevention and control of desertification.

(2) Further coordination should be strengthened between the Land Management Law and the Law of Prevention and Control of Desertification under the existing legal framework.

(3) The plan for the village town level should be constituted in the Law of the Prevention and Control of Desertification.

- *Administrative System*

(1) Separation of supervision and executive. An independent supervisory and administrative regulatory body could avoid the interference from the executive power into the implementation of law and policies.

(2) Coordination mechanism. The prevention and control of desertification requires supports from executive authorities and needs sound coordination mechanism.

- *Legal Rights and Interests*

(1) Improve the system of expropriation and make sure farmers can get fair compensation and reasonable care. In determining the compensation standards, the ecological value of the land should be taken into account.

(2) The ownership system and transfer mechanism should be further perfected.

- *Monitoring, Evaluation and Consolidation*

(1) Improve national system of land management and information sharing, especially the information collection of rural desertification; integrate existing information monitoring network; emphasize on land use monitoring and reporting responsibilities of village and town government.

(2) Strengthen the assessment of land use. Adopt IEM methodology with full consideration to local environmental, social and economic, historical and cultural factors.

(3) Specify rules on land consolidation. Attach importance to maintenance of ecological functions and environmental benefits.

- *Supporting and Incentive Mechanism*

(1) Supporting mechanism. Improve the funding mechanism, expand the scope of capital investment and strengthen the supervisions related to funds for prevention and control of desertification.

(2) Incentives mechanism. Take taxation and other measures to attract more people and enterprises to participate in the prevention and control of desertification.

- *Legal Responsibility*

Introduce new provisions on legal responsibility in the further legislation.

26. Improving China's Legislative Framework for Land Degradation Control

Zhou Ke, Cao Xia and Tan Baiping
Law School of China Renmin University

Abstract: The Chinese government has recently been attaching increasing importance to the application of effective legal tools to tackle land degradation (LD) issues. Based on the concept of sustainable development, China began developing and reaping the benefits of environmental and natural resources legislation including LD control regulations in the 1990s. In the past three years, some central-western provinces in China have been implementing a "People's Republic of China/Global Environment Facility (PRC/GEF) Partnership on LD Control of Dryland Ecosystems", which is based on an integrated ecosystem management (IEM) approach. IEM is designed to achieve a balanced, scientific and participatory approach to natural resources management, which creates the potential to improve the quality of Chinese environmental law and policy procedures. The paper studies the existing Chinese national laws and regulations pertinent to LD control within 9 areas covering land, desertification, soil erosion, grassland, forest, water, agriculture, wild animals and plants, and environment protection in detail, against IEM principles and basic legal elements. The main objective is to identify problems and provide feasible solutions and recommendations for the improvement of the existing laws and regulations. The authors conclude that the development of an improved national legislative framework is essential if LD control is to be successfully achieved. The paper is partly based on Component 1 — Improving Policies, Laws and Regulations for Land Degradation Control under PRC/GEF Partnership on Land Degradation in Dryland Ecosystems (TA 4357).

Keywords: Legislative framework; Land degradation control; Integrated ecosystem management; China.

Introduction

Land degradation (LD) is one of the severest environmental problems facing human beings. Frequent causes of LD include water erosion, wind erosion, land contamination, a reduction in soil fertility, water logging, salinization, lowering of the water table, loss of organic matter, invasive species, habitat conversion and aquifer degradation. Natural conditions are partly responsible for LD, but human-induced impacts are increasingly contributing to a larger share. It is estimated that in China 40% of land is affected by moderate to severe erosion (adb.org/projects/ PRC_GEF_Partnership). Water and wind erosion is common; water erosion which leads to soil erosion affects mainly the east, south and southwest regions of China, while wind erosion which results from desertification is a problem in the north and northwest regions (Du, 2004). During the last five decades, land affected by these erosion types throughout the country has increased between 20~30%. Areas affected by soil erosion increased from 1.53 million km^2 in the 1950s to 1.796 million km^2 in the 1990s (Huang, 2006).

The main anthropogenic contribution to LD in China has been land contamination. Chief pollution sources include industrial wastes (waste gas, waste water and waste residue); chemicals, such as pesticides and fertilizers; irrigation with industrial waste water or domestic sewage in some dry areas; heavy metals resulting from mining and other industrial activities; and acid rain (baike.baidu.com/view). Land contamination is characterized by concealment, accumulation and irreversibility, and its treatment is time-consuming, costly and more difficult than that of other forms of pollution. China is facing a very serious land contamination situation. According to the statistics released at the 2006 National TV Conference on the Land Contamination Survey (finance.sina.com.cn), at least 10% of total farming land — mostly in economically developed regions — is contaminated. Consequently, it is estimated that 12 million tons of grain are annually polluted by heavy metals, causing a direct loss of 20 billion RMB and countless health problems.

Land degradation in China has severely impacted poverty-stricken communities and has deteriorated the living conditions of millions in rural areas. Aware of this problem, the Chinese Government prioritized environmental protection in the development of its Western Development Strategy by recently launching, with the aid of the Asian Development Bank (ADB), a PRC/GEF Partnership on Land Degradation in Dryland Ecosystems. The partnership is to support the development and implementation of a 10-year (2003–2012) program to combat LD in West China. The partnership — whose aim is to identify and describe ecosystems in China that require different management approaches to preserve and restore functions and services, and to provide "a comprehensive framework to manage natural systems across sectors, and political or administrative boundaries within the context of sustainable development" (gefweb.org, 2000) — will be operating under an Integrated Ecosystems Management (IEM) approach seeking to enhance China's institutional capacity; to improve its regulatory and management systems; and to promote cleaner technologies and environmentally sustainable agricultural and natural resources management practices (adb.org/projects/PRC-GEF Partnership).

As part of this partnership, the authors of this paper are responsible for evaluating national laws and regulations pertaining to LD control in China. Applying the principles and basic legal elements for IEM analysis for LD control, the paper reviews and assesses existing LD laws and regulations in 9 areas: land, desertification, water and soil conservation, grassland, forest, water, agriculture, wild animals and plants, and environmental protection (Table 1). Some of the major problems of these legal tools are then examined to illustrate the challenges faced by the Chinese government and policy-makers in developing a coherent, consistent and responsive centralized legislative framework for combating LD in China. The paper concludes by outlining some key recommendations to improve this legislative framework.

2. Background: China's LD control legislative framework

Land degradation in China has traditionally been attributed to irrational anthropogenic activities vividly depicted as the "Six Excessive Acts": over-cultivating, over-lumbering, over-herding, herb over-collecting, over-hunting and overuse of water resources (Chen, 2003). However, the authors hold, a long-term deficiency in the regulatory framework of natural resources characterized by a lack of coherent, consistent and responsive laws and regulations has compounded the damaging effect of these activities.

China's LD-related national laws and regulations are numerous and of varying scope as they include articles in the Constitution, laws, administrative statutes, and departmental rules and regulations. These LD control provisions are scattered in sub-sectoral instruments and have not, in the past, been jointly considered as required by the IEM approach. Although beyond the scope of this article, it is worth mentioning that the focus of the instruments concerning LD at the local level varies greatly depending on geological and climatic differences. For example, not surprisingly, the legislation in some western provinces such as Inner Mongolia and Qinghai, which are rich in grassland, places more emphasis on grasslands while in Gansu and Ningxia, two provinces in the same administrative district as Qinhai Province but abundant in dryland, it places more emphasis on water and soil conservation.

China has typically managed natural resources disjointedly and indirectly through departments whose purpose is primarily economic development. As such, these departments may encourage the full exploitation of ecological elements within their jurisdiction, possibly leading to the deterioration of these elements, and contributing to the degradation of the ecosystem as a whole. Some of the departments involved in LD control are the Ministry of Land Resources, the Ministry of Agriculture, the State Forestry Bureau, the Ministry of Water Conservation, the State Environmental Protection Administration, the National Development and Reform Commission and the Ministry of Finance. LD control programs require concerted and integrated efforts from all these institutions which presents a challenge given that they have separate functions, they are at the same hierarchical level, and they function independently from each other. This complexity has resulted in a plethora of problems such as overlapping functions among various departments, lack of coordination and cooperation, and the issuance of fragmented and sectoral regulations with strong self-interest orientation (Lu & Liu, 2003). Not surprisingly, the overlapping cases tend to occur in resource-rich areas such as those areas suitable for forestry and husbandry, while the coordination deficits happen in resource-poor areas. For instance, the administrative departments responsible for land, agriculture (husbandry), and environmental protection are all empowered to exercise administrative powers over state-owned unreclaimed land or grassland. But in reality, none of these departments has ever done so. (Du, 2004)

China has traditionally employed administrative tools to control LD. However, economic incentives and involving interest-related organizations and the public in general can greatly contribute to the conservation of natural resources. Strengthening administrative powers and limiting the rights of land users, especially in a time when its economy was slow may have made sense in the past but not in the current economy. During the last decade, while seeking a win–win approach between protecting natural resources and increasing benefits to land users, the Government has initiated a mechanism of ecological compensation to encourage farmers to plant trees in farm land. In the reformation of collective forest rights, the State Forestry Bureau (SFB) has succeeded in contracting collectively-owned forest land on a long-term basis to forest farmers, who are held accountable for maintaining ecological integrity of the land. This contract-based approach has been adopted in the program for the reformation of state-owned forest rights.[1] These efforts are highly compatible with IEM ideas advocated in the PRC-GEF Partnership of LD Control.

3. Methodology: Integrated Ecosystem Management principles and elements

According to the Commission on Ecosystem Management of the World Conservation Union (IUCN), Integrated Ecosystem Management (IEM) "is a strategy for the integrated management of land, water and living resources that promotes conservation and sustainable use in an equitable way". (iucn.org) The PRC-GEF Partnership of LD Control has chosen IEM to evaluate China's current LD laws and regulations. This recently developed approach, which is consistent with the two major Rio conventions (Biological Diversity and Climate Change) and the UN Convention on Transboundary Waters signed in Helsinki (gefweb.org, 2000), requires interdisciplinary inputs and inter-departmental coordination and cooperation between all involved in forestry, agriculture, husbandry and fishing, water conservation, environmental protection, land resources, science and technology, finance, planning, and legislation. In other words, IEM will be used to streamline and build up relevant regulations, policies, and departments, social and economic systems on a comprehensive and cross-sectoral basis instead of a traditional sector-by-sector approach (gefweb.org, 2000).

IEM comprises 12 over-arching principles that were proposed by IUCN in 1995, and were first endorsed at the 5th Conference of the Parties to the Convention on Biological Diversity (CoP 5 in Nairobi, Kenya; May 2000) as the primary framework for action under the Convention (Table 2). These principles convey core ideas and values of the IEM approach, providing a fundamental tool for establishing and adopting detailed evaluation standards.

IEM also includes basic legal elements.[2] These elements embody the above 12 principles and serve as specific standards for assessing laws and regulations concerning LD control in the implementation of IEM. Originally, 17 elements were adopted in two technical assistance projects (3548 & 3708) developed by ADB with the former analyzing the capacity of the Chinese LD control policies and the latter analyzing the capacity of legal and policy framework in integrated management of the Yellow River basin,

---

[1] Implementation Guidelines of the State Forestry Bureau for Thoroughly Promoting Forestry by Law (in Chinese). the State Forestry Bureau, No. 196, issued on Nov. 5, 2004.
[2] The original 17 elements are: 1) general intent, 2) jurisdiction, 3) responsibility, 4) goals and objectives, 5) definitions, 6) duty of care, 7) hierarchy of responsibility, 8) institutional, 9) policy, 10) education, 11) research and investigation, 12) community participation, 13) land planning and IEM, 14) land management and IEM, 15) finance, 16) implementation and enforcement, 17) dispute resolution.

based on an analysis of both regional and global environmental regulatory frameworks in terms of sustainable management of land resources and the ecosystem. These basic legal elements play two roles: one is to help assess the functions of the existing instruments related to LD control; the other serves to guide possible revision of the existing LD-related instruments or to make new laws and regulations.

However, in order to accommodate China's LD control situation, 20 basic elements for IEM analysis have been jointly identified by the advisory groups and experts from both China and ADB. These amended 20 elements are: 1) Legislative purpose and basis; 2) Applicable scope and object; 3) Rights and obligations of social subjects over sustainable development, utilization and management of natural resources; 4) Security of rural land tenure and land quality protection; 5) Definitions of key terms; 6) Consistency of the legislative purpose with national and local policies and measures; 7) Joint duty of care; 8) Functions of government and administrative bodies; 9) Set-up of and roles of stakeholder institutions (organizations); 10) Administrative control; 11) Education, research and dissemination; 12) Investigation, monitoring, assessment and statistics; 13) Public and community participation; 14) Sustainable resources utilization or zoning and planning of ecological protection; 15) Resources ecology management and ecosystem management; 16) Financial inputs and market incentives; 17) Compliance and implementation; 18) Dispute resolution mechanism; 19) Legal liability; 20) Others.

These 20 elements constitute an indicator system for evaluating China's LD-related legal elements, a basis for evaluating the effectiveness of the relevant Chinese LD laws and regulations, as well as a basis for evaluating how IEM principles are embodied and applied in LD control. Specifically, the methodology applied in the evaluation is as follows: each instrument within the afore-mentioned 9 areas (land, desertification, water and soil conservation, grassland, forest, water, agriculture, wild animals and plants, and environmental protection) was evaluated against the 20 elements. To be exact, all the clauses of each and every instrument were classified according to content and were evaluated on a one-by-one basis from the dimensions of presentation, expression and application with five grades: A, B, C, D and E (Table 3). Presentation examines whether these 20 elements have been explicitly stipulated in the instrument in question. Expression measures whether and to what extent the content expressed in the law or regulation in question is compatible with IEM principles and element indicators. Application testifies if and to what degree the law or regulation in question is applied in practice.

4. Findings and summary

Using the methodology described above, the authors found the Chinese LD-related laws and regulations to exhibit, inter alia, the following characteristics in terms of presentation, expression and application.

4.1. Laws are comparatively more thorough than departmental rules and regulations

The 20 basic legal elements and 12 IEM principles are better represented in laws than in departmental rules and regulations. This can probably be explained by the fact that laws seek to adjust the relationships between interested parties as far as basic rights are concerned, which makes law-makers reassess what the core purpose of legislation on LD control is. The authors think that the essence of LD legislation should not focus only on management aspects as our present legislation mainly displays, but also on balancing the interests of various actors, resulting in more scientific and effective laws. Additionally, national laws have more clauses and a wider coverage than departmental regulations which are, more often than not, drafted by a particular ministry or commission to deal with a particular issue. For example, the Land Administrative Law (enacted in 1986 and revised in 2004) contains 8 chapters with 86 clauses covering basic principles, ownership and land use rights, land use planning, protection of farming land, land for construction, supervision and examination, legal liability and supplementary clauses, almost encompassing all 20 basic IEM elements. On the other hand, the Administrative Measures for the Pre-approval of Land for Construction Projects — a departmental regulation revised by the Ministry of Land Resources in 2004 — has only 16 clauses and is deficient in the expression and presentation of basic legal elements and principles, and is inadequately covered and supported by other instruments in content expression. This demonstrates that the type of regulation is a determinant factor in the level of the coverage. Therefore, China, in designing its LD regulatory framework, should focus its efforts on improved legislation in line with IEM ideas.

4.2. Recent laws and regulations better materialize the principles and elements than adopted ones

China started developing a sustainable development strategy in the early 1990s which serves as a source to LD-related legislation. Soon after 1979, China's priority shifted to economic development and at the time the environment was not of precedence on the Government's agenda. Some of the instruments adopted then are still in use and they differ significantly from those enacted since the 1990s. For instance, the 17 clauses in the Tentative Regulations on Management of Scenic Spots Areas (1985) include neither investigation, monitoring, assessment or statistics, nor resources or ecosystem management. Besides, provisions for legal liability are much too simple. Comparatively, the Grassland Law, first adopted in 1985 and revised in 2002, states that one of its purposes is to protect, develop, and rationally use grassland, to improve ecological environment, and to preserve biodiversity, which is basically in accordance with IEM ideas. The newly amended Grassland Law includes 75 clauses in 9 chapters covering basic principles, ownership of grassland, planning, development, utilization, protection, supervision and examination and legal liability. It incorporates almost all of the 20 basic elements.

4.3. Administrative control provisions are more detailed than other provisions

Out of the 20 basic elements, the following: 1) legislative purpose, 2) applicable scope, 5) definitions of key terms, 8) functions of government and administrative organs, 10) administrative control, and 19) legal liability, are expressed and illustrated in the LD-related laws and regulations. For example, among 38 clauses of the Environmental Impacts Assessment Law (EIA) — one of the most recently developed environmental laws in China — Clause 1 deals with "legislative purpose"; Clause 2 with "definitions of

key terms"; Clause 3 with "applicable scope"; 5 clauses (6, 7, 8, 9, 23) are about "functions of government and administrative organs". Altogether 11 clauses (10, 12,

13,14, 15, 16, 17, 18, 22, 25, 27) involve "administrative control". Another 8 clauses (28–35) relate to "legal liability". However, elements such as 3) rights and duties of social subjects over sustainable development, utilization and management of natural resources, especially women's status and rights, indigenous rights and interests, security of rural land tenure and land quality; 7) joint duty of care; and 14) sustainable resources utilization or zoning and planning of ecological conservation, are poorly reflected and presented. These elements are not even represented in the EIA Law.

Two reasons may explain this omission: One relates to the fact that China has long attached much importance to the protection of women's rights and interests, indigenous rights and interests, and special agencies are charged with relevant tasks of such protection. The problem is that these agencies are not involved in the management of economic resources and their rules and regulations have more of a policy nature than a law nature. That is to say that these special agencies have little dialogue with economic sectors when articulating policy-oriented regulations. Conversely, economic sectors tend to make provisions that exclude non-economic issues. As a result, a legislative vacuum is created. The second reason is that making provision for the aforementioned rights and interests entails reallocation of citizen's basic rights and social interests. Such provision is so important that it is difficult for any piece of administrative regulation to accommodate and fulfill the very task that only laws can do by directly coding the rights and interests in the book. (In China laws are traditionally more effective than departmental regulations in the perspective of legal status, thus important principles and institutional arrangements are usually explicitly set down in national laws rather than in departmental regulations.)

5. Recommendations

Combating LD successfully requires an enabling legislative framework with coherent and consistent legal tools that fully cover all the basic legal elements and principles with flexible and workable systems and mechanisms, and with applicable and enforceable procedures. In order to achieve such a framework, the authors of the paper propose that China's existing LD laws and regulations be streamlined in accordance with IEM elements, and that laws and regulations are systematically enacted, modified, abolished, and implemented as well. Although this is recommended for all 9 areas studied, in terms of legislation priority should be given to the areas of land resources and desertification control; and to the area of environmental protection as far as revision is concerned. Specifically, issues that need to be addressed in terms of either legislation or revision are as follows:

5.1 To expedite relevant legislation to bridge gaps

As has been noted, land contamination is among China's most serious LD issues. However, there are no laws or environmental standards for land contamination and, so far, no comprehensive and systematic land contamination surveys have been conducted for control measures to be based on. Additionally, most of the public is unaware of the seriousness and threats of this type of contamination. For these reasons, in 2007 the State Environmental Protection Administration determined that an overall programme for environmental protection of land, based on the survey and monitoring of the Status Quo of land contamination should be launched. The details of the program include:[3]

Conduct a national survey of the Status Quo of land contamination, establish and gradually improve a system of standards for environmental quality of land.

Establish systems for monitoring and assessing environmental quality of land;

Develop pilot projects in designated areas for the rehabilitation and integrated treatment of contaminated land;

Strengthen the supervision of land contamination of waste-water-irrigated areas, land for industrial purposes as well as areas around industrial quarters; and strictly control the irrigation by waste water in key grain and vegetable producing bases;

Develop actively ecological and organic agriculture; enhance strict control of the environment of cultivated areas where green, non-pollution and organic products are produced.

These initiatives will help combat LD in China. However, it is also essential to articulate the Law on the Prevention and Control of Land Contamination.

A few of issues need to be addressed in such a law. First, provisions should be thorough and contamination-specific, though land contamination issues are somewhat covered in China's Environmental Protection Law, in the Water Pollution Prevention and Control Law, in the Air Pollution Prevention and Control Law, and in the Law on the Prevention and Control of Environmental Pollution Caused by Solid Waste, their provisions are far from comprehensive or sufficiently specific. Second, a streamlined regulatory structure and clearly defined functions should also be addressed. China's regulatory structure is sort of a 'matrix muddle' of vertical and horizontal lines of authority and reporting (Andrews-Speed et al., 2003) involving numerous departments. They are all empowered to exercise their jurisdictional powers, but none of them could virtually carry out their powers in practice, leading to slow and poor implementation and enforcement of environmental laws and regulations. Third, given diverse sources of land contamination, each source should be specifically addressed, along with multiple tools and preventive provisions.

Another important deficiency can be found in the area of desertification prevention and control. Although China enacted the Law on the Prevention and Control of Desertification (PCD Law) in 2001, it lacks provisions concerning the protection of rights and interests, the coordination of interested parties, and the joint duty of care required by the IEM approach. In addition, the PCD Law lacks supporting administrative statutes that would enable its implementation. Traditionally, administrative powers in China have played a dominant role, and legislation by authorized administrative agencies can be binding. Hence, an administrative statute such as the Implementation Rules for the Law on the Prevention and Control of Desertification is imperative.

Besides the PCD Law, the Administrative Measures for Profitable Prevention and

---

[3] Suggestions for Further Enhancing Ecological Protection, the State Environmental Protection Administration, No. 37, 2007.

Control of Desertification (2004) explicitly stipulate the rights and obligations of units, individuals and private parties that conduct profitable desertification-combating activities. However, as valuable as these profitable activities are, more opportunities for desertification-combating programs for public interests need to be encouraged. The PCD Law includes provisions encouraging activities such as voluntary donations and provisions for a local competent authority to facilitate them (Clause 24). Unfortunately these provisions are too general to operate in practice.

5.2 To integrate IEM ideas into the revision of existing instruments

The study revealed that some instruments in the area of environmental protection need a thorough revision given that they were adopted in the 1970s or 80s. For example, the Environmental Protection Law (EPL), allegedly the basic environmental protection law in China, was originally enacted in 1979, amended in 1989, and has not been revised since then. Apart from being obsolete, the EPL has failed to lead environmental law for the following reasons. Firstly, Chinese basic laws are usually reviewed and passed by the National People's Congress (NPC), but the EPL was reviewed and passed by the standing committee of NPC; apparently its legal status is inferior to that of a basic law. Secondly, the EPL's principal clauses have already been reaffirmed or amended by those of newly enacted instruments. For example, protective provisions concerning the environment in the EPL are either non-existent or less specific and complete than those in the regulations.

Governing natural resources. Thirdly, the EPL's legislative purpose is narrow and sustainable development is absent from it. Fourthly, the law covers citizens' rights of accusation and complaint in very general terms, and provisions such as the rights of citizens to be informed about the environment, to make claims for environmental compensation, for public supervision and for the litigation in public interest cases are non-existent. Fifthly, due to lack of provisions for legal liability, when conflicts between economic development and environmental protection arise, the latter almost always gives way to the former (news.sohu.com, 2007). The EPL deficiencies described have greatly contributed to the poor performance of China's LD control activities. The EPL needs to be thoroughly revised so that it includes both pollution control and ecological protection.

This overhaul would benefit from the experiences of major industrialized countries that have accomplished similar transformations, by integrating IEM ideas, and by reconstructing principles and systems suitable for a basic law (Zhou and Zhu, 2005). To this end, the authors of this article propose an early revision of the EPL Law.

One more revision is needed in the present environmental protection law framework. Laws and regulations within this framework pay more attention to pollution-based provisions but neglect comprehensive adjustment of regional environmental integrity. The lack of fundamental legislation in the protection of natural resources coupled with undue regard to legislative planning for this field has constituted one of the reasons for ineffectiveness in LD control throughout the country. To remedy this deficiency, it is necessary for China, apart from an earlier amendment of the EPL, to promulgate the Ecological Protection Law, the Law on the Prevention and Control of Land Contamination, the Biological Safety Law and the Regulations for the Protection of the Rural Environment, and to improve the existing pollution control laws and regulations. The outcome of these measures would be an effective legislative framework with the Environmental Policy Law as the lead, ecological protection instruments and pollution control tools as mainstays.

Again within this framework, amendment must be made to the Water Pollution Prevention and Control Law, which was passed in 1984 and revised in 1996. Although it has been later supported by the Implementation Procedures for the Water Pollution Prevention and Control Law, a difference still exists when measured against IEM principles and basic legal elements given that several standards are out of date and need amending. In addition, as environmental protection in nature reserves gains importance, it is urgent to enact a Law on the Protection of Nature Reserves that would elevate the status of the Regulations on the Protection of Nature Reserves (1994).

5.3 To improve land use planning and zoning systems

China's existing land use planning system is composed of overall plans, detailed plans and specific plans, which are implemented at various levels including national, provincial, municipal, county and town. Unfortunately the successful implementation of this planning framework at different levels is hindered by the influence of an imperfect national economic system, as well as by the specific weaknesses and datedness of the plans per se such as unclearly defined functions, inflexibility of the planning system as a whole, and poor coordination among plans at various levels (Tang et al., 2004). The Land Administrative Law requires a close relationship between broader land use plans and specific ones, where the former should act as a basis for, or offer guidance to, the development of the latter. In practice, however, conflict is not uncommon when, for example, a higher level government agency, for lack of local information, develops a land use plan that hinders the development of a local land use plan that would fittingly address the development of the local economy. As modifications of higher government agency plans are a rare occurrence, lower level plans are implemented as a result of a lax enforcement process. Rather than offer broad strategic guidelines, China's land use system is a medley of policies, guidelines and goals that permeate every detail of country and provincial plans, leaving little autonomy for land administration at the local level where the plans are actually implemented. This has undermined the implementation of land use plans. In addition, there exists an apparent coordination problem between city planning and rural land use planning, between overall land use plans and special plans (e.g., Land Consolidation and Development Plans) (Tang et al., 2004).

An additional feature of the land zoning system is that it is largely human-centred. Land is categorized into three classes: land for agriculture (i.e., farming land, forest land, grassland, land for irrigation works etc.), land for construction (i.e., land for buildings, land for housing and public facilities, land for industry and mining, land for transportation and water conservation facilities, land for tourism and land for military purposes) and unused land. "Unused land" refers to land not used for agriculture or for construction. It includes weed infested areas, alkaline land, sandy land, uncovered land, nude rocky and stony land. These forms of land are unused because of their low economic value. The authors of the article contend that unused land should not be freely available to anyone who may be able to generate economic value from it since unused land may harbour a high degree of ecological value (Yue and Zhang, 2003). Hence, the "unused" category requires reassessment and reconsideration. One probable option is to adopt the term "ecological land" in lieu of "unused land" in the revision of the Land AdministrationLaw, which is nowadays a popular term in such countries as the US, European countries, Japan and Korea. Accordingly, the former three-class zoning system should be readjusted into land for agriculture, land for construction and land for ecological purposes (Xu et al., 2007).

The biggest administrative problem of land use zoning lies in the adoption of a planned economy in the context of a market economy. State planning that disregards market forces and local interests and needs is still prevalent. The existing legislation should be therefore revised accordingly. Further revision of the Land Administration Law should take into account warning zones according to ecological functions of used land, ecological protection requirements and obligations for the reclamation of farming land, and an emphasis on land quality data (including ecological quality) when land statistics and land monitoring are conducted (Du, 2005).

5.4 To enhance legal liability on LD control

Civil, administrative and criminal liabilities in China's LD control framework do not typically liaise and complement each other in the existing laws and regulations. For example, in 2002 revised Forestry Law, a number of clauses state that criminal liability shall be investigated and affixed by law when a grassland offence is proved; however, this provision is not present either in the criminal law or in judicial interpretations of the law, and cannot therefore be implemented in reality. Other coordination problems exist between the Grassland and Forestry Laws. In China both forests and grasslands are under the category of land for agricultural purposes. The actual use of the land is typically determined by the issuance of either a forest or a grassland permit. The Grassland Law stipulates that theState shall be responsible for protecting, developing and utilizing grassland through a unified planning system, but in practice, disputes about the ownership of forests and grasslands between the forestry department and the husbandry department exist. For example, the forestry and grassland departments are in dispute over the ownership of growing vegetation. To complicate matters, some local governments, looking to increase the rate of forest coverage, have rigidly required that a forest network be established linking farming land and grassland (Wang, 2004). This would undoubtedly add complexity and difficulty in departmental control, coordination and administrative enforcement of these departments.

Finally, China has placed a great deal of emphasis on the investigation and affixation of administrative liability rather than of civil liability, ignoring the significant role that the latter can play. Furthermore, lenient penalties slacken and undermine the performance of the legal liability functions of LD control.

6. Conclusions

China is facing serious land degradation problems mainly due to land contamination. The authors of the article contend that deficiencies in the regulatory framework of LD control have played a considerable role in exacerbating the problems. This study has sought to examine China's existing laws and regulations pertaining to LD control based on the IEM approach introduced by the PRC/GEF Partnership. The analysis has shown that China's existing legislative framework for LD control is characterized by:

Lack of linkages and coherence among LD-related instruments;

More emphasis on the control of single pollution elements, e.g., air pollution, water pollution, etc. with very limited attention given to integrated natural resources protection and ecosystem preservation;

A plethora of policy-like or conceptualized provisions characterized by more administrative control than rights-based incentives with weak implementation and operability;

Interlocking and overlapping management systems. From a legislative perspective, these are major constraints for China to effectively achieving LD control. To overcome them, laws and regulations should be revised and new instruments need to be designed in line with IEM principles and basic China-specific legal elements. This paper has proposed several improvement alternatives, including: the establishment of regulations for land contamination; the revision of the basic environmental protection law by emphasizing both pollution control and ecological conservation; and the enhancement of legal liability in LD control legislation.

Though developed to address the problems specific to LD-related laws and regulations at the national level, this IEM approach can also provide a starting point for analyses of instruments at the local level.

Acknowledgements

The authors are grateful to the Global Environment Facility (GEF) for funding the research. The support from Asian Development Bank (ADB), China's Central Office for Projects and Wuhan University of China is also acknowledged.

References

A Proposal for Speeding-up the Revision of Environmental Protection Law (in Chinese). Available at http://news.sohu.com/20070310/n248638052.shtml, last visited on July 15, 2007.

Andrews-Speed, P., Yang, M., Shen, L., Cao, S., 2003. The Regulation of China's Township and Village Coal Mines: a study of complexity and ineffectiveness. Journal of Cleaner Production 11:185-196.

Chen, X. Q., 2003-06-25. Experts Discussing Causes and Control of China's Desertification (in Chinese). Available at http://www.vos.com.cn/2003/06/25/8622.htm, last visited on Nov. 13, 2006.

Chen, Y. H., Mainland China Starts Land Contamination Control Projects" (in Chinese). Available at: http://news.phoenixtv.com/phoenixtv/83931275940855808/20061013/904120.shtml, last visited on May 22, 2007.

China Constitution (in Chinese), 2006-03-14. Available at: http://www.oefre.unibe.ch/law/icl/ch00000.html , last visited on Nov. 23, 2006.

Du, Q., 2004. Resource Legal Issues on Land Desertification Control and Counter Measures (in Chinese). Law Review1: 91-97.

Du, Q., 2005. Limitation of Major Territory-based Resource Laws in Ecological Conservation (in Chinese). Environmental Protection 6: 29-32.

GEF, Operational Program (OP) 12 on Integrated Ecosystem Management. Available at:http://www.gefweb.org/Operational_Policies/Operational_Programs/OP_12_English.pdf, last visited on

Nov. 21, 2006.

Huang, J. K., 2006-07-06. China's Land Degradation: water erosion and salinization (in Chinese). Available at: http://www.nmgland.cn/nmgland/zhjgd/zhjgd/3925.html, last visited on Nov. 16, 2006.

http://www.iucn.org/themes/cem/ourwork/ecapproach/index.html, last visited on Aug. 5, 2007.

Implementation Guidelines of the State Forestry Bureau for Thoroughly Promoting Forestry by Law (in Chinese). The State Forestry Bureau, No. 196, issued on Nov. 5, 2004.

Lu, Q., Liu, L. Q., 2003. Counter Measures for China's Desertification (in Chinese). Ecological Environment and Protection 5: 22-26.

2006 National TV Conference on Land Contamination Survey held in Beijing (in Chinese). Available at: http://finance.sina.com.cn/g/20060718/1233804158.shtml, last visited on Nov. 14, 2007.

People's Republic of China/Global Environment Facility (PRC/GEF) Partnership on Land Degradation in Dryland Ecosystems 2006-06-01. Available at: http://www.adb.org/projects/PRC_GEF_Partnership/, last visited on Nov. 21, 2006.

Suggestions for Further Enhancing Ecological Protection (in Chinese), the State Environmental Protection Administration, No. 37, 2007.

Tang, J., Zhao, X., Xia, M. 2004. Optimizing and Improving China's Land Use Planning System (in Chinese). Journal of Agricultural University of Northeast China (Social Sciences Edition), Vol. No. 3: 20-26.

Land Contamination (in Chinese). Available at: http://baike.baidu.com/view/786777.htm, last visited on Nov. 14, 2007.

Wang, R. 2004. An Investigation into the Truth of Enclosing Land for Forestry—Who are capturing grassland for planting trees (in Chinese). Available at http://finance.sina.com.cn/g/20041111/07581147034.shtml, last visited on July 2, 2007.

Xu, J., Zhou Y. K., Jin X. B., Yi L. Q. 2007. Discussion on Unused Land in Land Zoning System from the Perspective of Ecological Protection (in Chinese). Science of Resources, Vol. 29, No. 2: 137-141.

Yue, J., Zhang, X. 2003. A discussion on China's Land Zoning Issues (in Chinese). Arid Land Geography. Vol.26, No. 1: 78-87.

Zhou, K. 2001. On Ecological Environmental Law (in Chinese). Law Publishing House, Beijing, PR China.

Zhou, K, Zhu, X. 2005. Revision of the Environmental Protection Law and Its Historical Transformation (in Chinese). In Wang, S. Y (ed.): Sustainable Development and China's Environmental Law System—a monographic study of the revision of the Environmental Protection Law. Science Publishing House: 96-108.

Appendix

Table 1. PRC's laws and regulations related to LD control within 9 areas

Areas	Laws	Administrative Statutes	Departmental Rules & Regulations
Land	Land Administration Law (2004)	1. Implementation Regulations on Land Administration Law (1998) 2. Regulations on Land Reclamation (1988) 3. Regulations on Prevention and Control of Geological Disasters (2003)	1. Provisions on Public Hearings for Land Resources (2004) 2. Administrative Measures for the Pre-approval of Land for Construction Projects (2004)

Areas	Laws	Administrative Statutes	Departmental Rules & Regulations
Desertification Control	Law on Prevention and Control of Desertification (2001)		1. Administrative Measures for Profitable Prevention and Control of Desertification (2004) 2. Administrative Measures for Harnessing and Developing "Four Rural Wastes" (1998) 3. Administrative Measures for Monitoring National Desertification and Desert Land (2003) 4. Administrative Measures for Collection of Licorice Roots and Chinese Ephedra (2001)
Water & Soil Conservation	Water and Soil Conservation Law (1991)	Implementation Regulations on Water and Soil Conservation Law (1993)	1. Regulations on Water and Soil Conservation in Development of Bordering Areas within Shanxi, Shaanxi and Inner Mongolia (1988) 2. Administrative Measures for Water and Soil Conservation Schemes for Development Projects (1994) 3. Administrative Measures for Examination and Acceptance of Water and Soil Conservation Facilities for Development Projects (2002) 4. Regulations on Management of Compensatory Fees for Small-sized Farmland Irrigation Works and Water and Soil Conservation (1987)
Grassland	Grassland Law (2002)	Regulations on Prevention of Grassland Fires (1993)	1. Implementation Provisions for Prevention and Control of Grassland Insects and Mice (1987) 2. Administrative Measures for Balanced Grass and Husbandry (2005)
Forest Reserves	1. Forestry Law (1998) 2. Mineral Resources Law (1996)	1. Regulation on Prevention and Control of Forest Fires (1988) 2. Implementation Regulations on Forestry Law (2000) 3. Regulations on Arable Land Restoring to Forestry (2002) 4. Administrative Measures for Nature Reserves Characteristic of Forests and Animals (1985)	1. Administrative Measures for Forestry Management over Protective Construction Projects for Natural Forests (2004) 2. Administrative Measures for Ownership Registry of Forests and Forestland (2000) 3. Administrative Measures for Examination and Approval of Occupying and Requisitioning Forest Land (2001)
Water	Water Law (2002) Law on Prevention and Control of Floods (1997)	1. Implementation Measures for Water Intake Licensing System (1993) 2. Regulations on Compensation and Migrants Placement after Requisitioning Land for Large–Medium Irrigation and Power Works (1991)	1. Administrative Measures for Demonstration of Water Sources of Construction Projects (2002) 2. Administrative Provisions for Prevention and Control of Pollution in Protected Drinking Water Source Areas (1989)

Areas	Laws	Administrative Statutes	Departmental Rules & Regulations
Agriculture	1. Agricultural Law (2002) 2. Rural Land Contract Law (2002)	Regulations on Protection of Fundamental Farmland (1998)	Administrative Measures for Transfer of Use Rights to Contracted Rural Land (2005)
Wild Animals & Plants	Law on Protection of Wild Animals and Plants (1988)	1. Regulations on Protection of Wild Animals and Plants (1996) 2. Regulations on Protection of Wild Land Animals (1992)	Provisions for Protection of Wild Agricultural Plants (2002)
Environmental Protection	1. Environmental Protection Law (1989) 2. Environmental Impacts Assessment Law (2002) 3. Water Pollution Prevention and Control Law (1996) 4. Law on Prevention and Control of Environmental Pollution Caused by Solid Waste (2004)	1. Implementation Procedures for Water Pollution Prevention and Control Law (2003) 2. Regulations on Protection of Nature Reserves (1994) 3. Tentative Regulations on Management of Scenic Spots Areas (1985) 4. Administrative Regulations on Environmental Protection of Construction Projects (1998)	1. Tentative Procedures for Public Hearings over Administrative Permission in Environmental Protection (2004) 2. Administrative Provisions for Prevention and Control of Tailings Pollution (1992) 3. Administrative Procedures for Examination and Acceptance of Environmental Protection of Completed Construction Projects (2001)

Table 2. 12 principles of IEM

Principle No.	Content of the Principles

P1 The objectives of management of land, water and living resources are a matter of societal choice.
P2 Management should be decentralized to the lowest appropriate level.
P3 Ecosystem managers should consider the effects (actual or potential) of their activities on adjacent and other ecosystems.
P4 Recognizing potential gains from management, there is usually a need to understand and manage the ecosystem in an economic context. An ecosystem management program should:
i. Reduce those market distortions that adversely affect biological diversity
ii. Align incentives to promote biodiversity conservation and sustainable use
iii. Internalize costs and benefits in the given ecosystem to the extent feasible.
P5 Conservation of ecosystem structure and functioning, in order to maintain ecosystem services, should be a priority target of the ecosystem approach.
P6 Ecosystems must be managed within the limits of their functioning.
P7 The ecosystem approach should be undertaken at the appropriate spatial and temporal scales.
P8 Recognizing the varying temporal scales and lag-effects that characterize ecosystem processes, objectives for ecosystem management should be set for the long term.
P9 Management must recognize that change is inevitable.
P10 The ecosystem approach should seek the appropriate balance between, and integration of, conservation and use
of biological diversity.
P11 The ecosystem approach should consider all forms of relevant information, including scientific and indigenous and local
knowledge, innovations and practices.
P12 The ecosystem approach should involve all relevant sectors of society and scientific disciplines.

Source: http://www.gefweb.org/Operational_Policies/Operational_Programs/OP_12_English.pdf

Table 3. Evaluation criteria for 20 legal elements

Grades	Items		
	Presentation	Expression	Application
A	The law has a specific provision for the element.	Fuller coverage, and the content reflected is more compatible with IEM principles and element indicators.	Frequently applicable with good results.
B	The specific provision for the element is guided by other laws and regulations.	Fuller coverage, but there is slight difference between the content reflected and IEM principles and element indicators.	Generally applicable with good results.
C	The element is supported by other laws and regulations.	Some reference, and the content reflected is fairly compatible with IEM principles and element indicators.	Frequently applicable with moderate results.
D	The content of the element is not reflected in laws and regulations but in other regulatory instruments.	Some reference, but there is a slight difference between the content reflected and IEM principles and element indicators.	Generally applicable with moderate results.
E	The content of the element is not reflected in regulatory instruments at all.	No expressions at all.	Basically inapplicable.

CHAPTER IV
IEM Approach and Application

27. Developing a Global Strategy to Combat Land Degradation and Promote Sustainable Land Management

Michael Stocking
Vice-Chair, Scientific and Technical Advisory Panel of the GEF, Washington DC
Professor of Natural Resource Development
University of East Anglia

Extended Abstract

Land degradation is truly a global issue, requiring a global strategy not only to control it but also to promote sustainable land management. The Global Environment Facility (GEF) has recently (2007) undertaken a strategic review of support to the land degradation focal area, and shortly will undertake another to coincide with negotiations for a fifth replenishment of the GEF for the period 2010-2014. This paper sets out the leading considerations in the strategic review that led to the GEF-4 Land Degradation Focal Area Strategy, as well as the topics that should be taken forward to GEF-5. 'Sustainable land management' (SLM) is the defining approach, emphasising not only sustainability but also fully integrated approaches across landscapes and ecologies.

Translating SLM into practical action is, however, a conceptual and practical challenge. Two major considerations were taken into account in building strategic objectives for GEF-4: the policy and institutional environment conducive to prevention and control of land degradation; and how best to encourage effective actions on the ground. These factors were translated into two thematic topics: (1) the enabling environment that will place SLM in the mainstream of development policy and practice; and (2) up-scaling the mutual benefits for the global environment and local livelihoods through catalyzing SLM investments for large-scale impact.

Strategic Objective 1 emphasises integrated approaches to land degradation control that take the whole landscape and agro-ecosystems in order to influence policy, planning and regulatory frameworks. Strategic Objective 2 prioritises those areas where investment in SLM is most cost-effective in terms of global environmental and local developmental co-benefits. The GEF-4 Strategy then focuses on (1) projects and programs in critical agro-ecological zones and (2) innovative approaches to SLM that will inform the GEF about future priorities. This is done through three Strategic Programs that support (i) sustainable agriculture and rangeland management, (ii) sustainable forest management in production landscapes, and (iii) new and innovative approaches in SLM. This last Program has been the least used to date.

Reflecting the needs of financial donors and the imperative to show that investments are effective, the GEF now emphasises impact indicators that provide tangible evidence of attainment of project objectives and that will track the delivery of global environmental benefits such as total system carbon. Five indicators were chosen for the GEF-4 Strategy: (1) change in status of land degradation using global land degradation surveys now being provided by the LADA project; (2) change in land cover using remotely-sensed

NDVI surveys; (3) an indicator of well-being, the chosen one being prevalence of child malnutrition; (4) vulnerability to environmental disturbance on the grounds that human livelihoods are most precarious where environmental disturbance is greatest; and (5) vulnerability to climate change. It will be noted that these chosen indicators provide both direct and indirect measures of the impact of SLM investments and encompass topics of current concern such as climate change.

Through careful attention to global strategy, it is hoped that the GEF will provide the main financial mechanism to support investments in SLM well into the next decade that will complement support from other donors and agencies and bring substantial benefits to human society and the global environment.

Introduction

Since 1992, the Global Environment Facility (GEF) has been the main financial mechanism to support the implementation of the global environmental conventions for biodiversity, climate change and, since the Beijing GEF Assembly in October 2002, land degradation. To date, 32 donor countries have given US$7.4 billion to enable 160 developing countries and economies in transition to meet the incremental cost – that is, the additional cost over and above their national and domestic responsibilities – of implementing their obligations as convention signatories to conserve biodiversity, control climate change and reduce land degradation. This has generated over $28 billion in co-financing from other sources to support over 1,950 projects that produce global environmental benefits in the 160 countries.

A typical example of a GEF project in China is the granting of $10 million to co-finance a World Bank loan of $66 million for the Gansu and Xinjiang Pastoral Development project. The GEF has provided the financing for additional actions under this sheep pasture management project that will reduce land degradation, conserve biodiversity and sequester carbon (World Bank 2007). Approved in September 2003, the project aims to improve livelihoods of herders and farmers, through the establishment of improved grassland management, livestock production and marketing systems. It is the synergetic linkage provided by the GEF financing, not only between biodiversity, climate change and land degradation, but also between environmental protection and human development, that makes the GEF's contribution to sustainable development unique.

The GEF is now (November 2008) just over half way through its fourth four-year replenishment period, known as GEF-4. Planning for GEF-5 (2010-2014) is just starting. Notwithstanding the large financial investments by the GEF and other donors, unprecedented global environmental change is occurring that challenges not only the sustainability of ecosystems but also the very future of mankind. Recent assessments (Millennium Ecosystem Assessment 2005a; IPCC 2007; IAASTD 2008) all confirm the extremely rapid rates of loss of biodiversity, accelerating climate change, degradation of ecosystems, the critical need for sustainable agriculture, fisheries and forestry production and the consequent impact on livelihoods (GEF-STAP 2008). The Global Environmental Outlook – GEO4 is the most relevant current global assessment, and for land it predicts "the outlook to 2050 [with] dominating trends that are largely unavoidable, and caveats of risks that are very unpredictable, but which have serious implications for society." (UNEP 2007, p.110)

This paper is based charts the process towards GEF-4's strategic plan for land degradation (GEF 2007). It reproduces the thinking and the substance of the strategic plan that evolved. It also looks forward to the key topics that must inform GEF-5. In particular, the paper identifies current and future strategic priorities in order to deliver global environmental benefits through investments in land degradation control and sustainable land management.

Global Importance of Land Degradation Control

Land degradation is internationally recognized as a global process with global impacts (Stocking 2006). It is the primary way that damage to ecosystem functions and services occurs with consequent impact on livelihoods, economies and societies. Truly, it is a global environmental and developmental issue.

From a baseline in 1991 with the Global Assessment of Land Degradation (GLASOD) to a new global assessment as part of the GEF-funded LADA (Land Degradation Assessment in Drylands) project, the worldwide rate and extent of land degradation has significantly increased. The GLASOD assessment, while now recognised as somewhat flawed (Sonneveld and Dent, 2007), indicated that 15 percent of the world's land surface was degraded, whereas LADA identifies 24 percent as degrading. Furthermore, the new assessment identifies many new areas that are being affected, with areas of historical land degradation now so degraded that they are stable only at "stubbornly low levels of productivity" (Bai et al. 2008a, p. i). South China is identified along with other 'hot-spots' of land degradation, where a total of 1.5 billion people live under increasingly difficult and unsustainable circumstances (Bai et al. 2008b).

Therefore, through process drivers such as intensification of agriculture and opening up of steep land areas, land degradation is legitimately a global concern. However, if we add to the direct impacts of land degradation, the effects on other global environmental and developmental components, such as the goods and services provided by ecosystems, including storage of carbon as a buffer against climate change, then land degradation is magnified hugely as a threat to our planet. For example, taking just the soil and its below-ground living organisms (such as nematodes, bacteria, mycorrhizal fungi), the land and the world's soil biota provide free services to the estimated value of US1.5 trillion annually (Brussard et al 2007). At the same time, the soil stores at least 1500 Gigatonnes of organic carbon, far more than the atmosphere and all the world's vegetation combined (Dance 2008). The processes of land degradation deplete organic carbon, and hence are of massive global significance in driving environmental change.

Developing the GEF-4 Land Degradation Strategy
The Challenge

At an early stage in the development of the GEF-4 land degradation strategy, the purpose of the focal area was determined to be to "foster system-wide change to control the increasing severity and extent of land degradation in order to derive global environmental benefits." (GEF 2007, p.38). The 'tool' to achieve this would be Sustainable Land Management (SLM), investments in which would control land degradation cost-effectively in the landscape through delivering a suite of global environmental benefits

(GEF-STAP 2006).

An immediate general problem was that SLM is poorly understood, especially in regard to its integration with the wider landscape and building in sustainability. A working definition was adopted based upon impact monitoring guidelines developed for SLM. As Herweg et al (1999, p.15) succinctly put it, "Sustainability will remain an empty phrase if projects do not monitor their impacts". Sustainable land management (SLM) was therefore defined as the use of land resources (soils, forests, rangelands, water, animals and plants) for the production of goods to meet human needs while assuring the long-term productive potential. SLM was to be seen as the foundation of sustainable agriculture and land use, and a strategic component of sustainable development and poverty alleviation. It would address the often conflicting objectives of intensified economic and social development, while maintaining and enhancing ecological and global life support functions of land resources. Practicing SLM principles is one of the few options for land users to increase income without destroying the quality of the land as a basis of production. At the same time, GEF-4 would ensure that impact would be tracked and monitored, through sets of indicators that would verify whether investments in SLM were truly achieving global environmental and developmental benefits.

For the fourth replenishment of the GEF, US$ 300million was allocated to the focal area. Given the scale and intensity of land degradation, these resources cannot possibly meet the costs of prevention, control and reversal of land degradation in all affected areas. The Strategy, therefore, was to allocate the available resources selectively. Rehabilitation of already-degraded land or the development of control technologies would not receive money, as these either lack cost-effectiveness or are better supported by other agencies. The landscape approach, which embraces ecosystem principles, would instead be used to address processes that provide people with ecosystem goods and services at the local to global scales of operation. Priority would be given to areas (a) severely affected by land degradation but which have potential for the creation of an enabling environment for SLM, and (b) showing promising improvements that can be spread to neighbouring areas and other communities.

The Strategic Focus and Objectives

In the previous replenishment period (GEF-3), interventions in the land degradation focal area had focused on targeted capacity development and the implementation of innovative and indigenous sustainable land management practices. These priorities resulted in a diverse portfolio of proposals experimenting, for example, with programmatic partnership approaches or market-based financing mechanisms (e.g. payment for environmental services). An analysis of the GEF-3 portfolio resulted in the recommendation for GEF-4 to narrow the scope of interventions, in particular using the results of the then recently-published Desertification Synthesis (Millennium Ecosystem Assessment 2005b).

The Desertification Synthesis and subsequent analyses (e.g. Douglas 2006) had highlighted inter alia two principal barriers to effective action to combat land degradation: a weak policy and institutional environment that marginalises effective actions through a lack of national prioritisation of the issues of sustainable land management; a failure to appreciate the leverage on national development that could be achieved through targeted SLM interventions that would bring multiple benefits to a number of sectors simultaneously. Subsequent analyses have supported the finding that these barriers, along with deficiencies in the global institutions tasked with tackling the problem, are major impediments to controlling land degradation (e.g. Stringer 2008)

Two Strategic Objectives were developed to build a policy and institutional environment conducive to prevention and control of land degradation and to encourage effective actions on the ground (Table 1, Column 1)

Table 1: Land degradation focal area Strategic Objectives with their expected impacts and indicators (Source: GEF-4 Focal Area Strategy)

Strategic Objectives	Expected Impact	Impact Indicators
Strategic Objective 1: An enabling environment will place SLM in the main stream of development policy and practice at regional, national and local levels	Overall decrease in trend and/or severity of land degradation	% Increase in Net Primary Productivity (NPP) and Rain-use Efficiency (RUE)
	Protected ecosystem functions and processes, including carbon stocks in the soil, plants and biota, and fresh water	% Increase in carbon stocks (soil and plant biomass) and/or % availability of fresh water
Strategic Objective 2: Mutual benefits for the global environment and local livelihoods through catalyzing SLM investments for large-scale impact	A decrease in the vulnerability of local populations to the impacts of climate change	% decrease in mortality rates consequent upon crop failures and livestock deaths
	Improved livelihoods of rural (usually resource-poor) land users	% decrease in number of rural households below the poverty line
	Diversified funding sources for SLM	% increase in diversity of funding sources (e.g. private sector, CDM)

Strategic Objective 1: the Enabling Environment

Translating the rhetoric of SLM into practical action is not only a conceptual challenge but a practical difficulty. One of the principal global barriers is what is now called the 'enabling environment', that is the institutions, policies, legal frameworks and champions that are available to ensure that land degradation and SLM are taken seriously. Hurni (2000, p.83), for example, argues that "only a comprehensive, participatory approach involving stakeholders at all levels will have the potential to develop locally useful solutions within a favourable institutional environment".

Natural resource management issues involving land use are typically dealt with piecemeal; sectoral policies and regulatory frameworks are not harmonised, so there is no clarity in over-arching goals and no secure financing for SLM. National ministries tasked with 'land degradation' are often the weakest and most marginalised in government. Land degradation is most severe in countries where environmental issues are not in the mainstream of development policy and practice, and which lack sufficient institutional capacity. It is in this type of 'policy environment' that issues of poverty and disease affecting well-being are the drivers for further degradation. Policy reform, the GEF-4 strategy argued, is a priority.

This Strategic Objective #1, therefore, addresses the enabling environment for currently-recommended integrated approaches to land degradation control. Approaches that tackle the whole landscape and that include ecosystem principles to the management of natural resources, seeking to build institutional capacity for integrated management in the wider landscape, were to be given priority. The scope of the Strategic Objective was then to promote policy reform and build SLM competence and capacity in countries where the drivers of land degradation are potent, and the people most affected are poor and vulnerable.

Through addressing the 'enabling environment' and fostering mainstreaming of land degradation issues into national policies, the GEF-4 Strategy hoped to achieve the following outcomes that would make practical actions much more effective. Indeed, the Strategy argues these are pre-conditions for land degradation control:

Mainstreaming of SLM: SLM is fully supported by policy, regulatory and planning frameworks (e.g. institutional policies and programs, land tenure and water rights, and other incentives);

Institutional and professional capacity: Institutions have the capacity to support SLM at local, sub-national and national levels. Regional and trans-boundary institutions have the capacity to address and promote the management of transboundary resources (e.g. training, educational, monitoring and research capacities enhanced and extended to encompass ecosystem and other integrated approaches); and

Budget and financial allocations: Access to sustainable financing for SLM is facilitated (e.g. viable financing plans through national sector budgets, payments for environmental services, and access to small credit schemes).

The GEF-4 Strategy recognises that this Strategic Objective can only be achieved where the necessary institutions, even though weak, are currently in place. Therefore, while countries are prioritized according to need, project investments can only be made where the 'institutional architecture' is in place. So, countries for investment in SLM are identified through analysis of the drivers and impacts of land degradation - such as existing kinds and patterns of degradation, land use, poverty and well-being, and vulnerability to climate change – but finally chosen only if they meet the pre-condition of existing institutions with national and regional mandates in land resources management, including provision of services such as training and research. GEF investment would then seek to enable these institutions to fulfil their mandates by placing SLM in the main stream of public policy and by capacity building.

Strategic Objective 2: Up-scaling

Projects inevitably cover only limited geographical areas. Even major strategic investment programmes, such as TerrAfrica (http://www.terrafrica.org) in the land degradation focal area, are selective in terms of countries and places within countries for interventions. Funding constraints mean that projects are limited in terms of the spatial area of implementation, the number of people benefited directly and the thematic topics or technologies that can be covered. Many projects demonstrate what can be done in a pilot zone in the hope that the lessons learned will be applied more widely. Projects are also encouraged to develop a 'communication plan' whereby outputs are made more widely known. Indeed, the generic capability of a project output to be used outside the confines of the project envelope is one aspect that is considered prior to funding. If the outputs can demonstrate the rationale for action without additional GEF resources, then this

Table 2: Up-scaling: types and applications (Source: Gündel et al 2001)

Type of up-scaling	Description and application
Quantitative	Expansion' or 'growth' of impact; increase in number of people involved through replication of activities. Also known as 'scaling-out' or 'horizontal up-scaling'
Functional	Where projects or programmes expand the scale of their activities into new areas. For example, investment in land degradation control expanded to micro-credit and small business.
Political	Projects and programmes move beyond service delivery of, for example, technologies for land degradation control, towards wider change in policy and institutions.
Organizational	Improvement in organizational effectiveness and efficiency to allow for growth and sustainability of interventions. This is often achieved through leveraging co-finance, training, networking and capacity-building

makes the investment more attractive. This capability is what is known as 'scaling-up' or 'up-scaling' potential. Essentially, it measures the potential for impact from a project investment; the greater the impact, the more likely the project will be supported. A number of types of up-scaling have been recognised (Table 2), all of which are of interest to the GEF in achieving greater impact of projects and programmes especially in delivering global environmental benefits.

This Strategic Objective #2, therefore, prioritises those areas where investment in SLM will be most cost-effective in terms of benefits achieved for areas and populations both within and outside the immediate project area (Quantitative up-scaling); where there are mutual benefits for the global environment and local livelihoods (Functional up-scaling); where investments in SLM also achieve new policies and legal frameworks (Political up-scaling); and where the institutions for SLM are strengthened through, for example, training and additional financing (Organizational up-scaling).

The GEF-4 Strategy highlights the need to identify the most cost-effective investment that can replicate proven initiatives that are ready to be taken up widely; and tangible benefits to local livelihoods that will ensure that the initiatives are sustainable. This is in accord with guidance from the UNCCD Bonn Convention that emphasises the role of projects as catalysts for change in the land degradation focal area (http://www.unccd.int/cop/officialdocs/cop4/pdf/3add9(b)eng.pdf). It is also in accord with current scientific understanding of benefits achievable through integrated approaches. Synergies with other focal area objectives are also encouraged, including: adaptation to climate change, biodiversity conservation in production landscapes, and reductions in pollution and sedimentation of international water bodies.

Through addressing the interlinked aspects of up-scaling and maximising impact of project investments, this Strategic Objective is expected to achieve the following outcomes:

Up-scaling of SLM: Systematic large-scale application and dissemination of sustainable, community-based farming and forest management systems;

Developmental benefit: Communities benefit from applying and disseminating SLM practices; and Greater

application of integrated approaches: Sustainable financing for integrated approaches to SLM achieved.

The scope of this Strategic Objective encompasses actions of mutual benefit to the global environment and local people, through the adoption of best practices for the control and prevention of land degradation, and the measurable improvement in the delivery of ecosystem goods and services. The 'enabling environment' of Strategic Objective #1 needs to be in place in order to achieve these benefits, especially the people and institutions to handle integrated approaches to land resources management.

Priority Topics, Agro-Ecological Zones and Strategic Programs

The GEF-4 Strategy recognized that there are certain comparative advantages for GEF investments in a limited number of priority topics, and that other topics are more relevant to other agencies and funding systems such as the CGIAR. It was decided that GEF investment in the land degradation focal area should comprise: (1) projects and programs aimed at critical agro-ecological zones and, (2) innovative approaches to SLM that will inform the GEF about priorities beyond GEF-4.

The indicative list of kinds of interventions emphasises links between focal areas that will deliver global environmental benefits in the context of sustainable development. High-priority agro-ecological zones include:

Arid to semi-arid: cropland and rangeland issues, mixed land-uses, rainwater harvesting, small-scale irrigation, pastoral systems, traditional and local knowledge (cross-cuts with sustainable use and protection of dryland biodiversity, sustainable use of groundwater waters and; vulnerability to climate change and variability);

Semi-arid, dry sub-humid to temperate: mixed forest, rangeland and cropping, including subsistence agriculture, use of wood and non-wood resources, interactions with wildlife (cross-cuts with sustainable use and protection of biodiversity; sustainable forest management and vulnerability to climate change and variability);

Mountains and upland watersheds: including natural resources management to protect water sources and habitats, mountain communities (cross-cuts with protection of international water bodies, sustainable use and protection of biodiversity; sustainable forest management; and vulnerability to climate change and variability);

Humid forest margins: the forest/woodland mosaic in the wider landscape including crop and livestock production, protection of forest-margin biodiversity, management of highly-weathered acid soils and peat (cross-cuts with sustainable use and protection of biodiversity; sustainable forest management; and vulnerability to climate change and variability); and

Sub-humid to sub-tropical: rainfed agricultural zones, including issues of soil fertility, protection from soil erosion, sustainable use of groundwater (cross-cuts with climate change, biodiversity, and aspects of international waters).

In order to organize these priority zones and topics, the GEF-4 Strategy devised three Strategic Programs to act as global envelopes for both programmes and projects (see GEF 2007 and Table 3 for further details and descriptions of the Programs):

Sustainable agriculture and rangeland management;
Sustainable forest management in production landscapes; and
Innovative approaches in sustainable land management.

It is interesting to note that Strategic Programs #1 and #2 have received a large number of proposals, whereas #3 – innovative approaches to SLM - has had very few proposals to date in GEF-4. The reason for this disparity will be taken up in planning for GEF-5, with possibly greater guidance given to GEF agencies as to what constitutes 'innovation' and why it is important.

Table 3: Summary of Strategic Programs for GEF-4 (Source: GEF 2007).

Strategic Program	Expected Program Outcome	Program Outcome Indicators
1. Supporting Sustainable Agriculture and Rangeland Management	In intervention areas, an enabling environment for sustainable rain-fed crop production and rangeland management is created and natural resources (incl. dryland forests, water and energy) are managed in an integrated way	Each partner country has a new harmonised policy for each major land use type (agriculture, livestock) and/or has adopted a national land use policy % of extension programs offered by key institutions reflects ecosystem principles and concepts % increase in joint activities between specialized institutions % increase in allocation of resources to sectoral ministries dealing with natural resources Net and per caput access of rural land users to rural credit facilities and/or revolving funds % increase in area where SLM best practices are applied
2. Supporting Sustainable Forest Management (SFM) in Production Landscapes	Forest resources in humid forest margins, forest fragments and woodland resources in semi-arid and sub-humid ecosystems are managed sustainably as part of the wider landscape	Each partner country adopts a new harmonised policy for SFM and/or a national land use policy adopted % of extension programs offered by key institutions reflects ecosystem principles and concepts in wider landscape management, including forest and woodland resources % increase in allocation of resources to sector ministries dealing with forest and woodland resources % increase in net and per caput access of forest and woodland dependant land users to rural credit facilities and/or revolving funds % increase in area where SFM best practices are applied
3. Investing in New and Innovative Approaches in Sustainable Land Management	Enhance scientific and technical knowledge of emerging issues, facilitating the strategy discussion for GEF-5 and enhancing GEF operations in the LD FA	Newly created scientific and technical knowledge supports strategy discussion for GEF-5 % of designs of project to be financed in GEF-5 reflect new scientific and technical knowledge New knowledge assists % of GEF-4 financed projects in preparation and implementation

Impact Indicators

Environmental indicators are widely used for a number of purposes. For example, they may be used to assess environmental conditions and trends at global, national, regional and local scales; to forecast trends; to provide early warning information; to assess conditions in relation to goals and targets (Bakkes et al 1994). Indicators therefore feature throughout the GEF-4 Strategy with good reason since they provide tangible evidence of attainment of programme and project objectives, and, if carefully chosen, track the impact of an SLM investment in delivering global environmental objectives.

In this paper only global-level indicators will be considered, although in the GEF, indicators are needed at portfolio (i.e. focal area), strategic programme and objective, and project levels. The subject of indicators has been much visited in natural resource management, and several proposed sets of indicators have looked more like shopping lists of everything that might want to be known about a subject. This belies the purpose of indicators to present a simple, clear and rational selection of a number of variables for which it would be practically useful to know what change has been incurred as a result of the project. The purpose of a set of indicators is to steer the project in the direction originally intended. In order to be effective in guiding project progress, two requirements must be met.

First, an indicator must have a wider significance than its immediate role within the processes of change to be induced by a project. For example, change in the rate of soil erosion in tonnes per hectare is a common measure of the effectiveness of a soil conservation technology such as earth bunds. However, to be useful as an indicator of project progress, the chosen variable needs to serve a wider remit. Change in net primary productivity might be a better choice because it would integrate not only the conservation of soil on-site but also the increase in soil fertility and effects on production. An indicator needs to reduce a large quantity of data and processes down to a simpler form, while retaining the essential utility and meaning of the overall change that is being wanted by the project investment.

Secondly, an indicator must essentially be normative, that is, be comparable to an aim or reference value. This is really the difference between the desired and actual values during the progress of a project, so that the project may be steered back to its originally-designed purpose. The reference values are often the targets set by the project, but they may also be desirable outcomes in a wider sense. In the land degradation focal area, GEF investments are strictly targeted at delivering global environmental benefits such as reduced sedimentation of international waters and increase in stored carbon. However, there is a wider purpose in society for GEF investments, that is, to benefit human well-being. It is entirely reasonable, therefore, to include indicator variables that encompass, for example, increase in livelihoods on the assumption that the means of that increase is delivered through environmental benefits.

The GEF-4 Strategy for the land degradation focal area chose five global indicators, each represented by easily-accessible global mapping.

Indicator 1: Change in status of land degradation

This indicator is based upon the Global Land Degradation mapping, 1981-2003, undertaken by ISRIC in the Netherlands for the LADA project. The map (Figure 1) combines the trend of biomass production and trend of rain-use efficiency, over the 23-year period, at a resolution of 8km. The map shows areas where trends of both the biomass and rain-use efficiency are negative. For irrigated areas, only biomass trend is considered. Urban areas are excluded. The map highlights areas where land degradation has taken place over the reference period, as opposed to the total historical legacy of degradation. The map may be used to identify areas where GEF intervention is needed; also may be used to prioritize proposed project interventions.

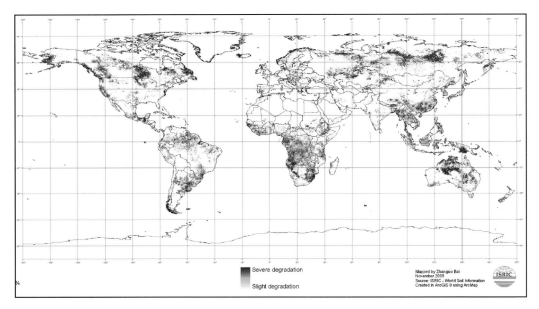

Figure 1: Indicator 1 – global land degradation, 1981-2003 (Source: ISRIC working document, February 2007)

Indicator 2: Change in vegetation land cover

This indicator is based on the land cover assessment for year 2000 from the work of the European Union Joint Research Centre based in Italy. The Normalised Difference Vegetation Index (NDVI) gives a measure of the vegetative cover on the land surface over wide areas. Dense vegetation shows up very strongly in the imagery, and areas with little or no vegetation are also clearly identified. The map (Figure 2) shows land cover categories at a resolution of 1km, mapped by interpretation of satellite imagery. The map may be used for comparison with the global land degradation map in order to

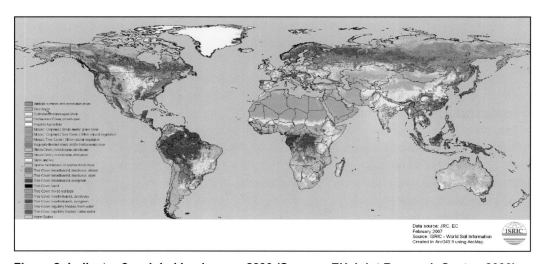

Figure 2: Indicator 2 – global land cover, 2000 (Source: EU Joint Research Centre, 2000)

assess which land cover categories are most affected by land degradation. Land cover categories are used as proxies for land use types and ecosystems.

Indicator 3: Prevalence of child malnutrition

This indicator was chosen in the GEF-4 Strategy to provide guidance not only on the beneficial impact of SLM investments on a key topic of human development but also to prioritise those areas that most need SLM in order to reduce developmental stresses, in this case, child malnutrition. The map (Figure 3) is essentially an indicator of poverty. Children are defined as underweight if their weight-for-age z-scores are more than two standard deviations below the median of the NCHS/CDC/ WHO International Reference Population. The map may be used to prioritize proposed project interventions and, also, to identify areas where land degradation and poverty are closely linked – and, therefore must be addressed simultaneously.

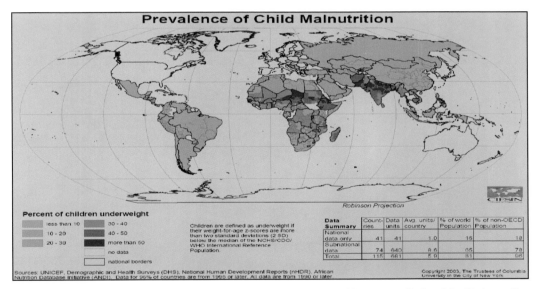

Figure 3: Indicator 3 – prevalence of child malnutrition (Source: Columbia University, 2003)

Indicator 4: Vulnerability to environmental disturbance

The map (Figure 4) presents the different grades of vulnerability of people to environmental disturbances. The Human Vulnerability Index is one of the five key measurements of the Environmental Sustainability Index. This component seeks to measure the interaction between humans and their environment, with a focus on how human livelihoods are affected by environmental changes. The map may be used to identify areas in which people are very sensitive to environmental changes and least prepared to absorb them. The map may be used to prioritize actions in proposed interventions on SLM on reducing the vulnerability of rural people to environmental disturbances such as land degradation.

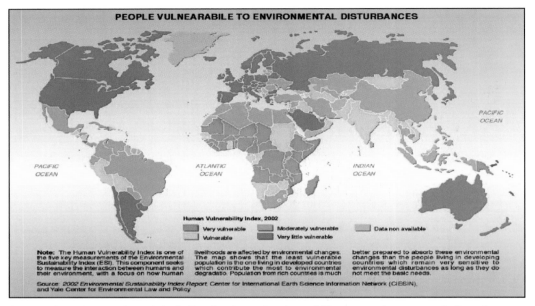

Figure 4: Indicator 4 – global distribution of vulnerability to environmental disturbances (Source: CIESIN and Yale Center for Environmental Law and Policy, 2002)

Indicator 5: Vulnerability to climate change

Rightly or wrongly, the threats posed by climate change have become perceived as the greatest current environmental challenge. There is, for example, proposals from GEF Council to make 'climate-proofing' mandatory for all projects, whether or not they explicitly refer to climate change. The map (Figure 5) presents the vulnerability index to climate change, which combines both national indices of exposure and sensibility. These indices are related to the variation of the annual mean temperature in 2100 equal to 3.3°C, calculated under the A2-550 ppm emission scenario (optimistic) and with climate sensitivity equal to 5.5°C (high value). The potential impacts of such a variation have been aggregated in the indices. The vulnerability spectrum ranges from modest to extreme vulnerable. The map may be used to identify areas that may be at future risk of land degradation due to impact of climate change. A comparison with the actual global land degradation map could help identify in particular those areas which are not at risk today, but which might be significantly affected by land degradation in the near future, so that preventative actions are undertaken.

Conclusion

Developing a global strategy to combat land degradation and promote sustainable land management was a complex process of selection, refinement and testing. Because the decision to create a new strategy came after the completion of the replenishment negotiations for GEF-4, the members of the Technical Advisory Group to the GEF in 2007 worked under intense pressure and had few opportunities to consult widely in finalising the structure and content of the Strategy. Nevertheless, the Strategy has been

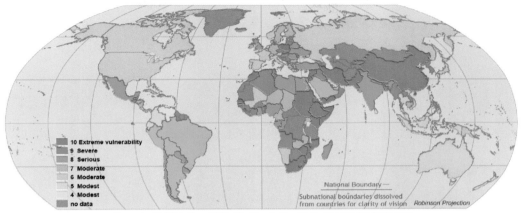

Figure 5: Indicator 5 – global distribution of vulnerability to climate change (Source: Wesleyan University and Columbia University, 2006)

effective in focussing GEF agencies' attention on aspects important to the GEF, specifically the delivery of global environmental benefits and the measurement and tracking of global impacts. For GEF-5, as already noted, the process of developing a new Strategy has started and will gather pace in 2009. The lessons learned in GEF-4 will be taken up, and a new agenda for the focal area of land degradation and the operational programme of sustainable land management will arise that will take us forward to 2014.

Acknowledgements

The core of this paper is based upon the GEF Focal Area Strategy for Land Degradation, for which the present author was one of a team of five on a GEF-4 Technical Advisory Group (TAG). The contributions of the other members of the TAG are acknowledged: Andrea Kutter (GEFSec); David Dent (ISRIC), Youba Sokona (OSS), and Goodspeed Kopolo (UNCCD Secretariat) The support of the Scientific and Technical Advisory Panel (STAP) and the GEF Secretariat is gratefully acknowledged.

References

Bai, Z.G., Dent, D.L., Olsson, L. & Schaepman 2008a. Global Assessment of Land Degradation: 1. Identification by remote sensing. GLADA Report 5. World Soil Information (ISRIC), Wageningen.

Bai, Z.G., Dent, D.L., Olsson, L. & Schaepman 2008b. Proxy global assessment of land degradation. Soil Use and Management 24: 223-234.

Bakkes, J.A., van den Born, G.J., Helder, J.C. & Swart, R.J. 1994. An Overview of Environmental Indicators: state of the art and perspectives. Report UNEP/EATR 94-01, United Nations Environment Programme, Nairobi.

Broussard, L., de Ruiter, P.C. & Brown, G.G. 2007. Soil biodiversity for agricultural sustainability. Agriculture, Ecosystems and Environment 121: 233-244.

Dance, A. 2008. What lies beneath. Nature 455: 724-725.

Douglas, I. 2006. The local drivers of land degradation in South-East Asia. Geographical Research 44: 123-134.

GEF 2007. Focal Area Strategies and Strategic Planning for GEF-4. Council Document GEF/C.31/10, Global Environment Facility, Washington DC - http://www.gefweb.org/uploadedFiles/Documents/Council_Documents__(PDF_DOC)/GEF_31/C.31.10%20Focal%20Area%20Strategies.pdf

GEF-STAP 2006. Land Degradation as a Global Environmental Issue: A Synthesis of Three Studies Commissioned by the Global Environment Facility to Strengthen the Knowledge Base to Support the Land Degradation Focal Area. GEF Council Document GEF/C.30/Inf8. Scientific and Technical Advisory Panel of the GEF, Washington DC.

GEF-STAP 2008. A Science Vision for GEF-5. Scientific and Technical Advisory Panel to the Global Environment Facility, Washington DC.

Gündel, S., Hancock, J. & Anderson, S. 2001. Scaling-up Strategies for Research in natural Resources Management: a comparative review. Natural Resources Institute, Chatham.

Hurni, H. 2000. Assessing sustainable land management (SLM). Agriculture, Ecosystems and Environment 81: 83-92.

IAASTD 2008. International Assessment of Agricultural Knowledge, Science and Technology for Development. Executive Summary of the Synthesis Report - http://www.agassessment.org/docs/SR_Exec_Sum_280508_English.pdf

Herweg, K, Steiner, K. & Slaats, J. 1999. Sustainable Land Management: Guidelines for Impact Monitoring. Centre for Development and Environment, Berne, 78pp.

IPCC 2007. Climate Change 2007: Synthesis Report. The Intergovernmental Panel on Climate Change, report approved at IPCC Plenary, Valencia, Spain - http://www.ipcc.ch/pdf/assessment-report/ar4/syr/ar4_syr_spm.pdf

Millennium Ecosystem Assessment 2005a. Ecosystems and Human Well-Being: Current State and Trends. Island Press, Washington DC.

Millennium Ecosystem Assessment, 2005b. Ecosystems and Human Well-being: Desertification Synthesis. World Resources Institute, Washington, DC.

Sonneveld, B.G.J.S. & Dent, D.L. 2007. How good is GLASOD? Journal of Environmental Management (in press – available online 20 December 2007, Science Direct)

Stocking, M. 2006. Land Degradation as a Global Environmental Issue: A Synthesis of Three Studies Commissioned by The Global Environment Facility to Strengthen the Knowledge Base to Support The Land Degradation Focal Area. Report for GEF Council, 5-8 December, Document GEF/C.30/Inf.8. Scientific and Technical Advisory Panel, Global Environment Facility, Washington DC, 17pp.

Stringer, L. 2008. Can the UN Convention to Combat Desertification guide sustainable use of the world's soils? Frontiers in Ecology and the Environment 6: 138-144.

The World Bank 2007. Global Environment Facility (GEF) Projects in China. - http://go.worldbank.org/CY9L4WRNA0

UNEP 2007. Global Environmental Outlook: GEO-4 – Environment and Development. United Nations Environment Programme, Nairobi - http://www.unep.org/geo/geo4/report/GEO-4_Report_Full_en.pdf

28. Using the Integrated Ecosystem Management Principles, Implementing Practice to Combat Desertification

Liu Tuo
Director General of Management Center of Desertification Prevention and Control, State Forestry Administration, China

Abstract: China's desertification area is 2.64 million km^2, accounting for 27.4% of the national total area, which is located in 18 provinces and autonomous regions, and has affected more than 400 million local people` livelihood. Desertification impacts China's economic and social development heavily, threatens to the survival of the human environment, restricts the development of the regional economy, widens the gap between regions, and act as one of the major reasons to global warming.

The Chinese government has paid great attention to desertification control. In recent years, a lot of integrated measures as well as international cooperation have been actively carried out to combat desertification in laws and regulations, engineering and scientific research. Five major principles were followed in undertaking desertification prevention and control work: (1) balance ecological and economical benefits; (2) take measures in key areas and undertake prevention in general area; (3) the government plays the leading role, and private corporate and the community are encouraged to involve and take up responsibilities; (4) comply to law, and insist on self-reliance; and (5) strengthen international cooperation. The effective results have been shown in prevention and control of desertification, that the desertification area has been in a reduction rate in 1283 km^2/a, which changed from growth rate of 3436 km^2/a. The local ecological situation has been improved and regional economic income has been increased, both of which have contributed to global Climate change mitigation and adaptation.

Desertification is one of the world-wide problems and a complicated engineering project to overcome. Taking its national situation into consideration, the Chinese government follows the scientific development concept and uses integrated ecosystem management approach in vegetation rehabilitation and reconstruction. It is expected that desertification should be under control by the middle of 21st century, and sustainable development of population, resources, and environment will be realized in desertification areas.

Key words: Integrated Ecosystem Management, desertification, practice

I. Situations and Influence of China's Desertification

China currently has a desertification area of 2.6362 million square kilometers, accounting for 27.4 % of its total land area, distributed in 18 provinces, regions and municipalities with more than 400 million people affected.

First, desertification threatens people's living environment. In China, desertification affects more than 50,000 villages, over 1,300 km of railways, 30,000 km of roads, thousands of reservoirs and 50,000 km of canals and ditches all year around. People experience serious problems related to sand- and dust-storms as they sweep across north China each spring and summer, in an average frequency of 8.8 times annually in recent years.

Second, desertification impedes economic development. The impediments mainly present in the following forms: decreasing of land productivity, reducing agricultural, forestry, and animal husbandry yield; leading to deep plowing and duplicate crop planting, along with waste of manpower, money and material resources; it also damages infrastructure, such as water conservancy, transportations, production and living conditions etc., greatly affecting economic safety and increasing the cost of development.

Third, desertification enlarges regional gaps. Sixty percent of the state-designated impoverished counties are located in degraded areas, leading to a vicious downward spiral of greater desertification and poverty, thereby leading to further un-balanced regional development.

Fourth, the process of desertification is a carbon-based process. The destruction of vegetation results in an increase in the surface reflection, leading to a hostile climate in part of the affected areas. Some scholars propose that the climate system and terrestrial ecosystems are corresponding to each other's changes, and the degradation of the ground vegetation, like desertification, has destroyed CO_2 balance in the atmosphere. A report by Massachusetts Institute of Technology in 2003 estimated that the amount of CO_2 loss by desertification in China in the last century was equivalent to 1.54 billion tons.

II. Basic Principles, Main Measures & Achievements in China's Desertification Control

The Chinese government attaches great importance to combating desertification. During the process, the government always adheres to the spirit of people-orientation, and respects the laws of nature and economy. Thus, it has adopted the following principles in combating desertification:

The first principle is coordination between ecological benefits and economic benefits. While driving forward combating desertification, great attention has been paid to people's livelihoods, regional economic development, and poverty alleviation of farmers and herdsmen.

The second principle is the combination of key control with general prevention. Control and prevention projects focus mainly on the eco-fragile areas, which significantly affect and thus might cast great damages on human settlement environment. The comprehensive desertification prevention and vegetation protection measures are also being concurrently implemented. All efforts orient to a harmonious development among agriculture, forestry and animal husbandry in combining natural forces and manpower together, emphasizing both on forest enclosure and treating, and on adoption of arbor-shrub-grass model.

The third principle puts emphasis on the leadership of governments at all levels, the participation of enterprises and the responsibilities by all social sectors in the process. Governments at all levels have earmarked funds for the control and prevention projects, and adopted preferential policies to mobilize enterprises and social sectors to participate in combating desertification.

The fourth principle is to implement integrated control and prevention in accordance to law and science. China has established a legal system with the Law of Desertification Control and Prevention as its mainstay, popularizing more than 100 technology models, and setting up a monitoring and evaluation system of desertification. Integration is guaranteed by the coordination of different parts and their corresponding measures.

The fifth principle is to strengthen international cooperation. We have set up cooperation and exchanges with dozens of countries and many organizations, through organizing and implementing a batch of foreign-aid projects like GEF-OP12 Program, LADA Program and Sino-Italy Afforestation at Aohan County, and introducing funds, technologies and new ideas via cooperative programs.

In the desertification control and prevention, we have employed the following measures:

(1) Afforestation and Grass Planting. Since 1978 many desertification projects concerning afforestation and grass planting have been implemented to increase vegetation coverage in the desertification lands, such as the Shelter Forest System in North, Northeast & Northwest China, the Desertification Prevention and Combating Project, Treating Sandstorm Sources around Beijing and Tianjin Project, and the Conversion of Cropland to Forest Project.

(2) Vegetation Protection. A strict vegetation protection system has been established to promote natural restoration of the ecosystems by implementing the Three Prohibition Policies of prohibiting collection of firewood, overgrazing and excessive reclamation.

(3) Water Resource Management and Water Saving. Water supply for industries, living and ecological areas is managed in accordance with the watershed conditions, and water saving technologies are popularized to improve water application efficiency.

(4) Eco-emigration. In uninhabitable and unproductive areas, an emigration strategy is adopted to improve people's living standard, restore vegetation and protect ecological systems.

(5) Reform of Production Model. Reforms are conducted to advocate a transformation from extensive to intensive management in agriculture, and to apply the policies of limiting the number of grazing animals by the carrying capacity of the pasture, and of pen raising (stall feeding), along with prohibiting and resting grazing systems.

With the above mentioned measures, remarkable progress has been achieved in combating desertification:

• Ecological conditions have been improved. Twenty percent of the nationwide desert land has been under control. Vegetation coverage in the key controlling sites has increased by over 20 %, and presents a positive evolving tendency. In the recent five years, the amount of national desert land shrunk by 75.85 million square miles on average annually, and sediment deposition in rivers decreased year after year. In areas of Treating Wind-blown Sandstorm Sources around Beijing and Tianjin Project, the amount of soil eroded by wind reduced one fifth under the same wind force, and the intensity

degree of sand- and dust-storms in some areas changed from strong/strong-plus to weak-plus/strong.
• The regional economic development has been promoted. Industries including characteristic plantation, breeding, processing and eco-tourism have developed continuously with a rising of a few backbone enterprises and famous brands. Farmers are enabled more ways to find jobs and increase of income, and their steps to poverty alleviation have been simplified. Some areas have witnessed a mutual promotion between ecology and economy and a harmonious relationship between man and nature. In the areas of Treating Wind-blown Sandstorm Sources around Beijing and Tianjin Project, over 16 million farmers and herdsmen profit from the project construction, with an increase of income near 50 % for the average per capita of farmers in 2005 compared to that in 2000.
• Sustainable development capacity has been improved. Taking the Project of Beijing and Tianjin as an example, through five years of continuous construction, social sustainable development capacity has improved by 22 %, the primary industry focusing on agriculture has reduced by 1.2 % annually, and industrial structure has been continuously optimized.
• Contribution has been made to mitigate and adapt to climate changes. Forest resources in China have increased at a relatively rapid pace by means of the afforestation strategy. The total area of forests has reached 175 million hectares, among which forest plantation exceeds 54 million hectares (equivalent to 800 million mu), ranking first in the world. The amount of CO_2 absorbed by forests in China has increased year by year as well. According to preliminary estimation by domestic experts, forests in China absorbed 500 million tons of CO_2 in 2004, equivalent to 8% of the total emission from all sources nationwide in the same period.

III. Desertification Combating Strategies in Next Stage

China, with the largest desert lands in the world, faces tremendous challenges in the process of combating desertification due to its national and desertification conditions:

(1) The desert lands are so large, taking up 27.4 % of the national total area, and therefore the control and prevention tasks are extremely arduous.

(2) With China's huge population and relatively slow economic development, some deserted areas, driven by over-zealous drive for economic benefit, have many problems of excessive reclamation, overgrazing and excessive excavation.

(3) Input for combating desertification is seriously insufficient, and it obviously does not match with the heavy tasks in desertification control and prevention.

Under the requirement of "strengthening combating desertification including rocky desertification" by the Chinese government, guided by outlooks of scientific development and policies of taking a priority in protection, actively involving in treatment and reasonable utilization, Chinese people will adopt integrated ecosystem management to improve the controllable desertification lands and realize coordinated development in economy, society, resources and environment in the desertification areas by upholding the principle of citizens-orientated and sustainable development. We will mainly employ four strategies:

• The strategy of natural restoration and rehabilitation of vegetation. The combating of desertification, mainly directed at vegetation restoration and rehabilitation, should abide by ecological principles and differential right treatments in different sites should be carried out, abide by principles of combining man-made plantation with natural recovery and of combining resources protection with reasonable utilization, as well as significantly increase the vegetation coverage rate so as to gradually create a positive cycle in the desert ecological systems.

- The strategy of promoting economic and social development. In desert areas, population increase will be adequately controlled, with purposes of gradually mitigating population pressures and improving people's living standard. Economic structure in the desert areas will be optimized, the linkages between agriculture, forestry and animal husbandry will be adjusted, and farm produce processing industry will be developed. Emphases will be put on alleviating poverty and increasing the income of farmers, striving to develop sand industry to promote economic development in desert areas and addressing poverty issues.
- The strategy of institutional innovation in combating desertification. Innovation in the desertification combating will be introduced to create a stable input mechanism with the public finance as a mainstay, improve an eco-compensation mechanism step by step, following the principles of endowing benefits to those who contribute to desertification control and prevention and of being compensated by those who benefit from the treatment, perfect the property mechanism to give a stable land-use right of farms and woodlands to farmers, explore the state eco-purchase system, reform the water and pasture management system, and many other positive outcomes.
- The strategy of improving the guarantee system for combating desertification. An organizational guarantee system will be perfected to establish a comprehensive inter-agency decision-making and consultation mechanism for more scientific decisions. A legal guarantee system will be set up to match with regulations and rules for combating desertification a profitable sand control system, an incentive and penalty system should be set up and implemented and become part of the assessment of local governments' work in the control. A supporting scientific system will be established to complete the monitoring and warning system.

Objectives: To improve the ecological condition in the key control and prevention sites by 2010; to make over half of national desertification lands under control with an obvious improvement of the ecological conditions in the wholly desertification areas by 2020; to form a stable eco-protection system with a highly efficient sand industry network and a completed guarantee system in ecological environment protection and natural resources utilization to take most desertification land in the country under control and realize a coordinative development in population, resources, environment and economy in deserted areas.

As a large developing country, China faces tough work ahead and a long way to go for combating desertification and achieving sustainable development. This is mainly due to population pressures, relative resource shortage, fragile ecological capacity, imbalance of regional development, and still tens of millions impoverished people. We are willing to actively carry out collaboration and exchanges with all partners in this respect and work for the realization of sustainable development.

29. Land Degradation and Sustainable Land Management in the Central Asia

Mr. Umid Abdullaev
Co-chair of the CACILM National Coordination Councils of Uzbekistan,
Director of Uzgip Institute, Uzbekistan

Abstract: This paper illustrated the status of land utilization and degradation in the region of Central Asian Countries, analyzed the environmental and social economic consequences of land degradation, concluded the experiences from SLM (sustainable land management), and introduced the GIS/RS approaches for land degradation assessment.

1. Baseline Information

Location:	Central Eurasia
Total land	3.882.000 km^2
Population	5.3 million people
Population density per 1 km^2: Kazakhstan - 6 Kyrgyzstan - 26 Tajikistan -45 Uzbekistan - 62	

2. Land Use in Central Asia

Mln.ha

	Total Area	Land Use (%)	Arable Land			Rangeland	Forest
			Total	Irrigated	Rainfed		
Kazakhstan	272.49	30.6	22.65	1.47	21.18	189.03	0.02
Kyrgyzstan	19.39	62	1.34	1.06	0.28	9.19	2.86
Tajikistan	14.31	5.1	0.73	0.50	0.23	3.74	0.55
Turkmenistan	48.81	81.3	1.73	1.70	0.03	38.15	2.21
Uzbekistan	44.74	63	5.10	4.30	0.80	23.00	2.81

Source: GM/UNDP (2007) based on the UNCCD National Reports, 2006

3. Current State of Land Degradation

Types of land degradation may vary depending on specific land use practices.

Mln.ha

| | Arable lands | | Salinization | Waterloggin | Overstocking |
	Water Erosion	Waterlogging			
Kazakhstan	272.49	30,6	22.65	1.47	21.18
Kyrgyzstan	19.39	62	1.34	1.06	0.28
Tajikistan	14.31	5,1	0.73	0.50	0.23
Turkmenistan	48.81	81,3	1.73	1.70	0.03
Uzbekistan	44.74	63	5.10	4.30	0.80

Source: GM/UNDP (2007) based on the UNCCD National Reports, 2006

4. Desertification of Coastal and Aquatic Ecosystems Related to Aral Sea Drying off

The current conditions of the natural ecosystems in the zone of Aral Sea crisis symbolize the largest problem resultant from the water mismanagement and agricultural management in the countries of region.

Annual sand and salt transport	million tonne	75
Usual precipitation in a form of sand and salt	kg/ha/year	520
Area of Aral region directly subject to deflation	km^2	42.000
Agricultural lands subject to erosion	million ha	2

5. Environmental and Socio-Economic Consequences and Threats

The condition of the environment has a direct impact on the living standard and health of the population, especially socially vulnerable groups. Major factors of such influence are: Significant reduction of food and cash crop yields (cereals – 48%; cotton – 39%; sugar beat – 52%; potato – 26% and vegetable – 34%); Decrease in efficiency of cattle breeding due to pasture degradation and reduction of fodder and agro biodiversity; Deterioration of quality of foodstuffs as a result of water contamination and soil pollution; Increase in the level of disease of the population, especially among women of childbearing age.

The Central Asia region losses USD 1.7 billions (or 3% GDP) annually due to inefficient water resources management. The annual decrease of the agricultural production in the region is estimated in the amount of USD 2 billions.

6. SLM Experience in Central Asia

Central Asia's SLM experience is to develop and implement projects oriented to sustainable land management through introduction of the conservation agriculture, projects oriented to build capacity and awareness raising on integrated water and land management, projects oriented to sustainable pasture management, projects oriented to build capacity of farmers and local communities through *Field Farmer School*, *Field Day* and others.

30. National Mechanism for Sharing Land Degradation Monitoring and Evaluation Information

Wu Bo
Professor, Research Institute of Forestry, Chinese Academy of Forestry, Beijing

1. Definition and types of land degradation

Land degradation refers to the reduction or loses of biological or economic productivity and complexity of rainfed cropland, irrigated cropland, or range, pasture, forest and woodlands resulting from one force or multi-forces. It include: 1) soil erosion caused by wind and/or water; 2) deterioration of the physical, chemical and biological or economic properties of soil; and 3) long-term loss of natural vegetation.

Land degradation is represented in reduction or loss of the biological or economic productivity and complexity. It behaves in many types including soil degradation, vegetation degradation, reduction or loss of biodiversity and land use value.

Land degradation processes can be divided in to three: 1) Physical process including soil erosion caused by water, wind, freezing and thawing, gravity and human factors, etc.. 2) Chemical process including reduction of soil fertility, soil salinization/alkilization and soil pollution, 3) Biological process including reduction of vegetation productivity and lose of biodiversity. These three processes are not acted individually and most of time, they are interacted, affected each other and take placed concomitantly.

Based on changes of characters, land degradation can be divided into four categories: 1) Degradation of land physical characters; 2) Degradation of land chemical characters; 3) Degradation of land biological characters; and 4) Degradation of land economic characters. Based on above classification, land degradation can be listed in following types: wind erosion, water erosion, gravity erosion, artificial erosion, reduction of soil fertility, soil salinization/alkalization, soil pollution, vegetation degradation, reduction of biodiversity, and land use conversion. These types of land degradation are crossed and overlapped. For example, while wind or water erosion takes place, vegetation degradation takes place at same time. When soil is polluted, vegetation on the soil will be also affected. When land use pattern changes, soil physical and chemical as well soil economic use characters will be also change.

2. Land degradation monitoring in China and its technical criterions

Following the function partition identified by the State Council, the State Forestry Administration (SFA), Ministry of Water Resources (MWR), Ministry of Agriculture, State Environment Protection Administration (SEPA), Ministry of Land & Resources (MLR), China Meteorology Administration (CMA), are related to land degradation monitoring and assessment.

Problems in land degradation monitoring and assessment in China include: 1) Lack of universal definition and classification system for land degradation. Each government department has its own definite land degradation, causing some overlap among government departments, while some of land degradation has been missed; 2) Lack of scientific and systemic technical criterions in land degradation monitoring and assessment. At present, most national and departmental technical criterions and provisions for land degradation monitoring and assessment were made by each government department. They are not well harmonized. There are differences in the adopted terms, indicators, investigation methods, grading criteria, and therefore, the data of land degradation among government departments are not comparable. 3) A coordinating mechanism is not well established among government departments in land degradation monitoring and assessment, which causes some overlapping and unnecessary waste of resource among government departments in land degradation monitoring and assessment. It also causes several sources of land degradation data, and leads users to loose ends and confusion.

The urgent issue in land degradation monitoring in China is the establishment of a scientific land degradation monitoring system with technical standard. The land degradation monitoring technical standard system should includes 5 types of technical standards: direction standard, universal basic standard, information technical standard, technical method standard and data management standard.

It is suggested a National Land Degradation Monitoring & Assessment Standardization Committee under the leadership of National Standardization Committee should be established, to coordinate, compile, examine and approve the national land degradation monitoring & assessment standards.

3. Status & the main problems in land degradation data sharing in China

In terms of information sharing worldwide at present, data has been shared in the following three models: 1) Data such as meteorological data, remote sensing data has been collected by national fund and shared freely by the whole society,; 2) Data with mass user has been collected by payoff institutions or other funders and shared by payment. 3) Scientific information is shared by contracts between data collectors and users from payment to free of charge gradually.

Trends for scientific data management and service are: 1) developed countries invest huge fund in data management infrastructure and data collecting in key scientific fields to promote data effective flow and lower-cost uses. 2) Law and regulation, policy and management system as well technical guarantees to ensure the scientific data management, sharing services orders. 3) Commonweal and basic scientific data management and sharing are regulated by government in a commercially operational model. Data release can be operated in three types: public visiting, cost reclaim and private-public cooperation. 4) Data services have become main indicators to evaluate national information technical standards and information development level. Policy supporting and huge fund investment from government have put into data management and development of the service system. As the data process industry and market experience become mature, more and more companies will participate into data services.

In land degradation, the specialized data sharing standards and regulations have not been established in China. Some government agencies including forestry, agriculture, water resources, environment protection and land resources departments which are related to land degradation monitoring have established a series of departmental standards and regulations within their own departments for data sharing and exchange. Some of these standards and regulations have become national standards or departmental standards.

Hardworking in recent years has resulted in great achievements in China in information resources and net work development, and great progress has also been made in land degradation monitoring data sharing. At present the problems existing in land degradation monitoring data coordination and sharing include: 1) lack of legal guarantee. International experience indicates that, legislation is the basic guarantee for information sharing. However, the situation in China is that there is not adequate legal guarantee in information sharing. Therefore, there are still many difficulties in land degradation monitoring data sharing. Only by legislation to regulate the responsibility, right, and obligation to the stakeholders related to land degradation monitoring data sharing, can the benefits be guaranteed. 2) Lack of necessary standards and regulations in data sharing. Data sharing in China has just started recent years and many technical standards and regulations are lagged behind. In data sharing of land degradation, there still are lacks of the consistent and necessary standards and regulations, such as assort code system, indicator system, database format regulation, data change standards, metadata standards disaccord. Many databases are only shared within the department and the network is not running. 3) Morbidity in data sharing quality and data updating, intellective property right are unclear. Parts of shared data have no metadata information or unsigned, some of data is not integrated or not calibrated. Most of data has not been updated in time and sustainability.

4. Proposal for design of a network for coordinating and sharing land degradation data

Establishing a national data sharing mechanism for land degradation monitoring and assessment to form a perfect information sharing system, which combine interior and outside information, online and offline information, national and regional data, thematic and integrated data. This sharing system could be divided into six layers: Management Layer, Technical Layer, User Layer, Service Layer, Data Layer, and Network Layer.

It is proposed that the following steps needs to follow to establish a national data sharing mechanism for land degradation monitoring and assessment: 1) Working-out policies, management regulations and agreements for data sharing; 2) Establishing coordinating organizations or groups; 3) Restructuring or reforming the existing organizations; 4) Establishing a standard building and setting down standards; 5) Building up unified data platforms of the fundamental geographic information; 6) Improving the existing thematic databases for standardization; 7) Development of a network for data sharing; 8) Establishing a network system at national and provincial levels:; 9) Extending the data application fields and users; 10) Setting up a mechanism in which the system can run safely and steadily.

Conceptual Structure of Network-Schema 1: Establish a new institution for disseminating, processing and sharing centralized data, in which data sharing without any obstacles can be realized.

Design: Establish a National Data Sharing Network Center for Land Degradation Monitoring and Assessment in one institution and equip it with powerful application server. At same time, establish distributed databases at the network center and all data points. All distributed databases should be connected to the application server and each data point only in charge of database establishment and maintenance, uploading data to network center or sending data by other means. The network center is in charge of the whole network operations and maintenance. Users visit the network center by browser to sharing network resources and services. For users, they can get data from one window (namely the network center), even the data belongs to different departments.

Characteristics of Schema 1: It is centralized data disseminating, management and providing whole information, i.e. sharing information, data reality, metadata, user information and network analysis and applications.

Conceptual Structure of Network-Schema 2: Adjust or reform some departments to establish a perfect general data sharing mechanism for centralized data disseminating and sharing, centralized and distributed processing to realize data sharing without obstacles.

Design: Use one institution's information center as the base to establish and maintenance a Main Center running the National Data Sharing Network for Land Degradation Monitoring and Assessment and equip the base with application server. At same time, establish distributed databases at the network center and data sub-centers and connect each other through the application server. For data sub-centers, they are not only in charge of database establishment and maintenance, but also in charge of establishment and maintenance of the application server connected with their local databases. Each data sub-center works and returns the results according the upper server requests. The main network center is in charge of whole network operations, maintenance, whole network control and management. Users visit the main network center by browser to sharing network resources and service.

Characteristics of Schema 2: Sharing information and data reality are managed distributively. Metadata, user information, network analysis and application services are managed centralized.

Conceptual Structure of Network-Schema 3: Based on present institutions, establish a practical and feasible data sharing mechanism. Data disseminating, processing and sharing are combined both in centralized and distributed manners. It can realize base data sharing.

Design: Gateway of the main center and sub-centers of the National Data Sharing Network for Land Degradation Monitoring and Assessment is established and maintained based on present information centers or network centers. Buy application server respectively and set up the network system for distributed data sharing between the main center and sub-centers. The user management is centralized at the main center. Every sub-center is structured independently but consistent with each other, i.e. each sub-center establishes its own web server, application server, database management system, while consistent standards are used for metadata collection and addressing issues. The main center and sub-centers store the metadata and data reality of their respective information, and install the unified metadata management software and search engine. The difference between the main center and sub-centers is that the main center provides the basic information for centralized sharing, such as the unified data platform of fundamental geographic information, thematic fundamental information and multi-thematic integrated data. Sub-centers provide distributed, shared and detailed thematic information.

Characteristics of Schema 3: Users information, fundamental geographic information, thematic fundamental and integrated information, as well as metadata are centralized managed. Detailed thematic information, data reality and related metadata are distributed managed. Network analysis and application services are also provided.

Compared the above three conceptual structures of network schemas, schema 3 is the most realistic and feasible one. It has the characteristics of: 1) with less technical restrict for each data sub-centers. The software for metadata collection, management and issue is centralized developed. It is easier for downloading, installation and use. It is also easily integrated. 2) It is flexible to establish the network. For different sub-centers, it has flexible methods in solving problems based on their situation to avoid lack of fund. It is much easier to realize at the provincial and local levels. 3) It is easier to keep consistency of data resources throughout the whole network system. When any data sub-center and its data content changes, it is not necessary to make more maintenance for the main center. 4) It could make the most of existing information sharing task from related institutions and extend the land degradation monitoring and assessment data and services. These not only meet the GEF project needs, but also enrich national information in sustainable development information sharing and scientific data sharing.

Table 1 Government departments related to land degradation monitoring in China

Government departments	Land degradation types	Operational organization	Implementation agencies
State Forestry Administration	Desertification/ sandstorm	Desertification combating center (CCICCD	China National Desertification Monitoring Center
	Forest resource	Department forest resources management	4 forest resources monitoring centers of State Forestry Administration
	wetland	Department of wildlife protection	Wetland monitoring center
	biodiversity		Wildlife monitoring center
Ministry of Water Resources	Soil erosion	Department of soil & water conservation	Soil & water conservation monitoring center
	Water resources	Department water resources management	Hydrological bureau

Government departments	Land degradation types	Operational organization	Implementation agencies
Ministry of Agriculture	grassland	Pasturage & veterinarian Department	Grassland supervisory center
	Soil fertility	Department of agriculture	National agricultural technical extension & service center
	biodiversity	Department crop management	Chinese academy of agriculture
State Environment Protection Administration	biodiversity	Department of nature	Nanjing Environment research institute
	sandstorm	Department of pollution control	China national environment mentoring station
	Soil pollution		
China Meteorology Administration	sandstorm	Department of forecasting and hazard reduction	China central Meteorology station, Climate center, satellite Meteorology center
Ministry of Land and Resources	Land use	Land management department	China academy of reconnaissance & land planning

Table 2 Website list for land degradation data sharing in China

Type of land degradation	Institution list for LD data sharing	Website for LD data sharing
Desertification	SFA homepage China desertification information Website North China desert database	http://www.forestry.gov.cn/ http://www.desertification.gov.cn/ http://sdb.casnw.net/bfsmh
Soil & water conservation (SWC)	MWR homepage China SWC monitoring Center Website China SWC eco-establishment Website	http://www.mwr.gov.cn http://www.cnscm.org http://www.swcc.org.cn/
Forest resources	SFA homepage China forestry information Website China SD forestry sub-center Website Science data sharing forestry data center	http://www.forestry.gov.cn/ http://www.lknet.forestry.ac.cn/ http://www.sdinfo.forestry.ac.cn/ http://www.lknet.ac.cn/ly/
Grassland	Ministry of Agriculture homepage China grassland information network Science data sharing grassland data center	http://www.agri.gov.cn/ http://www.grassland.net.cn/ http://www.grassland.org.cn
Soil	Ministry of Agriculture homepage National Agricultural tech center homepage CAS Nanjing Soil Institute homepage CAS science database-soil database	http://www.agri.gov.cn/ http://www.natesc.gov.cn http://www.issas.ac.cn http://www.csdb.cn/
Biodiversity	Ministry of Environmental Protection homepage Ministry of Agriculture homepage SFA homepage China natural reserve homepage Network of biodiversity & natural protection China biodiversity information exchange website	http://www.zhb.gov.cn/ http://www.agri.gov.cn/ http://www.forestry.gov.cn/ http://www.wildlife-plant.gov.cn/ http://www.biodiv.org.cn/ http://www.biodiv.gov.cn/
Sandstorm	CMA homepage China duststorm homepage Science data sharing weather data center China weather satellite information network	http://www.cma.gov.cn/cma_new/ http://www.duststorm.com.cn http://www.cdc.cma.gov.cn/ http://www.dear.cma.gov.cn/is_nsmc/
Water resources	Homepage of ministry of water resources	http://www.mwr.gov.cn/
Land use conversion	Homepage of ministry of land & resources	http://www.mlr.gov.cn/
Wetland	China wetland homepage	http://www.chnsd.com/

Table 3 Comparison of schemas for data coordinating and sharing network mechanism

Index	Schema 1	Schema 2	Schema 3
Technical advancement	good	good	good
Structure rationality	commonly	commonly	good
Capability according to policies, mechanisms and management regulations	good	good	good
Extensibility	not so good	not so good	good
Feasibility	not so good	not so good	good
Practicability	commonly	commonly	good
Costs of the network center	costliness	costliness	frugal
Management difficulty (outlay for construction, running and maintenance)	difficult	difficult	not so difficult
Time consuming for development	longer	longer	shorter
Staff for maintenance	more / centralized	more / centralized	less / decentralized
Applied efficiency of users	high / centralized	high / centralized	high / decentralized

31. A Legal and Policy Framework for IEM of Soil and Water—the New Zealand Model

David P Grinlinton, Kenneth A Palmer
Associate Professor of Law at the Faculty of Law, University of Auckland, New Zealand

Abstract: Over the last three decades 'sustainability' and the 'precautionary principle' have become internationally accepted guiding principles of human interaction with the natural environment. While these principles predominantly find expression in international treaties and agreements, it is far more difficult to incorporate them into domestic regulation in a meaningful, and enforceable, way.

Since the mid-1980's New Zealand has pursued a very active program of environmental and resource management reform. It has involved change at every level, from central and local government administrative restructuring, through legislative reform, to operational management at the local authority and municipal level. In the context of integrated environmental management ("IEM"), New Zealand has progressively introduced a number of policies and legal measures which attempt to apply an integrated approach to land and water management.

In 1991 New Zealand incorporated the principle of "sustainable management" as the statutory purpose of the Resource Management Act 1991 ("RMA"). The RMA repealed all pre-existing planning, water and soil, clean air and noise control legislation and was intended to provide an integrated approach to the use and management of land, air and water. Within the sustainable management purpose, the Act addresses the issues of inter-generational equity, environmental protection, and ecological integrity. It also provides a comprehensive policy-making and planning regime, and an integrated consenting and enforcement regime incorporating a specialist Environment Court. All policy-making, planning and decision-making functions are required to be undertaken in a manner that promotes the central purpose of sustainability. The precautionary approach is arguably implicit in this regime.

In respect of water and soil conservation, these principles are taken into account in the preparation of formal policy documents and planning instruments governing land and water use, and to decision-making on specific water and soil use applications. After 16 years of practice under this regulatory regime, a number of system failures have become apparent. On the other hand the regime has made consideration of sustainability issues a fundamental part of any soil and water use activities.

This paper will outline the IEM framework under the RMA in New Zealand. It is hoped this regime may provide a model or blueprint for the use of law and policy to achieve integrated and sustainable soil management in other jurisdictions.

Introduction

A brief summary of the main issues in New Zealand is useful to put the legal and

policy responses in context. The following particular problem areas have been identified (Ministry for the Environment, 1996; 2007: chapter 9 "Land",):
- Erosion, including surface, mass movement, fluvial and stream bank erosion;
- Loss of carbon and organic matter;
- Compaction and loss of soil structure;
- Nutrient depletion;
- Soil acidification; and
- Chemical contamination from industry and agriculture.

Geologically New Zealand is a relatively young country born of active tectonic movement in the subduction zone where the Pacific plate is forced beneath the Indo-Australian plate. Along with faulting and folding, glacial scouring and volcanic activity, this process has created a physically diverse country of high mountains, swiftly flowing rivers, heavily indented coastal topography, and highly mixed geology. Over two-thirds of New Zealand has slopes of greater than 12 degrees, and nearly one half of the country greater than 28 degrees. Three fifths of the country has a vertical elevation over 300 metres, and one fifth over 900 metres (Statistics New Zealand, 2008).

Prior to human settlement, 78% of New Zealand was under forest cover. Since Polynesian, and later European settlement, only 24.5 % of the natural forest cover remains, with a further 7.3% covered by planted production forest (Ministry for the Environment, 2007: Table 9.4, p 231). European settlement from the early 19th century resulted in the rapid conversion of forested land into cleared pasture and open farmland, which now comprises 51% of New Zealand's land area. Almost 75% of New Zealand is covered by easily eroded sedimentary rock and soils and has a relatively high annual rainfall of between 600 and 1600 mm, rising to 10-1200mm in the Southern Alps.

The most recent 'state of the environment' report for New Zealand sets out the current state and trends relating to land use and soils (Ministry for the Environment, 2007).

Soil erosion has long been a major problem in New Zealand (Memon and Perkins, 2000). Around 10% of New Zealand's land area is classed as "severely erodible", and erosion costs the country between $NZ100 million and $NZ150 million each year through loss of soil and nutrients (Ministry for the Environment, 2007: 4-5). The problem originates from the historical practice of clearing land, even steep erosion-prone land, for pastoral use. The problem is compounded in some mixed use and forestry areas by a reduced level of new forestry plantings, and felling of commercial forests without re-planting.

Loss of carbon and organic matter, and nutrient depletion, are further consequences of land use practices in New Zealand. In particular, intensive cropping, particularly where monoculture practices have prevailed, have led to depletion. Pastoral uses, while depleting certain nutrients and organic matter have contributed to very high levels of artificially added nutrient levels and altered acidity levels due to the use of lime and phosphate fertilizers to increase productivity. In some dairy farming soils, these levels are reaching saturation point leading to leaching of excess nitrogen into groundwater and rivers (Ministry for the Environment, 2007: 228, 237-239).

Compaction and loss of soil structure are further problems emanating from intensive agricultural and pastoral farming practices. Compaction is caused by farm animals, vehicle traffic and cultivation methods, and results in lowered macroporosity. Where macroporosity falls below a 10% threshold it will adversely affect growth and production. Half of all dairying sites in New Zealand, about 3.5% of New Zealand's land area, have fallen below this threshold (Ministry for the Environment, 2007: 231, 239).

Finally, chemical contamination from industry and agriculture is a persistent problem in New Zealand. A 1992 estimate put the number of contaminated sites at between 7,000 and 8,000, with 1,500 deemed to be at high risk to human health or to the environment. Currently there are no national standards establishing maximum levels of contaminants in soil, and authorities rely very much on self-reporting by either industry or by local authorities. While many of the contaminated sites are in urban and industrial areas, agricultural and rural areas are also affected. These include timber treatment sites where, historically, high levels of arsenic were used, and also an estimated 50,000 contaminated sheep dip sites where sheep have been dosed with chemical and biological agents for animal health and productivity reasons. Other areas include petroleum industry sites, gasworks and bio-solids (sewage and other industrial and domestic waste) (Ministry for the Environment, 2007: 248-251).

Early regulatory responses

The importance of preservation and proper management of forest cover was recognised very early in New Zealand with the passage of The New Zealand Forests Act 1874, the preamble to which stated:

Whereas it is expedient to make provision for preserving the soil and climate by tree planting, for providing timber for future industrial purposes, for subjecting some portion of the native forests to skilled management and proper control, and for these purposes to constitute State Forests. [Emphasis added]

In 1941 a "new era" of integrated soil and water conservation began with the passage of the Soil Conservation and Rivers Control Act (see Baumgart and Howitt, 1979). This measure was designed to promote soil conservation, prevent and mitigate soil erosion, and prevent damage by flooding. The system was aligned to natural catchment areas to better reflect the natural processes relating to water and soil. The Act established a National Soil Conservation and Rivers Control Authority, and Catchment Boards could be established to carry out the administrative and operational functions in water catchment districts.

The measure had mixed success due, in part, to parochial interests, the political strength of private landowners, and the multitude of different authorities involved in administering the Act.

The Water and Soil Conservation Act 1967 took the integrated approach a step further, by removing common law rights to water and vesting control of water "in its natural state" in the Crown. The Act essentially nationalized the use of water by requiring a person wishing to take, divert, dam, use water, or discharge wastes into water, to obtain a water permit from Regional Water Boards established under it. Existing water use was allowed to continue. Exemptions were granted in respect of taking water for domestic and stock use. No exemptions were granted in respect of the discharge of wastes into fresh water or ground where the waste could enter a natural watercourse: Water and Soil Conservation Act 1967, s 21 (see generally Palmer, 1983: 856-893).

An important feature of the water management system was that it reflected the desirability of regulation being based on a catchment area or region and this administrative imperative has been continued under the replacement legislation.

Integrated Environmental Management in New Zealand

In recent years New Zealand has attempted to create a policy and regulatory structure that reflects the complexity of environmental interactions in the broader context of land, air and water use. Such "integrated environmental management" must be applied not just to isolated statutory measures, but across the full spectrum of administration, regulation and implementation, including:
- administrative structures,
- policy-making and planning,
- legislation and regulation,
- processes of participation and decision-making, and
- operational implementation including environmental monitoring, impact assessment and enforcement of actions and responsibilities.

Administrative governance reforms 1986-89

The management of water and soil has long been regarded as a matter of national importance to central government, and the operational management of such resources the province of regional and local government. The following administrative reforms took place in the late 1980s – early 1990s:

Central government restructuring

The Environment Act 1986 was part of central government reforms to clarify the policy making functions of central government departments as against the work functions which had formerly been combined. Until that time, the dominant Ministry of Works and Development had responsibilities for planning and delivery of public works throughout the country.

The Act established the Ministry for the Environment, and the separate Parliamentary Commissioner for the Environment (PCE). The PCE was to be an independent "system guardian" for the environment: Environment Act 1986, s 4. For the first time, the term "environment" was given an expansive meaning, encompassing ecosystems and their constituent parts, including all natural and physical resources, and the physical, social, economic, cultural and aesthetic aspects of an area: Environment Act 1986, s 2. Further, the Act recognised in the management of natural and physical resources that a full and balanced account should be taken of the intrinsic values of ecosystems, all values placed by people on the quality of the environment, the rights of the Maori (the indigenous people of New Zealand), the sustainability of natural and physical resources, and the needs of future generations: Environment Act 1986, Preamble. This purpose represented a statement of sustainable management, which was reflected in later legislation.

In the following year, the Conservation Act 1987 was passed establishing a new Department of Conservation, to have responsibility for administration of national parks and public (Crown) conservation lands. The Department has particular functions in advocating conservation and sustainable management of those lands which cover approximately 30% of New Zealand's land area (see Department of Conservation, 2008; Nolan (ed), 2005: paras 2.25-2.29). In addition, other Government ministries were established to take over functions relating to promotion of agriculture and forestry, fisheries, transport, health, civil defence and emergency management. A Department of Building and Housing was later established to focus specifically on promotion of sustainable housing (Nolan, 2005: paras 2.30-2.37).

Local government restructuring

The reform of central Government policy structures, to provide transparency in strategy and policy making, was complemented by reform at the local authority level. Between 1988 and 1989, the Local Government Commission reviewed all existing local authorities, resulting in a major reform and reduction in the number of authorities. The statutory guidelines required recognition of the existence of different communities, but also identified the need for local authorities to be able to efficiently and effectively exercise functions, duties and powers, and to provide for effective accountability for the delivery of services.

The end result was to substantially reduce the number of existing public bodies and divide the country into 12 regions, and 74 districts. The regions were to be presided over by elected regional councils. The districts would be controlled either by elected city councils or district councils. In drawing up the boundaries for the regions, the Local Government Commission deliberately defined boundaries which followed catchment areas, with the intent that comprehensive integrated management of water and soil conservation could be achieved. The functions in respect of river control and land drainage were allocated to regional and district councils. Regional Councils were given the responsibility for regional planning, to provide broad policy directions for land use planning, which should be implemented at the district level (see Palmer, 1993: 7-10).

Environmental management law and policy reforms 1988-91

In the late 1980s and early 1990s the government, through the newly created Ministry for the Environment, proceeded to develop and implement a range of new policies and legislation. Underlying these environmental reforms was the desire to incorporate the normative principle of "sustainability" under a single integrated system of resource management (Grinlinton, 2002: 19-46; Williams, 1997: chapters 2 & 3). The concept was consistent with the Brundtland Report produced by the World Commission on Environment and Development (1987), which gave general recognition to the objective of sustainable development as an essential condition for future survival of the earth planet, to the extent of recognising intra-generational equity by redistribution of wealth, and inter-general generational equity through maintaining the viability of the ecosystem for the benefit of future generations.

The Resource Management Act 1991

The Resource Management Act 1991 ("RMA") was central to the reforms. It attempted to integrate into one statute the law relating to the management of land, air and water and replaced over 50 other Acts dealing with these matters. The overriding thrust of the legislation is to provide for integrated resource management, to ensure that decisions made in respect of particular environmental issues are not made without regard to consequences in respect of other issues. The RMA requires an holistic approach to planning and administration. It recognises the balance required between environmental objectives, social and cultural objectives, and economic objectives. The statute does not impose a deep ecology philosophy on administrators but provides a relatively shallow pragmatic ecological standard as a baseline for activities (see Grinlinton, 2002: 19-46).

The purpose and principles of the RMA

The RMA has as its central purpose " the sustainable management of natural and physical resources" (s 5(1)).

The pivotal definition of sustainable management is then stated in section 5(2).

(2) In this Act, sustainable management means managing the use, development, and protection of natural and physical resources in a way, or at a rate, which enables people and communities to provide for their social, economic, and cultural wellbeing and for their health and safety while—

(a) Sustaining the potential of natural and physical resources (excluding minerals) to meet the reasonably foreseeable needs of future generations; and

(b) Safeguarding the life-supporting capacity of air, water, soil, and ecosystems; and

(c) Avoiding, remedying, or mitigating any adverse effects of activities on the environment.

All functions and decision-making carried out under the Act must be guided by this purpose, and must actively promote it. In this sense the Act itself provides a powerful statement of government policy.

The "sustainable management" purpose is possibly unique in domestic legislation. However, the definition has given rise to some difficulties in interpretation. The balance between the "management purpose" of providing for the wellbeing of communities appears to be qualified by so-called ecological "bottom lines" in s 5(2)(a)-(c). However, the courts have taken the view that the words should be given a wide meaning of purpose and principles, rather than strictly subjugating the "management purpose" to the ecological "bottom lines". The prevailing view was stated in North Shore City Council v Auckland Regional Council (in 1997) as follows:

The method of applying s 5 then involves an overall broad judgment of whether a proposal would promote the sustainable management of natural and physical resources. That recognises the Act has a single purpose …. Such a judgment allows for comparison of conflicting considerations at the scale or degree of them, and their relative significance or proportion in the final outcome.

This pragmatic view of the purpose of sustainable management, which recognises the necessity in many instances of evaluation and decision making to reach an "overall broad judgment" has been broadly endorsed. The concept, purpose, or ethic of sustainable management has been seen as the prime objective, rather than a direction to take a narrow legalistic approach to the particular words. The purpose of sustainable management can be seen as a constitutional statement recognising the intrinsic value of the environment, and the need to safeguard for future generations the state of the environment.

The RMA also includes certain other supplementary purposes as "matters of national importance", under section 6 of the RMA, and other matters to be considered by decision-makers in s 7. Many of these have direct relevance to water and soil conservation.

The policy and planning structure under the RMA

The RMA creates a vertically and horizontally integrated structure for environmental management. It provides for central government policies, regional government polices and planning instruments, and territorial (city/municipal) level planning instruments. Each level of government has differing, but sometimes overlapping resource management responsibilities. Vertical integration is achieved by the requirement that lower level plans and policies must "give effect to" higher level policies and plans: RMA, ss 67(2), (3), 75(3). Lateral integration by the requirement to consult with neighbouring councils, central government agencies, some NGOs and other interest groups when preparing such instruments.

National level policies and standards

Under the Act central government may promulgate "National Policy Statements" (NPSs) and "National Environmental Standards" (NESs) pertaining to various aspects of environmental protection and natural resource management. There are as yet no NPSs or NESs specifically on water or soil conservation produced under the RMA. This is something of a failing of the RMA system. There are, however, a number of other statements and documents that provide guidance to regional councils and others.

A "Water Conservation Order" (WCO) may also be made under the Resource Management Act, by the Governor-General by Order in Council, on recommendation of the Minister for the Environment: RMA, s 214. The purpose of a WCO is to provide for the preservation as far as possible in its natural state of any water body that is considered to be outstanding, and to protect in particular a habitat for aquatic organisms; a fishery; the wild, scenic or other natural characteristics; scientific and ecological values; and protect recreational, historical, spiritual, or cultural purposes; protect other characteristics considered to be of outstanding significance to Maori (indigenous people): RMA, s 199. At all times, the Water Conservation Order process is subject to the "sustainable management" purpose of the RMA.

Very few WCOs have been made, with opposition mainly coming come from public bodies responsible for electric power generation, which prefer the waters remain available for dams, or other utilization (see Nolan (ed), 2005: paras 8.62-8.85).

Regional and territorial (municipal) level policy and planning

Strategic planning and operational management of land air and water resources is largely devolved to regional councils and "territorial" authorities (city & district/municipal councils) (see Palmer, 1993: 564-568; Grinlinton, 2002: 19-20).

Part III of the RMA contains a number of enforceable "duties and restrictions". Section 9 prevents the use of land in a way that breaches the Act itself, any Plan or any rule in a regional or district plan unless a resource consent is obtained. Even more onerous are the prohibitions in ss 12, 14 and 15, on coastal activities, water use, or discharges of contaminants into water or into the atmosphere unless permitted in a plan, or a resource consent is obtained.

Furthermore, s 17(1) of the RMA provides:

Every person has a duty to avoid, remedy, or mitigate any adverse effect on the environment arising from an activity carried on by or on behalf of that person, whether or not the activity is in accordance with a rule in a plan, [or] a resource consent,

This general duty can be enforced through "abatement notices" issued by a regional council or territorial authority, or "enforcement orders" issued by the Environment Court. Failure to comply with these orders constitutes an offence under the Act leading to the possibility of imprisonment or heavy fines.

Specific duties of Regional councils

Regional councils now have primary responsibility for managing water use and discharges into water, and uses of land that have regional significance. This includes soil conservation and erosion control measures. Under s 30(1)(a) of the RMA, Regional Councils are required to establish and implement measures to achieve "integrated management of the natural and physical resources" of their region. Section 30(1)(c) specifically mandates them to "control ⋯ the use of land for the purpose of … soil conservation".

Before preparing a proposed plan or any rules in a change or review of a plan, the Regional Council must consider alternatives, benefits and costs. Specifically it must carry out an evaluation which takes into account the "risk of acting or not acting if there is uncertain or insufficient information about the subject matter of the policies, rules or other methods": RMA, s 32(4)(b). This procedural step endorses a precautionary approach in dealing with land management issues.

Water allocation plans

Another function of regional councils is to include policies and rules in relation to the taking or use or other allocation of water in significant waterways: RMA, s 30. In practice, the powers to make policies and plans, envisaged in the RMA 1991, have not been significantly implemented. This state of affairs is partly due to the existence and continuation of irrigation schemes, and authorised activities preceding the 1991 Act continuing. For example, many of the major power stations on rivers for hydro power electricity generation have been established at earlier dates. The validity of these existing use rights has been protected under the RMA and recognised by the Courts.

It is acknowledged that in special circumstances, central Government may override these principles in the national interest, to provide a priority for one particular type of use above the needs of other competing users. This outcome would require special legislation, which may or may not reflect community acceptance.

Where a Regional Council fails to prepare a Water Allocation Plan, the Minister has a recently acquired power of direction to require this to be undertaken: RMA, s 25A.

Territorial (municipal) authorities

District and city councils have primary responsibility for land use and subdivision, air use and atmospheric discharges of localised significance.

Coastal management

Coastal management policy is primarily the responsibility of central government through the Department of Conservation although some management of the coastal area is delegated to regional councils. Coastal erosion is managed through this Department of Conservation/Regional Council management structure.

The "resource consent" permitting system

People wishing to undertake activities with environmental effects are required to apply for "resource consents". Often a number of different resource consents may be required for a particular activity. For example, a factory may require a range of land use permits, water use permits and discharge permits to operate.

The integrated nature of the system is illustrated by the resource consent application procedure. Applications for resource consents may be made on a publicly notified or non-notified basis in accordance with statutory notification criteria. The Council hears the application and, where notified, "any person" may make submissions. The decision must be made in accordance with the statutory purpose of "promoting sustainable management" and in accordance with the objectives and criteria in the Plan. The Plan, in turn, must give effect to higher level regional or Government Policy statements, and is also subject to the sustainable management purpose.

Further integration of decision-making is providing for by "joint hearing committees" made up of representatives of the various consent authorities, and which can conduct hearings and grant all resource consents required in one hearing and decision-making process.

When publicly notified, hearings for applications are open to objections and submissions by any person without the need to have locus standi ('standing').

In considering an application, the consent authority must have regard to the purposes and objectives of the RMA under sections 5, 6 and 7, as outlined earlier. They must also have regard to the actual and potential effects on the environment of allowing the activity, and be guided by the relevant policy statements and plans which have been put in place. In the evaluation, these individual matters are not determinative, and an overall broad assessment of all relevant issues will be considered: RMA, s 104. Where a consent is granted, conditions may be imposed to remedy and mitigate adverse effects: RMA, ss 108, 222.

In respect of applications for a discharge permit, the consent authority may in addition have regard to the nature of the discharge and the sensitivity of the receiving environment to adverse effects, and possible alternative methods of discharge into other receiving environments: RMA, s 105.

Further, concerning a discharge of a contaminant into water, or onto land which may result in the contaminant entering into water, the consent authority must consider, whether after reasonable mixing, the discharge will result in the production of conspicuous oil and grease films, a conspicuous change in color or visual clarity of the water, emission of objectionable odours, rendering the water unsuitable for farm animals, and any significant adverse effect on aquatic life. In these circumstances, the consent authority may not grant a discharge permit, unless it is satisfied there are exceptional circumstances or the discharge is of a temporary nature, or related to temporary maintenance work: RMA, s 107.

Decisions at the council level can be appealed to the specialist Environment Court by both the applicant and any objectors. This appeal can be on both law and merits issues. The Environment Court is also bound by the sustainable management purpose of the Act. Further appeals to the High Court and Court of Appeal can only be on matters of law, including judicial review: RMA, ss 299, 301.

Conservation land

The RMA has only limited effect in National Parks, Reserves and other public lands administered by the Department of Conservation. These lands comprise collectively around 30% of New Zealand's land area. However, the Conservation Act 1987 sets out a very similar system of land management policy-making and planning as under the RMA. People wishing to undertake activities on DoC land are required to comply with these policies and plans, and to apply for licences or "concessions". These applications are rigorously assessed for compliance with the conservation principles in the legislation, and any relevant polices and plans.

Other measures promoting water and soil conservation

In the global context, New Zealand is a signatory to a number of international statements of policy and agreements which include references to the prevention and management of soil degradation. These include the World Conservation Strategy (1980), the Brundtland Report (1987), the Rio Declaration on Environment and Development (1992) and Agenda 21 (1992).

The New Zealand government has also produced a number of reports (see, eg, Ministry for the Environment, 1997; 2007), strategies (eg, Ministry for the Environment, 1995; 1996), handbooks (eg, Ministry for the Environment, 2001) and databases (eg, Landcare Research, 2008).

There are also a number of quasi-governmental initiatives and non-governmental organizations involved in promoting soil conservation awareness and implementing practical measures.

Conclusion

There is no simple answer or universal model to address the challenge of sustainably managing and conserving the world's soil and water resources. International agreements, conventions and strategies such as the Rio Declaration and Agenda 21 provide States with some normative guidance. However, the difficulty is to implement these global themes at the national level in an effective way that recognizes the complexity of the problem, and in particular, the geophysical, ecological and sociological interactions and influences that contribute to it.

An "integrated" problem requires an "integrated" solution. Integration must occur at a number of levels. First, and probably foremost, any system for the management and conservation of soils must have strong normative guiding principles. "Sustainability" and the "precautionary approach" provide these. Secondly, these principles must be fully integrated into every level of administration, policy-making, regulation and implementation of the system. Thirdly, the system itself must be part of an integrated environmental management structure reflecting the interrelatedness of soil health with other aspects of the biosphere.

New Zealand has implemented such a system. The pre-existing regime required significant reforms, including the integration of administrative structures at central and local government levels, and integration of environmental and resource legislation. The RMA, while not without some flaws, provides an interesting and reasonably effective example of integrated environmental and natural resource management based on sustainable development principles.

References
Baumgart, I. L. & Howitt, P. A. 1979, 'Trends in Law Relating to Conservation and Preservation of Natural Resources', New Zealand Journal of Ecology, Vol 2, p 68.

Department of Conservation 2008, [Online] Available at: www.doc.govt.nz.

Grinlinton, D. P. 2002, 'Contemporary Environmental Law in New Zealand' in Environmental Law for a Sustainable Society, ed. K. Bosselmann & D. P. Grinlinton, NZCEL Monograph Series, Vol 1, NZCEL, Auckland.

Landcare Research 2008, New Zealand Land Resource Inventory (NZLRI), [Online] Available at: http://www.landcareresearch.co.nz/databases/nzlri.asp.

Memon, P. A. & Perkins, H. 2000 Environmental Planning and Management in New Zealand, Dunmore Press Ltd, Palmerston North, pp 148-149, and 152.

Ministry for the Environment 1995, Environment 2010 Strategy: A Statement on the Government's Strategy on the Environment, MfE, Wellington.

Ministry for the Environment 1996, Sustainable Land Management Strategy, [Online] Available at: http://www.mfe.govt.nz/issues/land/soil/strategy.html.

Ministry for the Environment 2007, Environment New Zealand 2007, [Online] Available at: http://www.mfe.govt.nz/publications/ser/enz07-dec07/index.html.

Ministry for the Environment 2001, Soil Conservation Technical Handbook, MfE, Welington), [Online] Available at:

   http://www.mfe.govt.nz/publications/land/soil-conservation-handbook-jun01/index.html

Nolan, D. (ed.) 2005, Environmental and Resource Management Law, LexisNexis, Wellington.

Palmer, K. A. 1983, Planning and Development Law in New Zealand, Vol 2.

Palmer, K. A. 1993, Local Government Law in New Zealand, Law Book Co, Wellington.

Statistics New Zealand 2008, New Zealand Official Yearbook on the Web, Para 16.3 Environmental and resource management [Online] Available at: http://www2.stats.govt.nz/domino/external/PASFull/pasfull.nsf/b45013b35df34b774c2567ed00092825/4c2567ef00247c6acc25697a0004407a?OpenDocument

Williams, D. A. R., et al, 1997, Environmental and Resource Management Law in New Zealand, 2nd ed,

World Commission on Environment and Development, 1987 Our Common Future (Gro Brundtland, Chairperson, referred to herein as "the Brundtland Report").

Notes
See Glenmark Homestead Ltd v North Canterbury Catchment Board [1978] 1 NZLR 407.

The PCE was to be an officer responsible to Parliament, and not subject to direction by a particular minister.

Local Government Amendment Act (No 3) 1988, schedule.

The Brundtland Report was subsequently adopted by the United Nations Convention known as the Rio Declaration 1992, and the Agenda 21 Statements of Implementation.

For example, in New Zealand Rail Ltd v Marlborough District Council [1994] NZRMA 70 at 86, Grieg J in the High Court upheld a consent to construct an export wharf in a natural part of a coastal area, as more important than conservation of the coastline.

In North Shore City Council v Auckland Regional Council [1997] NZRMA 59 at 94, Sheppard J presiding. Here, the Environment Court approved restricting the metropolitan urban limit around Auckland City in a location to protect an estuary from pollution.

Matters in s 6, RMA, include preservation of the coastal environment, wetlands lakes and rivers, and their margins, outstanding landscapes, and indigenous flora and fauna. Matters in s 7, RMA, include the efficient use and development of natural and physical resources, the intrinsic values of ecosystems, and maintenance and enhancement of environmental quality.

Regional government is the middle level of government comprising some 12 regional councils covering the entire country. Municipal ("Territorial") authorities (City and District councils) are the lowest level of government comprising some 72 municipalities.

"Environment" is defined in s 2, RMA, to include "ecosystems and their constituent parts", and "all natural and physical resources". The latter is further defined to include "land, water, air, soil, minerals, and energy, all forms of plants and animals … and all structures".

The RMA provides very strong enforcement mechanisms, including imprisonment for up to 2 years, fines up to $NZ200,000, strict liability and vicarious corporate liability. To date, there have only been 3 prison sentences imposed on environmental offenders: In Franklin District Council v McCollum (District Court, CRN 3057005960, 14 February 1994), a pig farmer was convicted, fined $5,000 and sentenced to 6 months imprisonment for allowing pollution, but the sentence was suspended; in R v Borrett [2004] NZRMA 248 the Court of Appeal upheld a sentence of 3 months for illegal earthworks and removal of protected native vegetation); and in R v Conway [2005] NZRMA 274 the Court of Appeal upheld a sentence of 3 months for pollution of waterways with oil and fuel. In all cases the offences were at the serious end of the spectrum and involved almost complete disregard for the law.

See Aoraki Water Trust v Meridian Energy Ltd [2005] NZRMA 251.

"Resource consents" include land use consents, subdivision consents, water permits, coastal permits and discharge permits under the RMA: ss 2, 87.

Traditionally under the common law, "standing" requires the litigant to have a property interest or some special interest greater than the general community. While "any person" can theoretically make submissions and objections to proposed plans, and also to resource consent applications that are publicly notified, in reality less than 5% of resource consent applications are notified, so "open participation" is an illusion!

[1] The Environment Court (previously the Planning Tribunal) is a specialist judicial body set up to arbitrate and adjudicate on environmental disputes (Part XI, RMA).

[2] Agreed at the United Nations Conference on Environment and Development, 3-14 June 1992, Rio de Janeiro, Brazil.

---

[1] Defined by the IPCC as "lands used for agricultural production, consisting of cropland, managed grassland and permanent crops including agro-forestry and bio-energy crops"

[2] The Pew report: http://www.pewclimate.org/docUploads/Agriculture%27s%20Role%20in%20GHG%20Mitigation.pdf

32. Application and Practice of Integrated Ecosystem Management (IEM) in Gansu GEF Pastoral Development Project

Hua Limin
Environment Officer, the World Bank Financed Gansu Pastoral Development Project Management Office

Abstract: Integrated Ecosystem Management (IEM) is a new approach and concept in sustainable natural resource management. It integrates ecology, economics, sociology, and management science into one management with a purpose to set up a transdepartmental and transjuristictional management framework. In 2003, the GEF Gansu Pastoral Development Project, which is being implemented by World Bank (WB), first introduced IEM approach and concept into sustainable pastoral development in Qilian Mountains and the Loess Plateau in Gansu Province. This paper discusses the environmental issues and their causes in the Project Areas. It also provides a list of project activities, the outputs of these activities tailored for the Project objectives after a participatory rapid appraisal (PRA) was undertaken, and a summary of the experience of the Project.

1. Background

Land degradation and biodiversity loss are major environmental problems in West China. They are also main causes for poverty of the local farmers /herders and the main constraints to the local economic development. The Government of China has paid a great attention to these problems. Obvious results have been achieved in land degradation control in West China after great efforts being given to rehabilitation and protection of these areas, in particular the implementation of China's West Development Program. Relevant sectors have undertaken studies on land degradation and biodiversity loss in both traditional and modern ways. However, those studies have been undertaken within single sectors with no input from other sectors, which results in biased and inconsistent policies lacking efficiency in implementation. Thus, land degradation remains severe. In 2003, following its operational plan 12 whose focus is IEM, GEF approved to fund the Gansu Xijiang Pastoral Development Project. The commencement of the Project was a valuable trial addressing the worsening land degradation in China.

Grassland is one of the main landscapes in West China. Guansu is located in a transitional region where the Loess, the Qinghai-Tibet Plateau and the West Qinling Mountains join, which indicates its complex topography and diverse climates. Gansu has a grassland area of 17.90 million ha and is one of the 6 largest pastoral areas in China. Its rare and valuable livestock and pasture species include Tian Zhu White Yak, Tan sheep, alfalfa which covers a large area, ormosia, Minshan Red Clover and Minshan Timothy. It is also abundant in wild life resources.

The main GEF Project areas in Gansu Provinces include the Qilian Mountains and part of the Loess. The Qilian Mountain grassland is an area of water input into the Shi Yang, Heihe, Shule River Systems and 56 other inland rivers, and is the only water source for the farming area in Hexi Corridor Oasis. Therefore, the ecological situation of the Qilian Mountain Grassland is directly linked to farming and livelihood of Hexi Corridor Oasis. Qilian Mountain Grassland contains a variety of landscapes including desert, temperate grassland and mountain meadows with very rich biodiversity. However, in the past 30 years, the glacier has shrunk and grassland has degraded which has become challenge to biodiversity conservation. Part of the Project areas is located in the loess with a large population. Due to the problem of land being over developed, rainfalls which mainly happen in summer have reduced. The lack of vegetation cover has caused severe soil erosion.

2. Main Environmental Issues and Causes in Project Areas

The Gansu GEF Project Areas are of vulnerable ecosystems with outstanding environmental problems including grassland degradation, soil erosion, biodiversity loss, and decreasing carbon sequestration capacity. These problems have become constraints to local farmers' and herders' income and economic development. They are also threats to local cultures and social stabability. Precondition for implementing the Gansu GEF Project is to identify the causes of grassland degradation and measures to be taken. Pre-project surveys and analysis indicate the following are the main problems in the Project areas.

2.1 Pasture-Livestock Conflict and Livelihood

It is well known that the direct cause for grassland degradation is the conflict between pasture and livestock. However, what causes imbalance between pasture and livestock? It varies all over the world. In the Gansu GEF Project areas, analysis shows the conflict is 'little grass for many animals'. 'Little grass' means inadequate pasture supply in the livestock production system. The first main cause for this inadequacy lies in the nature of the natural grassland in Gansu. The grassland in Gansu is of either alpine-cold grassland, or alpine-cold meadow or desert grassland, which indicates low productivity. In addition, harsh climate and limited land area available for pasture development add difficulties to livestock feed production. The main problem in livestock in pastoral areas on the grassland is the shortage of pasture in winter and spring that affects the benefit of the livestock system. The second is the pressure from the large population and influence from food security policies in the mixed farming-grazing areas and the farming areas, where livestock feed production is subjected to food production which has been placed first. Although greater efforts have been given to restructuring of plantation in the rural areas in China in recent years, forage grass and feed production hardly meet the demand of increasing development of livestock. Therefore, inadequate livestock feed supply is the main cause of the pasture-livestock conflict in both pastoral and farming-grazing areas. 'Many animals' refers to the problem in which the number of livestock exceeds the carrying capability of the grassland. The main cause for this is that local communities have few options for income generation and the extensive livestock production system. They have to increase the livestock quantity to maintain livelihood. An example is that in Gannan County, one of the Project areas, livestock is the only income for the local community. However, their expenses in education, health care and transportation are higher than in the communities in other areas due to their harsher natural environment. Therefore, addressing the imbalance of pasture and livestock must be taken as the first measure in grassland degradation control. In the GEF Project areas in Gansu, an increase in livestock feed supply can be achieved by expanding pasture area where it is suitable for developing planted pasture and introducing improved pasture species to increase productivity. At the same time, reduction of livestock number can be achieved by introducing improved livestock species and improving livestock management to improve livestock quality.

2.2 Policy Impact

Previous studies have shown that the most severe grassland degradation occurs in the farming-grazing areas where agricultural cultivation has great influence while grassland area is small and not valued by the local government. Although the farming-grazing areas are adjacent to pastoral grassland, grassland management has been very weak. No policies for grassland household contract and balance for pasture and livestock have been developed and implemented by the local government due to the pressure from the population. In Yongchang County and Liangzhou District, Project areas in Gansu Province, income from pastoral production takes up a large proportion in total household income, accounting for 65-75%. However, the distribution of grassland area is only 135 mu/household in Yongchang and 149 mu/household in Liangzhou. With such large population and small household grassland area, the County/District has not been able to implement a grassland household contract system as which has been adopted in other pastoral areas, neither has an effective grassland management been implemented, resulting in severe overuse and degradation of the grassland. The lack of overall grassland management policies has been one of the main causes for the grassland degradation.

2.3 Threat to Biodiversity

Located in the area where the Loess, the Qinghai-Tibet Plateau and the Inner Mongolia-Xinjiang Plateau join, Gansu Province is rich in biodiversity. Here live the Tianzhu White Yak which is a unique species well adapted to the alpine-cold climate and the *Cymnocarpos przwalsii* Maxim, a relic plant from the Mediterranean. The plateau wetland in the west part of Qilian Mountains and the low-lying meadow in Hexi Corridor are important habitats for migrant birds. However, the biodiversity in Gansu has been affected by human disturbance and inappropriate development of the natural resource, in particular the Qilian Mountain Grassland where overgrazing and mining have resulted in reduction of plant species, increase of poisonous weeds and impact on wild animals' habitat and breeding. In accordance with the reality of pastoral industry in the GEF project areas in Gansu Province, effective conservation efforts must be given to agricultural biodiversity.

2.4 Lack of Capacity Building for Farmers /Herders

Although it has been stated that the direct cause for grassland degradation is an imbalance between pasture availability and livestock carrying rate, the underlying cause is related to people, the users of the grassland. The lack of capacity building for farmers/herders has been the main factor that constrains sustainability of pastoral development. Constrained by transportation and training facilities, input for capacity building for farmers/herders in pastoral areas in Gansu and even in China has been far less than input for facilities for grassland conservation works for two reasons. Firstly, capacity building for farmers and herders has not been highly emphasized by decision makers in the government. Capacity building is not included as one of the main activities in government funded grassland conservation projects, e.g the Grazing Ban Project. Secondly, outflow of labor force and off-farm workers in the farming-grazing areas become great pressure for effective training. Statistics of the second national agricultural mass survey shows that a rural population of 2.478 million, which accounts for 20% of the total labour resource of Gansu, have migrated from the rural areas. 63.6% of these migrant workers completed junior or senior high schools and 81.6% of them are aged between 21 and 51. Migration of the young people from the rural areas means great challenge to capacity building for farmers and herders.

2.5 Small Household Production Model

Household-based business is the main model in China's pastoral industry, which is different from the commercial pastoral farm business in Canada and Australia. In the household-based business model, use of grassland is extensive and most households have several livestock species. Such a 'small but all-round' production model is not helpful in increasing economic gains. Traditionally, the household property is split into shares when the children grow up or when the sons set up their own families. In Sunan County, the number of households increased from 1508 in 1954 to 6818 in 2007, which

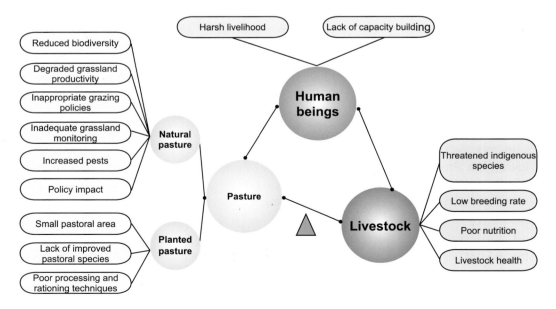

Figure 1: Main problems in pastoral development in Gansu GEF project areas

means that the grassland area per household has become smaller while the total grassland areas remain the same. With such small grassland area, it is difficult for individual households to implement measures to improve management such as zoning for rotational grazing and grassland improvement. Such a business model has resulted in low household economic income and impact on livelihood, leading to unsustainable development.

2. Project Policies and Approaches

Analysis of environmental problems in the Gansu GEF Pastoral Development Project areas shows that the factors causing an imbalance between pasture and livestock can be classified into two groups: one that can be controlled by human beings (degraded grassland, decreased livestock productivity and lack of management techniques) and the other that cannot be controlled by human beings. The Gansu GEF Pastoral Development Project focuses on those that can be controlled by humans. Increasing pastoral feed supply (including at technical and management levels) has been identified as a point for breakthrough. At the same time, other approaches including improving livestock breeds and management are adopted as technical supports. The final purpose is to increase farmers /herders' income and improve the capacity of the local managers and technical personnel, which will provide experience in grassland degradation control for other areas with similar natural conditions.

Participatory methods are used in implementation of the project. Attention is given to the status and the role of the community in conservation and sustainable development of the grassland. Based on participatory surveys and assessment, the design of the project activities takes the GEF Project objectives, the need of the community and comments of other stakeholders into account. During the implementation process, technical assistance was given to farmers and herders in undertaking various project activities (including training and application studies). The Project Coordinator works with Project households to undertake evaluation at completion of annual activities to develop an implementation plan for next year. Such an implementation approach is good complement to and provides applicable experience for the government-led environmental management projects.

3. Project Activities

In accordance with the causes of grassland degradation in the Project areas and following the Project objectives, using participatory surveys and evaluation as the base, the Gansu GEF Project identified the following as project activities:

3.1 Activities identified for increasing pastoral feed supply and improving grassland productivity

(1) Development of planted pasture: increasing grassland area and introducing improved pasture species where it is suitable for planted pasture to increase pastoral feed supply and quality.

(2) Pastoral rationing and processing: to address nutrition loss in the course of storing and feeding, the Project provides mowers, cutting machines and silage pits for farmers/

herders, changes the traditional pasture processing and usage model to improve pasture quality, with more attention given to addressing the lack of feeding stock in winter.

(3) Rotational grazing, grazing rest, and grazing ban on natural grassland: various grassland management measures including rotational grazing and grazing rest are taken in accordance with degradation levels of different grasslands to enable vegetation rehabilitation by enforcing intervention.

(4) Rat and insect pest control: in areas with severe rat and insect pest problems, studies on the causes and levels of rat and insect pest are conducted and ecological approaches including establishment of eagle racks and biochemical control of rat and pest insect are applied.

(5) Collection and sowing of wild pasture seeds: natural pasture seeds are collected from the natural grasslands where pasture develops well for re-propagating or sowing.

3.2 Activities for improving livestock productivity

(1) Introduction of improved species: introduce improved species including wild yak, Tibetan sheep and table-purposed breeding sheep to improve livestock productivity, with specific conditions of different project areas taken into account.

(2) Building warm sheds: cold winter is one of the main factors affecting sheep fattening. It increases feed demand while decreasing farmers' income. The Project supports farmers' to set up warm sheds which is a cost-effective way to address this problem.

(3) Development of grazing management policies: communities develop their own grazing management policies to regulate grazing practices of each member, following discussions and consultations with community members, e.g. setting times for grazing and grazing rest and roles of community members in grassland management etc.

(4) Construction of livestock water supply points: it is important that livestock has clean water. Protecting and improving livestock water is essential in improving livestock productivity.

3.3 Rural energy and environmental protection related activities in support of livestock production

(1) Provision of solar stoves for farmers/herders: Farmers/herders in West China collect shrubs or grass sods for fuel. The Project provides inexpensive solar stoves for farmers to help address this problem.

(2) Provision of solar panels for farmers/herders: due to the complex topography in the rangeland pastoral areas and nomadic grazing, herders have difficulties in access to electricity supply and lack information transferred through TV and radios. The Project provides solar panels as a mean to improve living standard which is also an approach to expand project impact.

(3) Construction of biogas pits: increases in livestock quantity and construction of warm sheds lay good foundation for construction of biogas pits. The Project provides funds (farmers provide labor force) to support farmers to build biogas pits in the warm sheds. It helps to improve farmers' living standard and the residue of the pits is good fertilizer which is good support for eco-agriculture.

3.4 Grassland management activities related to biodiversity conservation

(1) Protection of endangered plant species: following the catalogue of national endan-

gered plant species, *Gymnocarpos przewalskii* Maxim and Ephedra etc. in project areas were identified for protection and fences were built. Protection results are monitored and protection approaches are studied for sustainability.

(2) Water source conservation and wetland management: surveys, monitoring and protection of wetland resource have been undertaken in source area of the inland river Yulin River and in the Oasis wetland in Hexi Corridor. At the same time, protection and development plans have been developed to enhance wetland management.

3.5 Reform on policies for grassland property right in farming-grazing areas: grassland as part of collective property has been identified as one of the reasons for poor grassland management in two project counties. The Project introduced two models: a joint-household contract system and a price bidding contract system, which shall provide experience in grassland management in the farming-grazing areas in China.

3.6 Conservation of local livestock species

In Gansu, conservation of Tianzhu White yaks and Jingtai Tan sheep have been funded by GEF. The purpose for conservation of the local species is to prevent them and the existing indigenous livestock genes from extinction. The main measures adopted include maintenance and expansion of the number of the existing indigenous species, selective breeding, purification and rejuvenation to ensure the purity of the local species.

(1) Establishing core groups: To preserve the purity of the local species, the White Yak and Tan Sheep, it is essential that selective breeding is undertaken and core groups of breeders are established following national standards. The GEF Project has established household-based core groups in Tianzhu and Jingtai.

(2) Using biotechnology to preserve local species: preservation of the quality local genetic resource such as blood and sperms is critical in biodiversity conservation. The Project supports collection and refrigeration of sperms of quality breeds of White Yaks to conserve the genes and to enhance artificial insemination and increase the impact of the quality male yaks.

(3) Establishing information centers to preserve local species: GEF provides funds to the Project areas for purchasing computers to set up white yak conservation and breeding database to record accurate and comprehensive preservation data, which is to guide the production. A database for preservation of the white yak has been established.

(4) Livestock management activities

• Increase in pastoral feed production: promoting improved pasture species suitable for alpine and cold areas to increase pastoral feed production, and improving farmers' processing techniques, e.g. hay modulation and silage.

• Monitoring of natural grassland productivity: monitoring the productivity of grassland is very important in obtaining data for implementation of scientific grazing. Technicians working on Tianzhu White Yak Farm selected representative monitoring points and are undertaking monthly monitoring to provide technical assurance for balance pasture-livestock management.

(5) Development of a disease control and monitoring system: the Project provided

necessary equipments and medicine for emergency rescuing, trained local technical personnel and farmers/herders, which strengthened the local animal disease monitoring and control system.

3.7 Training, consultant services and applied studies

(1) Training: one of the key tasks of the GEF Project is to enhance capacity building. The Gansu GEF Pastoral Development Project conducted training need analysis and then organized training sessions for different target groups. Training courses were organized at international, national, provincial, county and farmer levels. Currently, farmer/herder training is the focus. Participatory demonstration by technical personnel for farmers/herders is adopted in farmer and herder training.

(2) Environmental education: Environmental education targeted at school children in the project areas was undertaken when providing technical training. Environmental education curriculums and practices were organized to increase children's environmental awareness and their care for their homesteads, and through them to influence and increase their family members' environmental awareness.

(3) Consultant services: The GEF Project values the role of consultant services (which takes up 40% of the total investment of the Project). In accordance with the technical issues raised from implementation of the project and the needs from farmers/herders, the Project designed and delivered 19 application studies and technical

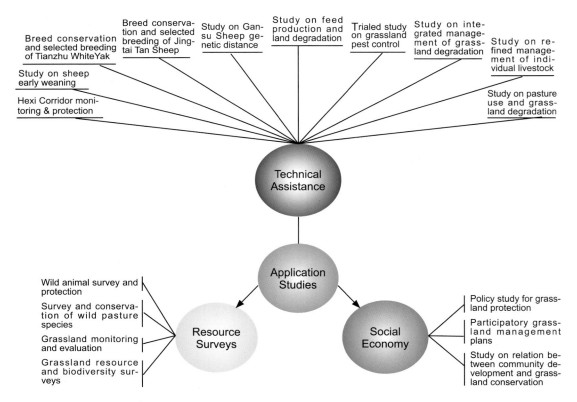

Figure 4: Application Studies of Gansu GEF Project

services involving (i) resource surveys including grassland and wild animal resource surveys, for the purpose of obtaining information on the natural resource of the project areas to lay foundation for other efforts; (ii) social economy, as it is understood that studies on grassland policies and community development are necessary as grassland degradation will not be addressed only by science and technology, and that social and humane science are needed in addressing some of these problems; and (iii) application studies needed to address the problems in practice and to direct and improve pastoral development management activities. To date, 10 papers on these application studies and technical services have been published. It is estimated 50 papers will be published by the end of the Project. The following figure shows the consultant services provided by the Gansu GEF Pastoral Development Project.

3.8 Project management, Monitoring and Assessment

(1) Project management: The Gansu GEF Project is a component of the Gansu Pastoral Development Project Funded by the World Bank (WB). Project management follows WB requirements set for pastoral development such as institutional, financial and procurement management. However, different attention is given to different activities. In the GEF component, more attention is given to capacity building including training and technical assistance. Investment in capacity building takes up 55% of the total fund for the GEF component.

(2) Monitoring and assessment: there is a need for monitoring and assessment to increase implementation quality. The Project Coordinator of the Gansu GEF Pastoral Development Project visits farmers to collect information on project progress, results and issues raised from implementation every year to support improvement of implementation and management. Only the project activities that meet farmers/herders' need and project objectives are approved when developing implementation plans at the beginning of the year, using results of the monitoring and assessment undertaken for the past year. Monitoring and assessment is undertaken both internally by the GEF Project and by a third party, a Canadian company Agriteam.

4. Project Outputs

(1) Evident results of the participatory village grassland management plans: A number of villages in each project county were selected as pilot sites. Community members and other relevant stakeholders worked together to develop grassland management plans catered for local conditions, which has resulted in evident effect. An example is that grazing ban has been implemented in some of the GEF Project villages. The Project supports the community to expand pastoral feed area, provides funds for cutting machines and building silage pits to increase pastoral feed supply, assists farmers/herders to improve breeding and productivity, and provides training for every project activity for farmers and herders. These are all helpful in reducing grazing pressure, conserving biodiversity and increasing farmers/herders' income. The Project is progressing towards achieving sustainable development.

(2) Improved infrastructure: Livelihood environment in Gansu Province has been harsh. The Project integrated the need of the communities, project objectives and

expert advice together to identify where it needs to be improved for integrated development. Assistance was given to the community to set up planted grassland, silage pits and warm sheds, supported by introduction of improved species to increase integrated productivity. Farmers' livelihood has been obviously improved.

(3) Strengthened capacity building: Although improvement of infrastructure is an urgent need by farmers/herders and the local government, experience shows it is more important that capacity building is provided to change thinking and improve capability of local managers, technical personnel, farmers and herders. In particular, capacity building provided for farmers and herders has changed from classroom lecturing to technical demonstration trials, i.e. improving technical personnel's skills in addressing problems through 'learning by doing' approach. Such an approach combines 'soft and hard wares' and is embraced by managers, technical personnel, and farmers/herders.

(4) Community-based reform for policies for grassland property right: The absence of clear policies for grassland property right has been part of the causes for grassland degradation in West China. The Project selected two farming-grazing areas, Yongchang County and Liangzhou District, as demonstration sites for grassland property right policy reform, in pursue of setting up trials and demonstration to improve management at the policy level. A joint-household contract system was trialed in Yongchang County, while in Liangzhou District price bidding auction was trialed. To support the reform, the Project assists farmers/herders to develop planted pasture and improve livestock species to increase pasture supply and livestock productivity. The result of the trials is that grazing pressure has been obviously reduced without compromising farmers' income. The reform of grassland property right in farming-grazing areas is a valuable trial to address grass degradation in the mixed farming and grazing areas in China.

(5) Development of appropriate training materials: At the beginning of the Project, published books were used as training materials. However, it was found that the books were not practical and too difficult to understand. The Project then employed a professional to identify training need which was followed to develop different training materials for different trainers. To date, 12 training manuals have been developed for technical personnel and managers. For farmers/herders, simple folders, posters and videos have been developed. The development of training materials is playing an important role in capacity building.

(6) Environmental education campaigns: Technical training for farmers/herders was emphasized as a key point at the beginning of the Project. However, a review of the results of the training and experience from other international assistance projects show that improving environmental education among school children and establish a long term mechanism in which children influence their parents to increase their awareness is good complement to farmers/herders technical training. Taking the specific natures of the grassland environment and schools into account, the PMOs at all levels, consultants and school teachers worked together. They started with environmental education review, using experience of other projects. To date, an instructive manual for environmental education for school teachers, environmental booklets and VCD have been developed. Some schools have integrated environmental education into official curriculums.

5. Discussion

The Gansu GEF Pastoral Development Project followed IEM approach at all aspects. However, constrained by the legal and administration systems it has a long way to go before meeting the IEM approach. It is difficult for the project management and implementing agencies to achieve 'trans-department' cooperation or management. Project management authority remains limited within one department. Inter-departmental cooperation was tried at the beginning of the Project. However, it was much too difficult and time consuming and project progress was affected. So more attention has been given to multi-disciplinary, integrative and comprehensive project activities, e.g. grassland management is integrated with livestock production, woodland or wetland management. Social sciences including policy studies and community development were integrated into technical studies for grassland management so that problems are solved in a systematic framework. Although it is the beginning of IEM, it provides experience useful for other projects being implemented or to be implemented in China.

6. Conclusion

(1) A set of integrated interventions aiming at increasing grassland productivity and reducing grazing pressure is needed in addressing grassland degradation. At the same time, sociological issues including farmers/herders' livelihood and community development must be considered. The Gansu GEF Project integrated individual activities into livestock production and social development activities, which enables integration and systematization to be emphasized.

(2) Participatory approaches are essential for project implementation. Adoption of participatory methods enables project decision making more democratic and project activities fit better with the local situation. The results of participatory monitoring provide better rationales for adjusting project activities.

(3) Capacity building guarantees long term impact of the Project. The Project provided a large number of training courses for project managers, technical personnel and farmers/herders. Training improves management and production skills. With long term impact, it is more important in changing people's thinking.

(4) Attention must be given to the household-based small business model. Household-based business remains the main model in China's pastoral industry, which is different from the commercialized farmers' business model. The Gansu GEF Project cares for development at the household level. More attention has been given to farmers' livelihood and applicability of project activities.

(5) Long term impact of policies. China's grassland degradation is a complex problem. It is not only a matter of technical restructuring but a matter of policy impact including land policies, economic policies for pastoral development and cross-sectoral cooperation. Therefore, constraint from policies must be removed before addressing grassland degradation. Only in this way, technical interventions will have sustainable impact.

7. Acknowledgement

Many thanks to Madam Sari Söderström for giving me this opportunity to attend this international workshop and for her support to my work. My efforts would not have been effective if there had not been support from her. Thanks must be given to the PMO of PRC-GEF Partnership on Land Degradation in Dryland Ecosystems for offering me the opportunity to deliver this presentation so that project experience has been shared. I must thank Mr. Victor Squires and David Michalk for their support and assistance in fulfilling my role in the Gansu GEF Project. My most sincere thanks also go to all technical personnel, managers and farmers/herders in the Gansu GEF project areas. Their hard work and humbleness has become a life memory.

References

[1] Cai Shouqiu: *On Integrated Ecosystem Management*, Proceedings of Gansu Political and Legal University, *2006(5): 19-20*

[2] Jiang Zehui, *Integrated Ecosystem Management*, China Forestry Publishing House, Beijing, 2006

[3] Gansu Grassland General Station: *Gansu Grassland Resources*, Gansu Science and Technology Publishing House, Lanzhou, 1999: 15-18

[4] Hua Limin, *Barriers to using feed balance systems for range livestock production – Case study at Dacha Village, Sunan County of Gansu Province*, Proceedings of the World Grassland/Steppe Conference, 2008

[5] Chengxu: *Advancing Front Problems in Modern Ecology in Study of Farming-Livestock Fringe*, Resource Science

[6] Main Data Announcement of the Second National Agricultural Survey in Gansu Province, http://www.gstj.gov.cn/doc/

33. An Introduction of Ningxia Integrated Ecosystem and Agricultural Development Project of Asian Development Bank

Ma Minxia
Deputy Director General of the Department of Finance, Ningxia, China

Abstract: This paper introduces the establishment background, general situation and some highlights of the Integrated Ecological and Agricultural Development Project in the eastern Helan Mountains in Ningxia Hui Autonomous Region. The project area is located to the east of Helan Mountain and to the west of the Yellow River, occupying 3,655 square kilometers and covering three districts and two counties in Yinchuan City. The area also includes nine state-owned farms of the farming system and a total population of 1.1 million. Its natural environmental conditions are complicated with grassland ecosystem as the major component. Environmental factors are comparatively poor, with a relatively poor natural ecosystem function. The environmental capacity is small with a vulnerable ecosystem structure. The strong interference from human's economic activities has resulted in issues like desertification, soil and water loss, and natural grassland degradation.

The ecological environment protection in the eastern Helan Mountains was listed in the Eleventh Five-year Plan of the provincial Party Committee and government of Ningxia Hui Autonomous Region, which is very significant to the protection and restoration of Yinchuan's ecological environment, water sources and biological diversity. After the program officers' repeated investigations, consultations and analyses under the GEF OP-12 Framework, the Integrated Ecological and Agricultural Development Project in the eastern Helan Mountains has been proposed in accordance with the laws of natural and ecological development and the requirements of the provincial party committee and government. The project incorporates four types of items, including IEM Capacity Building, Management of Water and Soil Resources, Improvement of Livelihoods in Rural Areas, and Protection of Ecological Systems, covering twenty-seven sub-items. The total investment is RMB 1.66 billion yuan with US$100 million loan coming from the Asian Development Bank, US$4.6 million donated by GEF, and RMB 850 million yuan as local matching funds. The overall goal of the project is to improve the environmental management, restore the ecological system and increase farmers' incomes in the project area.

The specific objectives are:

(1) To improve the ecosystem management capability with IEM demonstration examples through the reform of related policies, regulations and institutions;

(2) To increase farmers' incomes and solve their livelihood problems through project activities such as animal husbandry and cash crop farming;

(3) To reduce water and pesticide use in cultivated land and increase efficiency of water resources utilization through integrated water management;

(4) To decrease the loss of agricultural water use to further improve water resource deployment by the nine major lakes and marsh systems;
(5) To protect the biodiversity and fifteen threatened wildlife species.
The project building tasks mainly incorporate:
1. The ecological shelter-forest system construction in the west of Yinchuan City;
2. Wetlands rehabilitation of 104,800 mu (1 ha = 15 mu) through building fences and cultivating aquatic plants;
3. Integrated water resources management;
4. Alternative livelihood projects.

1. Establishment Background of the Integrated Ecological and Agricultural Development Project in the eastern Helan Mountains

(1) Origin of GEF OP-12 Project
• In 2002, the Chinese Government reached an agreement with GEF on IEM project.
• In August 2002, the Standing Committee of National People's Congress sent a group of experts to Ningxia for an investigation.
• In January 2003, China and GEF established a project for strengthening the institutional capacity building.
(2) Observations after the Project Investigation and Communication
• Existence of protectionism, "offside", and "absence" in some industries and departments;
• Defection in policies and laws with various loopholes;
• Duplication of efforts for not sharing information resources.
(3) Driving the Operation of IEM on the Basis of the Project
• Collaboration among different departments
• Linkages among different policies and laws;
• Design of multi-function demonstration projects;
• Effective combination of different sectors and industries in related areas;
• Establishment of industry chains among agriculture, forestry, water conservancy, animal husbandry, plant protection, and processing, etc.
(4) The ecological environment protection in the eastern Helan Mountains was listed in the Eleventh Five-year Plan of the Party Committee and government of Ningxia Hui Autonomous Region, which is very significant to the protection and restoration of Yinchuan City's ecological environment, water sources and biodiversity.
(5) After the program officers' repeated investigations, consultation and analyses under the GEF OP-12 Framework, the Integrated Ecological and Agricultural Development Project in the eastern Helan Mountains has been proposed in accordance with the laws of natural and ecological development and the requirements of the provincial party committee and government.

2. General Introduction to the Ecological and Agricultural Development of the Asian Development Bank Loan Project

(1) General Information of Ningxia Hui Autonomous Region
Ningxia is located in the Northwest of China with a total land area of 66,400 square

kilometers and a total population of 6.10 million. Ningxia Hui Autonomous Region is one of the only five autonomous regions in China. Its terrain is high in the south and low in the north, with loess hilly areas in the south, the Erdos Terrace in the middle, and the alluvial plain of the Yellow River in the north. The project area of Asian Development Bank is settled in the northern plain, which is an important ecological shelter in the west of Yinchuan Plain.

(2) Scope of Ecological and Agricultural Development of the Asian Development Bank Project

The project area is located to the east of Helan Mountains and to the west of the Yellow River, occupying 3,655 square kilometers and covering three districts and two counties in Yinchuan and nine state-owned farms of the farming system with a total population of 1.1 million. The project incorporates four types of items including IEM Capacity Building, Management of Water and Soil Resources, Improvement of Livelihoods in Rural Areas, and Protection of Ecological Systems, covering twenty-seven sub-items.

(3) Characteristics of the Ecological System in the Eastern Helan Mountains
- Its natural environment condition is complicated with grassland as the major component;
- Environmental conditions are comparatively poor, with a relatively poor natural ecosystem function;
- The environmental capacity is small with a vulnerable ecosystem structure;
- The strong interference from human's economic activities results in problems like land desertification, soil and water loss, and natural grassland degradation.

(4) Classification of Ecosystems in the Eastern Helan Mountains
- Mountainous ecosystem
- Diluvial plain ecosystem
- Yellow River oasis ecosystem
- Wetland ecosystem
- Sandy land ecosystem

(5) Main Ecological Problems in the Project Area
- Climate warming and rainfall decreasing brought about environmental deterioration;
- Estrepement. Blind reclamation led to the destruction of primary vegetation.
- Coyoting. Disorderly coyoting sand and stone for architecture use damaged the vegetation coverage.
- Irrational grazing. Overgrazing caused serious degradation of grassland.
- Deforestation. The area of bushes decreased by 38% due to uncontrolled felling.
- The groundwater has been decreasing by a rate of 1 m per year because of the over use of groundwater resources.

(6) Investment of the Project

The total investment is RMB 1.66 billion yuan with US$100 million loan coming from the Asian Development Bank, US$4.6 million donated by the GEF, and RMB 850 million yuan as local matching funds.

(7) Goals of the project

The overall goal of the project is to improve the environmental management, restore the ecological system and increase farmers' income in the project area. The specific

objectives are:
- To improve the ecological system management capability with IEM demonstration examples through the reform of related policies, regulations and institutions;
- To increase farmers' incomes and solve their livelihood problems through project activities such as animal husbandry and cash crop farming;
- To reduce water and pesticide use in cultivated land area and increase the efficiency of water resources utilization through integrated water management;
- To decrease the loss of agricultural water use to further improve water resource deployment by the 9 major lakes and marsh systems;
- To protect the biodiversity and 15 threatened wildlife species.

(8) Main Tasks of the Project
- Building the ecological shelter-forest system construction in the west of Yinchuan;
- Wetlands rehabilitation of 104,800 mu (1 ha = 15 mu) through building fences and cultivating aquatic plants;
- Integrated water resources management;
- Alternative Livelihood projects.

3. Highlights

(1) A consistent effort in exploration and accumulation is needed to raise people's consciousness and understanding of the protracted and arduous nature in promoting IEM.

(2) The management institutions should be at a higher level of integrated coordination capacity, and all executive sectors can share information with each other.

(3) The concept of human resources is of great importance. Equal emphasis should be placed on proper management staff and experts.

(4) Long-term support from international organizations such as GEF, ADB and WB is of significance to achieve our common goals.

34. Implementing GEF Objectives in a Systems Framework in Western China

Victor R. Squires
GEF international consultant, WB/GEF Gansu&Xinjiang Pastoral Development Project
e-mail dryland1812@internode.on.net

Synopsis The World Bank/GEF project operates in Gansu and Xinjiang. It deals mainly with grazed rangelands and has several objectives relating to improving livelihoods and achieving sustainable use of rangelands. It is the first of the large scale projects under the PRC/GEF partnership. Lessons drawn from this project are outlined.

Preamble

The Gansu and Xinjiang Pastoral Development Project (GXPDP) is the first large-scale demonstration project under the PRC/GEF Partnership, complementing the smaller ADB-financed Capacity Building to Combat Land Degradation Project. It is intended that these two projects, together with a further five, will provide an array of lessons and models on integrated land management for replication in the western region and more widely across the China. The GEF component of the IRBD GXPD project is being implemented in both Gansu, and Xinjiang.

As the counterpart project of the IBRD GXPDP, the GEF funded activities that are focus of this Workshop are closely linked to rural development. The GEF project gives an opportunity for mainstreaming biodiversity conservation and ecosystem management into efforts to improve the lives and livelihoods of herders and farmers and contribute to China's Great Western Development Strategy of environmental protection and economic development in its western regions.

GEF Project Objectives

The overall GXPDP development objective is to improve the lives and livelihoods of herders and farmers in the project areas, through establishment of improved grassland management and livestock production and marketing systems, while sustaining the pastoral resources. The project will build the capacity of provincial institutions down to the level of relevant township bureaus and farmers' associations through the provision of works, equipment, materials, Technical Assistance (TA) and training. Communities will contribute their labor. This constitutes the Baseline Scenario.

The global environmental objective of the project is to maintain and nurture natural grassland ecosystems to enhance global environmental benefits. The project aims to mitigate land degradation, conserve globally important biodiversity, and enhance carbon sequestration through promotion of integrated ecosystem management in the grassland, desert, and forest ecosystems of the Qilian Shan, Tian Shan, and Altaishan mountain ranges in Western China.

The Baseline Scenario consists of five components, to be implemented over a six-year period: (i) Grassland Management; (ii) Livestock Productivity Improvement; (iii) Market Systems Development; (iv) Applied Research, Training and Extension; and, (v) Project Management, Monitoring and Evaluation. Activities started in 2004. The GX-PDP design provides for significant overlap between its five components, reflecting the integrated nature of the project. Strong links between grassland management/forage development, livestock breeding and production, and market systems underpin development activities.

As an integrated pastoral development project, livestock production and grassland management is at the core, with the GEF component emphasizing the conservation of key mountain grassland eco-systems and their biodiversity and carbon storage capacity in selected sites of global environmental significance. This is to be achieved by: (i) designing and implementing global environment-friendly participatory grassland management plans, including management of investments with demonstration purposes; (ii) demonstration and dissemination of good practices on landscape-level grassland eco-systems management; and (iii) increasing public awareness regarding linkages of biodiversity conservation and sustainable grassland ecosystem management The global environmental objective is to maintain and nurture natural grassland ecosystems to enhance global environmental benefits, specifically: mitigate land degradation; conserve globally significant biodiversity; and, enhance carbon sequestration. The relevant GEF Operational Programs (OP) are OP 4 on Mountain Ecosystems, OP 12 on Integrated Ecosystem Management and OP 13 on Conservation and Sustainable Use of Biological Diversity Important to Agriculture

The GEF activities should also build environmental management capacity within the local Animal Husbandry Bureaus and Grassland Monitoring Stations, and other relevant local government entities, and would provide the means to integrate biodiversity conservation objectives into their activities.

The GEF component of this major World Bank funded project aims to maintain and nurture natural grassland ecosystems. Specifically to:

(1) Mitigate land degradation.
(2) Conserve globally important biodiversity.
(3) Enhance carbon sequestration.

The specific intervention sites for the GEF activities are located in areas that are defined as national priorities in the National Environmental Action Plan (NEAP) (1998) and the Biodiversity Strategy Action Plan and are regionally significant biodiversity corridors in Qilian Shan, Tian Shan and Altai Shan.

Implementing GEF activities in a systems framework

One stated objective of the GEF project is to assure sustainability of the rangelands. This means keeping options open for future generations. The approach must is based on scientific principles—understanding of *why* rather than *how*. Certainly we need to know how to grow better and healthier livestock, how to prevent and control soil erosion, how to get more rainfall into the soil to grow more grass. But to achieve sustainability of rangelands these 'how-to' tasks have to be done as subheadings in the larger framework of getting what people want from the land without violating the ecological limits of carrying capacity.

The ultimate global objectives of the GEF project in Gansu and Xinjiang are to conserve mountain grassland eco-systems and their biodiversity and carbon storage capacity in selected sites of global environmental significance.

There are three major elements in most counties (people, grassland and livestock). The project revolves around 3 intersecting sets of activities that are related to each of these elements. Each activity-set has the potential to reduce the pressure on the grasslands. In other words to manage the vast areas of grassland in a way that achieves the global objective of conserving biodiversity and capture carbon while at the same time contributing to livelihoods of millions of people.

The project's activities focus on the village level to maximize the participation of the community in the management of the grassland resource A lesson learned from the project implementation to date is that there is a certain minimum number of interventions (or actions taken) that need to be applied together. Single-factor approaches are unlikely to be very effective. The idea of "sets" of linked interventions or activities is central to the success of the project. This realization has had profound impacts on the way in which the seemingly unrelated activities and inputs such as Training, Applied Research, Livestock production and grassland improvement are now seen as part of an integrated pastoral grazing system. This realization and the subsequent action plans are bringing us closer to achieving Integrated Ecosystem Management (IEM).

Management of grasslands in a sustainable way is a process to balance the productive use of land, water and vegetation with the needs of the people whose livelihood depends on them. Plans to improve grassland management and achieve closer integration of animal husbandry must have support of the land users (the village or herder community) as they involve trade-offs between economic, social and environmental benefits derived from the grasslands. Additionally, regulatory actions required to share and manage these grassland resources cannot be efficiently implemented without the support of the affected communities. This support derived from understanding of the management process necessary to achieve sustainability and the agreement of the land users.

No progress can be made in developing sustainable grassland management and meeting the other project objectives without an understanding of how the grassland/livestock/people system works. The choice of GEF interventions from an extensive menu of possibilities depends on matching the proposed technical or policy intervention to the perceived need.

There are constraints that may reduce the impact of efforts to introduce the proposed community-based grassland management plan (CBMP). The proposed strategy framework that underlies the CBMP requires that land users (herders and farmers) be sensitized to sustainable environmental management and accept restrictions on the use of renewable natural resources within a specific area. This framework goes somewhat against historic and traditional understandings that these resources (grazing land and water) are a "free" gift of the global commons. Over coming this barrier in mind-set is particularly important on a project that relies on the cooperation of the whole village in the formulation of integrated grassland management plans and which is based on demonstration areas that represent a miniscule fraction of the entire grassland area.

GEF incrementally-financed activities to be conducted under the GEF component are shown in Box 1:

> **BOX 1 GEF-FINANCED ACTIVITIES**
> - Piloting the development and implementation of participatory resource management plans by village communities in the globally significant ecosystems of the Qilian Shan, Tian Shan, and Altai Shan of Western China.
> - Training herders and provincial/county bureau staff in the core competencies required for integrated ecosystem management (IEM) and achieving effective trade-off between maximising livestock numbers and environmental conservation.
> - Designing and implement a system to assess and monitor the project's global environmental impacts.
> - Preserving local livestock breeds.

There is a clear need to tackle the causes of the land degradation problem and not just deal with the consequences. Most effort in the past have been aimed in "solving" minor problems such as "how to get more forage from each hectare" rather than deal with the underlying causes of lower productivity such as insecure land tenure, unclear boundaries for the assigned grazing user rights, lack of clear policy on how to balance livestock numbers and feed supplies. The project has made progress though over the last two years in gaining a better understanding among both land users (herders and farmers) and the county and provincial level technical staff of the keys to sustainability and the realities of the market economy.

There is recognition now too of the fact that there are few management options available to the land users. Those that do exist fall into two categories:

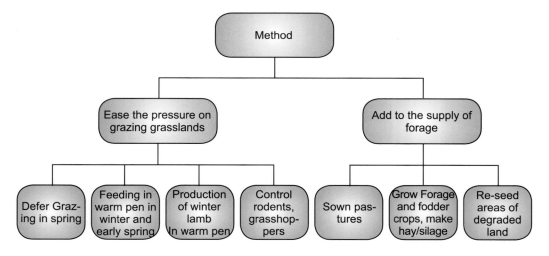

Figure 1: There are only two main options open to herders in western China

Reduce total grazing pressure (from livestock, from mammalian competitors such as rodents and wildlife; and from grasshoppers and other invertebrate pests); and by reducing herd/flock sizes through heavier culling and through adoption of precision management to cull unproductive animals. Breed improvement also falls into this category as a longer term strategy but it is not a panacea. Improved breeds will not perform well unless they get better feed!

Increase feed supply and/or utilization efficiency (by planting sown pastures, and fodder crops, by utilizing crop residues in a better way e.g. urea treatment, by conserving fodder as hay or silage). Better ration formulation for penned animals helps to make better use of the available feed and allows the tailoring of ration to the specific animal's need. Of course reducing the competition from rangeland pests like rodents and grasshoppers should be part of the strategy to reduce grazing pressure.

But the main message from the project is that "more means less". By this I mean that more livestock often means less income - not more. Research and demonstration has shown that it is possible to (i) increase profit at the current stocking rate by precision management that culls unproductive animals; (ii) maintains profit at lower stocking rates.

There is a clear need to tackle the causes of the land degradation problem and not just deal with the consequences. The World Bank/GEF project has no mandate to deal with policy and regulation but we note that this is key part of the ADB/GEF project that is the main focus of this Workshop. We look forward to reforms that arise from this work.

35. An IEM Approach to the Land Degradation Control and Conservation of Biodiversity in Dryland Ecosystems

Zheng Bo
Project Officer, International Fund for Agriculture Development

Abstract: This article introduces the development of agricultural projects of the International Fund for Agricultural Development in China, reviewing the history, the policy background of the project, the project establishment standards, the aim of integrated ecosystem management (IEM) and the procedure and process of the project implementation. In addition, introduction also includes project background, objective and activities in relation to each project area and the sub-projects.

The International Fund for Agricultural Development (IFAD) entered into China in 1981. It has offered loans to 21 rural development projects, which have amounted to $528 million. The projects have included areas such as soil improvement, irrigation, soil and water conservation, as well as grain production, cash crops, and animal and aquatic products. On-going projects include the Rural Development Project in Inner Mongolia, the Agricultural Development Project in Xinjiang, Poverty Alleviation Project in the southern Gansu region, Rural Finance Project, and Environment Conservation and Poverty Alleviation Project in Ningxia and Shanxi. Each project area is representative in view of geographical distribution and ecological environment and socio-economical conditions. Furthermore, the aim and the activity content of each project and its subprojects are systematically designed, and the experience of the project will be extended and applied to other IFAD projects as well as other similar domestic projects.

1. The International Fund for Agricultural Development (IFAD)
(1) Entered China in 1981.
(2) Offered loans to 21 Chinese rural development projects, which have amounted to $528 million.
(3) Project regions: remote areas, mountainous regions, ethnic minority regions, and comparatively poor regions.
(4) Target groups: populations of relative poverty, rural women.
(5) Project areas include soil improvement, irrigation, soil and water conservation, as well as the production of grain, economic crops, animal and aquatic products.
(6) Approximately 30 million people in 21 provinces in China have benefitted from the IFAD project.

2. On-going Projects
(1) The Rural Development Project in Inner Mongolia;
(2) The Agricultural Development Project in Xinjiang;

(3) Poverty Alleviation Project in southern Gansu region;
(4) Rural Finance Project;
(5) Environment Conservation and Poverty Alleviation Project in Ningxia and Shanxi.

3. The Process

Main Milestone and Dates	Related Key Outcomes
Project identification (March, 2003)	Compilation of the founding precepts for land degradation
Organizing group and seminar (May, 2006)	Preparation for project launch
Local project preparation (June to December, 2006)	Completion of 20 studies (draft)
Provincial IEM seminar (August to September, 2006)	Identification of the project areas
Suspension of GEF financial aid (October 2006)	Project suspension
GEF–PRC's agreement on re-orientation of the project (January 2007)	Tightening of the project budget and addition of biodiversity conservation
GEF Project Identification Framework (PIF) submission and approval	30 million dollars (GEF 4.5 million + 0.5 million) reduction in budget / incorporating conservation area into the project
Preparation of group meeting (September 2007)	Determination of the aims / objectives / influences
The seminar of logical framework (November 2007)	Confirmation of the budget / final seminar
The last preparation group meeting (February 2008)	

4. Policy Background

(1) GEF:
• Integrated ecosystem management (OP # 12);
• Sustainable land management (OP # 15);
• Conservation and sustainable use of the biodiversity significant to agriculture (OP # 13);
• Arid and semi-arid ecosystems (OP # 1)

(2) GEF-PRC plan:
• PRC–GEF partnership on land degradation (Country Program Framework [CPF] and project # 1: to create an adequate environment and to strengthen the capacity building of the stakeholders);
• China Biodiversity Partnership and Framework of Action (CBPFA)

(3) PRC country plan:
• Conversion of cropland to forest;
• Grazing ban;
• Afforestation

(4) Provincial plan: the Eleventh Five-year Plan of three provinces/regions

(5) IFAD project:
• Poverty relief in southern region of Gansu province (Gansu);
• Environmental conservation and poverty alleviation (Ningxia and Shanxi)

5. Aims and Objectives in Application of IEM to Preserve the Biodiversity of Dryland Ecosystems and to Combat Land Degradation

Project Aim	Development Objective
"...to significantly mitigate the loss of biodiversity in relevant dryland ecosystems influenced by land degradation..."	"... to make and implement IEM plan specific to each project site, so as to protect and rehabilitate existing conservation areas being threatened by land degradation... "

Project area
The standards adopted for setting up the project.
• With reference to the work already-done by the OP12 project.
• The maximum degree of overlap between GEF and IFAD projects has been considered on the bases of eco-functional zoning of State Environmental Protection Administration (SEPA) and agro-ecological zoning of the Ministry of Agriculture.

Taizi Mountain Project (Gansu)
• Position: Qinghai-Tibet Plateau and the Loess Plateau.
• Ecological zone: the Loess Plateau.
• IEM project area: 2,138 km^2 (the area of Taizi Mountain conservation zone [847 km^2], marginal area [1,291 km^2])
• Landforms: grassland (34%), farmland (30%), natural and man-made forest (22%), wasteland (6%), wetland (1%), and others (7%).
• Land degradation: seriously degraded with an estimated 80% of desert expansion
• 337,330 people, 179 villages (95 IFAD villages), 18 townships, 2 counties.
• Livelihood: rain-fed agriculture, animal husbandry, Chinese herbal medicine, fruit trees.
• Economic status: poverty county (net income per capita of farmers being RMB 1,300 per annum).
• PA objective: strict natural conservation (protection of wildlife).
• The main threats to PA: excessive cutting, firewood collection, digging Chinese herbal medicine.

Haba Lake IEM Project Area (Ningxia)
• Position: Mao Us Desert.
• Ecological area: the transitional area from desert steppe to arid grassland .
• IEM project area: 5,400 km^2 (Haba Lake [840 km^2], marginal area [4,560 km^2]).
• Landforms: grassland (58%), forests (19%), farmland (9%), and others (14%).
• Land degradation: seriously degraded with the desertification area being 80% of the total area.
• 83,432 people, 58 villages (32 IFAD Village), 5 townships, 1 county.
• Livelihood: animal husbandry, rain-fed agriculture, labor outflows.
• Economic status: poverty county (net income per capita of farmers being RMB 2,480 per annum).
• PA objective: strict natural conservation (protection of wildlife).
• The main threats to PA: grazing, harvesting non-timber forest products.

Luya Mountain IEM Project Area (Shanxi)
• Position: Lüliang Mountain.
• Ecological area: mixed deciduous and coniferous forests in mountain area and shrub ecological zone.
• IEM project area: 1,147 km^2 (Luya Mountain [215 km^2], marginal area [932km^2]).
• Landforms: forests (46%), farmland (19%), grassland 36.1 km^2, water area 20.7 km^2, construction area 22 km^2, other agricultural land 3.7 km^2, unused land 311.9 km^2; grassland (34%), farmland (30%), forests (22%), wasteland (6%), wetland (1%), and others (7%).
• Land degradation: seriously degraded (with 40% of land experiencing soil and water loss at different degrees).
• 65,000 people, 213 villages (189 IFAD Villages), 8 townships, 3 counties.
• Livelihood: rain-fed agriculture, animal husbandry, Chinese herbal medicine, agricultural processing, labor outflows.
• Economic status: poverty county (net income per capita of farmers being RMB 1,100 per annum).
• PA objective: habitat/ species management area.
• The main threats to PA: grazing, harvesting of medicinal herbs, firewood collection.

Sub-projects and Activities
• Sub-project #1: Planning, policy and institutional capacity building.
• Sub-project #2: Community-based eco-planning and restoration; alternative/sustainable livelihood.
• Sub-project #3: Protection of biodiversity and conservation area.
• Sub-project #4: Raising public awareness.
• Sub-project #5: Project management, monitoring and evaluation, knowledge sharing.

Sub-project 1: Planning, Policy and Institutional Capacity Building
• Objectives:
"To promote the improvement of the policy environment, strengthen the institutional capacity, and support the implementation in the three provinces/regions…"
• Activities:
- (Planning): the eco-planning of the project area, to guide the follow-up project activities, and to protect and rehabilitate related ecological areas;
-(Policy): to study and evaluate the impact of the existing policy, and to support the promotion of the sustainable land management measures and the preservation of the biodiversity during the future policy-making process;
-(Institutional capacity building): to improve the capacity of the project office and cross-project office, technology department, and farmer association, in particular the capacities of institutions at county-level, township-level and village-level in project areas, for adoption and application of ecological conception during their work.

Sub-project 2: Community-based Eco-planning and Rehabilitation, for Alternative/ Sustainable Livelihood Activities
(1) Objectives:
• Incorporate eco-planning and rehabilitation activities during the process of the

village-level development planning;
• Focus on the specific factors that threaten the holisticity of the conservation area, to support the development of field activities in villages in marginal areas.

(2) Activities:
• At the village-level, to integrate IEM into the process of village assessment and/or village-level developmental planning; to support the development of incremental activities (for example, increasing the number of different tree species during tree-planting activities)
• At the township/county-level, to facilitate the adoption of more sustainable activities (for example, planting Chinese herbal medicine species), and to adopt alternative livelihood (for example, adopting biogas technology)

Sub-project 3: The Protection of Biodiversity in both the Conservation Area and Outside the Conservation Area

(1) Objectives:
• Enhance the conservation of the project area, and protect the ecosystem and significant biodiversity;
• Restore the degraded land outside the conservation area, thus partially restoring some ecological processes and functions

(2) Activities:
• Support: the compilation of the overall plan and management plan;
Personnel training;
Boundary demarcation;
Adoption of community co-management based on participation of the local communities;
Purchasing of equipment;
Technique studies;
• Support: the plantation of local grass species to revegetate the natural grasslands;
Planting local tree species to rehabilitate the degraded forests

Sub-project 4: Raising Public Awareness

(1) Objectives:
"…to raise local community, decision-maker and the public awareness, to promote the improvement of the environment, and to understand the profit brought by the adoption of IEM in the arid and semi-arid regions…"

(2) Activities:
Focusing on the environmental problems, China has already begun to implement an effective, large-scale and formal environmental education plan in primary and middle schools. The present sub-project is meant to supplement these formal educational activities, to develop public education and communication activities, and to support the village-level land-users to choose alternative development objectives as their priority. The second target group is those in charge of land policy management, including those decision-makers in townships and counties.

4. To Integrate the GEF-supported Activities into Other IFAD- and GEF-supported Projects in China

IFAD activities (baseline)	GEF activities (increment)
• Establish VIGs • Compile VDPs • Identify priority activities • Support the implementation of the priority activities (examples of incomplete activities) - develop irrigable land - improve arid land agriculture - control desertification - encourage to grow cash crops - plant trees - encourage development of animal husbandry	• Raise IEM awareness • Develop village level eco-plan • Establish standards for choosing activities that can be listed for GEF support • Support the implementation of the priority activities (examples of incomplete activities) diversify tree species (to increase biodiversity and carbon sink) cultivate plant zones (to increase ecological services) promote alternative livelihoods (to mitigate the pressures on ecosystem and agricultural biodiversity) rehabilitate village public lands (habitat rehabilitation and carbon sink)

Application and Extension

• The experiences will be extended and applied in the projects of the International Fund for Agricultural Development.

• The experiences will be extended to other similar national projects.

36. CPRWRP's Approach and Challenges to Slow down Degradation in Mountainous Areas of the Upper Yangtze River Basin.

Piet van der Poel
Soil Conservation Specialist, EU-China River Basin Management Programme.
No.1863 Jiefang Avenue, Wuhan, 430010, P.R.China
Tel: +86 27 82865716
Fax: +86 27 82416125
E-mail: pipoelcn@gmail.com
Website: www.euchinarivers.org

The WB-EU supported Changjiang and Peal River Watershed Rehabilitation Project (CPRWRP) started implementation in late 2006. It receives technical support from the EU-China River Basin Management Programme. Its approach is based on the results of the WB Loess Plateau projects, adjusted for the differences in ecological and socio-economic conditions between the two areas.

The project area consists of hilly or mountainous areas in 37 counties of Yunnan, Guizhou, Hubei and Chongqing. Many of the selected sub-watersheds have serious degradation, such as rocky desertification. Many of the sub-watersheds are inhabited by minority ethnic groups and the project counties belong to the poorest in the region. In order to promote participation of poor rural households in the project activities, the EU provides a Grant for the Poor.

Major challenges for the approach:
• Soil conservation alone is not attractive for rural households, as the effect of reduced soil loss is long-term and virtually imperceptible for local farmers. Consequently, the program combines livelihood improvement measures with soil conservation measures. Only livelihood improvement measures that have a direct link to soil conservation are supported.
• Poor farmers have little or no money to invest in conservation measures and have no incentives to change their land use, if this increases the risk of not producing sufficient food for the family. Consequently, these farmers should be provided with long-term loans to enable them to invest in changing their land use.
• Soil conservation projects in the past have often been characterised by inadequate long-term impacts. A participatory approach should ensure the active involvement of the population in project planning and implementation. This should ensure that the selected measures and species are well adapted to the local biophysical and socio-economic situation and lead to more sustainable results.

Major challenges for the implementation
Although the rating of the last WB supervision mission was moderately positive,

subsequent progress has reportedly been slow. Various causes can be indicated.
- PMO staff have too high workloads;
- Serving the designated main target group, the poor households, and the wish of many PMO staff to work more with large households are hard to reconcile;
- The unit prices for project measures are unrealistically low, causing measures to become unattractive for both P/CPMOs and farmers;
- Procedures have not been clear because until recently the project has operated without a complete Project Implementation Manual;
- RBMP support to the project had a slow start and its support to the county level staff has been limited;
- The MIS was designed based on incomplete specifications and has not functioned satisfactorily. It has increased the workload of PMO staff and has failed to deliver the management information that it should produce;
- Participation seems to consist mostly of consultation and information without sufficient active participation of the stakeholders in planning.

1. Introduction

The WB-EU supported Changjiang and Peal River Watershed Rehabilitation Project (CPRWRP) started its implementation in late 2006, after several years of project preparation.

The WB provides a loan of 100 million US$ to support the provinces to implement the project. The EU provides a grant of 10 million Euros, mainly aimed at subsidizing land use changes and conservation measures that also increase the income of rural households and a special Grant for the Poor to reduce the total loan for the poorest households in the intervention area.

The CPRWRP receives technical support from the EU-China River Basin Management Programme. This support consist of one long-term advisor, short-term advisors and support for workshops, training, study tours, publications, studies and surveys.

The approach of the CPRWRP is based on the results of the WB Loess Plateau projects, adjusted for the differences in ecological and socio-economic conditions between the two areas. Shallow, often rocky and infertile soils in an area with 800 mm to 1400 mm of annual rainfall require different conservation measures than deep and fertile loess soils in a much drier climate. For example the use of tractors and bulldozers to reshape the land is quite common in the Loess Plateau area, but is of little use in most of the small watersheds of the CPRWRP area. Other changes include more emphasis on monitoring and an Management Information System.

The project area consists of hilly and mountainous sub-watersheds in 37 counties of four provinces / municipalities: Yunnan, Guizhou, Hubei and Chongqing. Many of the selected sub-watersheds show serious degradation, such as rocky desertification in large areas of Guizhou and Yunnan, and somewhat less extreme degradation in other areas. Many of the sub-watersheds are inhabited by minority ethnic groups and the project counties belong to the poorest in the region. In order to promote participation in project activities of a main target group of the project, the poor rural households, the EU provides a Grant for the Poor.

The CPRWRP approach is in principle participatory and based on collaboration between villages and the PMO in developing a Preliminary Sub-watershed Design and implementing it. To assist in the planning of the sub-watershed activities a Participatory Design Manual was prepared. It has recently been improved and up-dated with support of the RBMP. The design consists of four phases:

(1) A project preparation phase, including the selection of sub-watersheds. This phase has been mostly completed, except that due to a forthcoming restructuring (due to unexpectedly high increases in prices of materials and labour and to the large fall in the US dollar to the China RMB exchange rate) of the project the number of sub-watersheds will have to be reduced by eliminating the least suitable sub-watersheds. Due to socio-economic development in some counties some of the originally selected sub-watersheds no longer meet the selection criteria;

(2) A sub-watershed preparation phase in which available information is collected, pamphlets and posters about the project's objectives and approach are distributed, village meetings to discuss the project are organised and villagers vote on acceptance of the project, and if so select village planning teams, which include representatives of the farmers and of the various stakeholder groups;

(3) A communication, awareness raising and planning phase, in which the details of the project activities and support are discussed, PRA's and field surveys are conducted and a Preliminary Sub-watershed Design Report is prepared and approved;

(4) A hamlet-level detailed planning phase, in which the measures proposed by the village planning teams are discussed and agreed upon at plot level with the farmers directly involved. This phase is followed shortly by the start of the implementation of the project activities. This phase is not included in the planning for the Preliminary Sub-watershed Design Report in order to reduce the period between the agreement with farmers on the measures to implement and the actual start of the implementation. In the past this period was much longer and problems occurred due to farmers changing their mind in the meantime.

The Preliminary Sub-watershed Design should be well adapted to the local environment and to the socio-economic situation of the local population. Involvement of the local population and of the relevant line agencies should help achieving this. Flexibility is required as some changes may need to be made to adjust the plan to the preferences of individual farmers. This requires the auditing agency to show similar flexibility.

2. Major challenges for the approach

Unattractive cost–benefit ratio of the proposed soil and water conservation and livelihood improvement measures

Soil and water conservation measures in general have two main impacts:
- Reducing soil erosion which in the long run will improve soil fertility, organic matter content, water holding capacity, and stop further soil degradation
- Increasing infiltration, which may lead to an increase in crop production, especially in dry areas where water is the limiting factor for crop productivity

In dry areas, there is an almost immediate positive effect of soil and water conservation measures. In more humid areas the effect of soil conservation is long term and often not perceptible to the local farmers in the short term. Consequently, in humid areas the challenge of introducing soil and water conservation is much bigger. Introduced measures should have additional social or economic benefits to make them attractive to the farmers. A good example from semi-arid Mali is the adoption of hedges by the local population, even in villages not covered by any soil conservation project. Research showed that the main reasons for adopting the hedges were that it helped in stopping powerful neighbour from extending their land each year a bit by ploughing up a little bit of their poorer neighbour's land. Hedges also reduced crop damage by free roaming livestock. Soil conservation was mentioned as a third reason only.

Recent calculations have shown that the benefits of stone-wall terraces in the project area are marginal. Stone lines along the contour lines in fields in dry areas of West Africa became very popular because the farmers could easily notice that crops just above the lines were taller and greener and produced higher yields than those halfway between the stone lines.

One way in which the project aims to address this unattractive cost-benefit ratio is by also supporting measures which have mainly economic benefits, but are directly linked to the soil and water conservation measures to be introduced. For example, the project assists farmers in buying calves, which can be sold or kept to provide milk and manure or reproduce, but these cattle should be kept in stables and not be allowed to roam freely in order not to degrade the grazing land / vulnerable vegetation cover. The project also supports a change from sloping land annual cropping to orchards. However, if species are poorly chosen or if maintenance of the seedlings or trees is poor, the impact on farmers' income may be marginal or even negative.

Conclusion: Concentrate on soil and water conservation measures with clear and short-term economic or socio-economic benefits or a combination of conservation and livelihood improvement measures. In dry areas water conservation should be emphasized.

High investment costs

The CPRWRP mainly proposes changes of land use e.g. from the traditional sloping land with annual crops to orchards. Orchards allow the cultivation of annual crops in between the planted trees for the first couple of years, but as the trees grow bigger the crop yields decline, often before the trees start producing. This loss of income will create serious problems, especially for the poor households that already have problems to make ends meet. The investment cost of orchards is also beyond the capacity of many of the rural households. Poor farmers have little or no money to invest in conservation measures and have no incentives to change their land use if it increases the risk of not producing sufficient food for their family. In order to promote the proposed change of land use, a concessional loan should be available for the households to make the investments. A grace period before the repayment of the loan starts should make it possible for all households to make the transition from annual crops to perennial crops. Due to the non-availability of some of the counterpart funds and the project not requesting an advance on the loan from the WB, farmers and contractors often had to pre-finance project activities. Especially for poor farmers in some areas this has created problems.

Conclusion: the availability of counterpart funds or funds from other sources is an absolute requirement at the start of project activities. Pre-financing of project measures by households or contractors should be avoided at all costs.

Loans to farmers

Many projects and government programmes provide seedlings and infrastructure to farmers at no costs, the farmers only contributing their labour during planting or construction. Thus, farmers may be reluctant to sign loan agreements indicating that they will have to refund part of the costs. In Yunnan, where the Finance Bureau (FB) required farmers to sign a new loan agreement based on actual rather than planned work, farmers appear very reluctant. The reluctance to sign loan agreements is also due to the fact that many of the "loans" are pre-financed by the farmers and reimbursed from the WB loan, after certification of the measures. If the farmer pre-finances there may not be a need for a loan. If the farmer has taken out another loan, e.g. from the Village Committee, the WB loan (usually referred to as reimbursement by the PMOs) merely replaces the local loan with a WB loan. Often, the farmers do not see any money as the reimbursement goes directly to e.g. the seedling supplier. In this case they hardly consider it a loan. Besides they often think that it is up to them to decide how to spend the loan rather than it being determined by a project. The small amount of many of the loans for farmers also appears to constitute a high workload for the Finance Bureaus.

Conclusion: One can debate the pros and cons of free supply of seedlings and other goods or services to farmers, but different policies by projects and government programmes should be avoided, except for pilot studies. Clear policies on the repayment of farmers' loans should be adopted and enforced.

Inadequate long term impact

Soil conservation projects in the past have often been characterised by a lack of long-term impact. In Indonesia areas treated under the government Regreening Programme a few years later could not be distinguished from non-treated areas, due to the lack of maintenance. For example from the planted grass strips only occasional clumps could be found. The exception was a village where there was a milk cooperative and thus a market for cut grass. All terrace risers in this village were protected by grass strips. Part of this failure may be due to the poor selection of measures, e.g. measures that were well adapted to the local biophysical situation but did not fit into the local farming systems. The participatory approach adopted by the project should ensure that the population gets actively involved in the project planning and implementation, including the choice of measures and species. This should guarantee that suitable measures and species are selected and that the villagers consider the measures as their own and consequently will invest time an effort in maintaining them.

Conclusion: Propose measures that fit in the socio-economic conditions and farming systems of the villagers, address real or perceived problems and use a participatory approach for the selection of measures and species and for the planning, implementation and monitoring of work progress and financing.

Project not addressing main problems of an area

Solving the main problems of a sub-watershed may be beyond the reach of the project as it can spend only a limited amount of funds on activities such as drinking water and little or none on health and education. CWMP and the county governments have addressed this by "resource integration", having various county departments collaborate to address the problems of a number of sub-watersheds in an integrated way. This resource integrations would also solve the problem of lack of support from higher levels reported by PMO offices and would make collaboration with other line agencies easier as there is no longer only a WR Department Soil Conservation project but a county level effort to address the problems in some of the sub-watersheds.

Conclusion: Make certain that important problems of the villagers are addressed and explore the possibility to join forces with other project or programmes for an integrated approach.

Disregarding the impact on the environment

In a few cases measures or species selected may have unintended side effects. This can include:

- contour cultivation measures leading to the concentration of run-off and the formation of gullies or to landslides;
- the introduction of species that are invasive, such as e.g. Leucaena glauca;
- Planting of monoculture forest, which may reduce soil erosion and may produce wood for construction or firewood, but may reduce biodiversity e.g. by replacing natural grassland with a one-species forest.

Conclusion: Evaluate site conditions and the possible impact of proposed measures on the environment before the selection of measures/species. Try to restore the original climax vegetation where possible.

3. Major challenges for the implementation

The WB supervision mission in April 2008 raised the rating of the project to moderately satisfactory, but since then progress has been slow. Various causes can be indicated for this.

- Lack of support from upper levels

The project is considered to be a Water Resources Department/Bureau project and collaboration with other line agencies and with township staff suffers because County Leading Groups reportedly have been ineffective. The lack of interest could also be linked to the lack of incentives for the collaboration of line agency and township staff with the project. The China Watershed Management Project CWMP indicated that through "Resource Integration" they managed to agree with other agencies involved on a common and integrated approach to solve the problems in priority sub-watersheds.

Recommendation: Some degree of "Resource integration" should be considered, but transparency should be maintained.

- Lack of authority of the CPIO

The Central Project Implementation Office (CPIO) has not has not officially been authorised by the Ministry of Water Resources (MWR) to manage the project. Thus, it has limited authority in relation to the Provinces, which at times ignore recommendations from the CPIO. The logic is that the World Bank loan is to the Provinces, but a more clear management structure is advisable.

- PMO staff have a high workload.

This has been a regular complaint. Two main causes identified by PMO staff are the participatory approach and double work due to an ineffective MIS system. The participatory approach requires more time of the staff than the traditional top down approach. It is believed that the end results will be better and more sustainable, if real involvement of villagers is achieved. CWMP reports that for certain activities, such as tree planting, the villagers may take more responsibilities in a real participatory project, thus reducing the workload of PMO staff.

More efficient planning of activities could make many of the tasks less time consuming and more effective. For example in one sub-watershed CPMO staff planned meetings with villagers of all major natural villages. Although the effort is to be appreciated, it may be more effective to only have meetings with each of the usually three administrative villages. To involve farmers and promote discussion between them and the farmers' representatives, the CPMO staff should post information and hand out leaflets in the natural villages and hamlets a week before the meeting with the farmers' representatives and village leaders. This should stimulate discussions between farmers and their representatives, leading the latter to come to the meeting prepared for discussions on what this project is about. Moreover, the meeting should take the main problems existing in the village as its point of departure and indicate where the project could help the villagers solving or reducing these problems. This would be more effective than giving a long monologue about objectives, activities and conditions of the project. It should make the villagers feel that the project will address some of their main problems. Some tasks may be simplified while others could possibly be skipped. Several CPMO staff consider farmers' labour input promises a rather meaningless requirement.

The MIS, notwithstanding efforts of CPIO and RBMP to make it operational, has not functioned adequately up till now. A major complaint is that the system requires substantial double data input and does not help the CPMO staff in the preparation of plans, planning reports and progress reports. Consequently, many of them have recently ignored CPIO requests to complete data entry for the sub-watersheds presently under implementation.

Recommendation: strive for real participation, not just consultation, keeping the focus of the project on the interests of the main target groups.

• Township lack of interest.

Because the project works with provincial, prefecture and county staff and the latter often cannot carry out all day-to-day work in the sub-watersheds, this is often left to the township staff, with a major involvement for the township SWC technician. However, there are no operating funds for the township staff and as the participatory approach is very labour intensive, township staff in several counties reportedly show a lack of interest in carrying out CPRWRP work.

Recommendation: CPMOs should try to find ways to motivate township staff to fully support the project

• Farmers lack of interest.

Farmers' interest in the project seems to be waning. This is partially due to the farmers ending up having to pre-finance the project loans rather than receiving project loans or goods and having to repay some of the goods/services provided to them. Reimbursement of investments at unrealistically low unit prices has further reduced the interest.

Many farmers fail to see the benefits of the project for them. Pre-financing of project measures by farmers is often very difficult for the target group, the poor. For similar reasons there is also a lack of interest from contractors to carry out project interventions. The availability of funds could have been solved by an advance on the WB loan, but for unknown reasons this has not been requested. Also the timely availability of counterpart funds could have eliminated or reduced this problem.

- Conflict between the designated main target group: the poor households and the wish of many PMO staff to work more with large households. Moreover, the project supports small households to acquire a couple of cows or calves, while at the same time the government promotes large household livestock raising operations.

It is believed that most of the PMO staff finds the more progressive larger farmers easier to work with and more responsive to new ideas. This is quite logical, as these farmers can afford to invest in potentially profitable long-term activities. Poor farmers, whose main concern may be to produce enough food for their family this year, can not afford to experiment with new measures that in the short run may lead to a reduction in yield. For them, a somewhat lower but reliable yield is preferable over a less reliable but potentially higher yield. Risk avoidance is their main strategy. Moreover, working more with larger farmers would also reduce the workload of the staff.

It should be borne in mind that there is little point in investing heavily in keeping poor farmers in rural areas a few years longer. The RBMP employed Farmer's Association team pointed out that the project should probably concentrate on those poor, small and medium sized-farmers, who either are going to stay in the rural areas because there is no attraction for them in the big cities (which may be the case for certain minority ethnic households) and/or those that have a reasonable chance of becoming sustainable farmers. The richer large farmers do in general not need the support of the project as they are powerful enough to develop links with middlemen or buyers to sell their products at reasonable prices or to get the technical support needed. In Chongqing, the project objectives appear to be in conflict with the policy of promoting poor farmers to move to the cities by providing housing and job-training.

Large households farming marginal and eroding sloping land should be included in the project in order to reach the project's conservation objectives.

Recommendation: Search for ways to combine targeting the intended beneficiaries and existing policies.

- Participation seems mainly to consist of consultation and information rather than serious involvement of the stakeholders in the project planning and decision making. The approach appears to be stuck in between a real participatory approach and a traditional top-down approach. Most of the project staff are engineers and many have little experience and training in participatory methodologies. Some have indicated that this is a problem and more training is required. Management at the CWRC level considers technical inputs more relevant than soft science inputs such as participation and farmers association training. Efforts to increase the capacity of the staff in understanding and applying participatory methodologies should continue. Raising the awareness of management level project staff on the importance a good understanding and knowledge of participatory theories and techniques of technical project staff should assist in changing the prevailing reluctant attitude towards participatory training.

Recommendation: Training for all levels of project staff in the principles and techniques of participatory approaches is required

- The unit prices for project measures have become unrealistically low, causing the measures to become unattractive to implement for both PMOs and farmers. This problem is due to the fast increase in prices of materials and labour in China and the long period of devaluation of the US dollar against the China RMB. The Province Project Management Offices (PPMOs) and CPIO have proposed updated unit prices, which recently have been forwarded by the MoF to the WB for approval. Acceptance of the new unit prices will require the project to be restructured by scaling down the expected outputs. This is currently under discussion.

Recommendation: Use realistic unit prices and design an easy way to adjust these without the need for a time-consuming restructuring process.

- Shortcomings in the project design and implementation

Because the project employs mainly engineers, certain social aspects of the approach appear to get insufficient attention. Occasionally, some technical aspects of the measures that are being introduced get overlooked or alternative measures are not being considered.

Some of the project supported plant species are recommended to be planted as orchard or plantation crops with more or less equal spacing between rows and plants. However, several of these species will be more effective for soil conservation when planted as hedges along contours or along farm boundaries. The inclusion of hedges is presently being discussed.

Moreover, in the context of Chinese government responsibilities the emphasis is often on achieving planned quantities and spending the budget rather than on sustainable results. In order to build stone wall terraces cheaply, one county reduced the walls to stone lines near the edge of existing earth terraces. It was pointed out to them that these terraces would be ineffective and subsequently proper stone-wall terraces were built.

Post management of project introduced measures has proven difficult and lack of maintenance/management may negatively affect the productivity of e.g. project supported orchards. Project supported farmers associations, such as used in CWMP, would not work for CPRWRP under the present conditions because no budget for support to the running expenses of such FAs is available. Farmers Associations should have clear socio-economic benefits for their members. The possibility of promoting viable FAs may need to be reconsidered. These should probably be dealing with marketing of products and management of resources.

- Procedures of the project have not been clear or are too complex or too strict.

The project has tried to operate until recently without a complete Project Implementation Manual (PIM). This has caused considerable confusion as often staff was not quite clear on what procedures to follow. Only recently has the CPIO with assistance of an expert provided by the World Bank finalised most of the PIM. The RBMP has assisted the CPIO in revising the PDM and is presently assisting in producing the Monitoring and Evaluation framework and M&E operational manual. The CPIO will in collaboration with the RBMP produce the PDM for farmers. When these parts of the PIM are completed, there should be less confusion about procedures and responsibilities. However, County Project Management Offices' (CPMO) complaints, such as excessive documentation requirements for reimbursement, will probably continue.

At present, it looks like the PIM may end up at some 1500 pages, including 455 pages procurement and management procedures, 600 pages safeguard documents, 100 pages PDM, 100 pages M&E and more to come. This appears excessive as shown by repetitions and contradictions between different documents. Especially some of the safeguard documents are unnecessarily voluminous and it is hard to extract the relevant guidelines and recommendations from these documents. As many documents have been prepared by different organisations differences and contradictions exist.

Recommendation: develop clear, concise, complete and consistent procedures, guidelines and operational manuals before or right at the start of the project. Limit safeguard documents to a brief situation overview and clear and concise guidelines and instructions.

- The MIS was designed based on incomplete specifications and has not functioned satisfactorily. While it should have reduced the workload of PMO staff, it has actually increased it without providing the management information that it should produce. One could probably say that the MIS design was premature, but there was pressure on the CPIO to have a management system in place at the start of the project. However, the system did not produce the needed outputs, required double data entry, proved to be extremely slow and unstable and is presently largely ignored by many of the counties. Efforts of the RBMP and CPIO to solve the existing problems have lead to a considerable improvement of the MIS, but some of the major drawbacks still exist and many counties are reluctant to enter all their data, which presently mainly serves management purposes for the upper management levels without much benefit to the CPMOs. Discussions on how to solve the problems are continuing.

Recommendation: in order to be useful to the CPMOs, the MIS should assist them in producing the Preliminary Sub-watershed Design Report and planning maps as well as the project progress reports and maps. Starting with a simple MIS for the most important functions and extending it as and when required should be considered.

- Support to the project by the RBMP has been slow due to:
  - Slow start-up, including staff changes
  - Complicated bureaucratic procedures
  - Some of the hired short-term experts produced only marginally acceptable reports.
  - It has proven difficult to identify and hire adequately qualified Chinese short-term experts.

In hindsight, it would have been beneficial if a long-term technical advisor had been available during the start-up of the project. Simplified procedures and more flexibility would also help.

4. Other challenges/issues

Plant species not recommended for the project are introduced by other projects
Other projects in the area are introducing species not recommended by the project.

For example, the province of Yunnan is said to be introducing one million hectares of physic nut (Jatropha curcas). The WB eliminated this species from the list of potential species, mainly because the yield per hectare is low and harvesting the nuts is a very labour intensive job. Consequently the economic outlook for this species is not good. However, the effort placed by other projects on this species may confuse staff.

Species introduced by other projects do not follow conservation principles

Other projects or farmers and to some degree even CPRWRP are introducing measures in ways not promoting soil conservation. For example, in several counties in Yunnan mulberry (Morus alba) is planted to produce feed for silk worms. However, most of these plantations are made using (almost) even spacing between plants and rows, while mulberry would be a very suitable species to plant as hedges with wider spacing between the rows and narrow spacing between the plants and with the rows running more or less at right angles to the slope. Also the plantation of physic nut (Jatropha curcas) in Yunnan could be better done as hedges around existing fields than as plantations, covering the whole field. Promotion of hedges has been discussed during the recent workshop and will be given more attention in the project.

Other projects introducing the same activities but with different support measures

Lack of integration of the activities with those of similar projects within an area is often a problem. Within the project area there are for example different projects or government programmes introducing biogas pits, some offering more favourable conditions than the CPRWRP. Consequently, farmers may become selective and collaboration between projects may turn into competition. Project staff have had a hard time introducing measures with conditions that the farmers know are less favourable than those of government programmes. This difference is very hard to explain to the farmers because the project is supposed to target poor counties and households.

Other activities in area causing heavy soil erosion

Other economic or development activities in the area may counteract all the positive effects of the project's erosion control measures. For example road construction may cause more erosion in a sub-watershed than the project measures manage to reduce. In most countries in Asia road construction is still done in a way minimizing investment, but leading to high maintenance costs due to high erosion rates, especially by land slides of the cut slopes and of the overburden dumped over the edge of the road under construction.

37. Application and Promotion of Integrated Ecosystem Management in China[1]

Cai Shouqiu [2]
Professor, Wuhan University

Abstract: This paper provides detailed explanation of fundamental meaning and content of IEM, introduction and conclusion of basic situation of IEM application in the field of policies and laws in foreign countries, summary and generalization of the IEM practice regarding legislation and policy-making in China. Also, four proposals are put forth as to adoption and Popularization of IEM in the field of policies and laws. (1) The prospective basic law on Environment and Resources, such as the Basic Law on Environment and Resources of the People's Republic of China, should define the ecological methods including IEM. (2) The revision of the Forest Law of the People's Republic of China should define the ecological methods including IEM, and also turn the IEM guidance rules, fundamental principles and measures to or let it be operable forest rules and legal systems. (3) The establishment of relevant standard documentation and other policy oriented documents as regards resources and environment, such as laws, administrative rules, local rules, departmental regulations, regulations of local government in the fields of forest, grassland, soil, land water, ocean, wildlife and biological diversity, the ecological methods including IEM method should be applied and popularized to the maximum, as well as implementation in light of the real situation in China. (4) It's recommended to establish specific IEM law criteria. Still, prior establishment of governmental rules such as Measures for Forest IEM, Measures for Wetland IEM are proposed.

Key Words: Environment and Resources Law; Prevention & Control of Land Degradation, Ecological Method, IEM

I. Basic Meaning & Content of IEM

The ecology is an applied science which studies the correlations between organic object and the environment, which constitutes the major theoretical basis for establishing environmental policy and law. The core conceptions of ecology are zoology and ecosystem. The ecological approach, also known as (Eco-approach), ecological method and ecological adjustment mechanism, stand for methods for application of ecological principles and rules. The ecological approach covers specific methods for research of resources and environment oriented policies and laws, which fully embodies the characteristics of research of resources and environment oriented policies.

---

[1] Refer to the General Situation of IEM Approaches Development, written by Cai Shouqiu, published in the Tribune of Political Science and Law, the 3rd issue, 2006.
[2] Caldwell, L 1970. The ecosystem as a criterion for public land 2 . policy. Natural Resources Journal 10(2):203-221.

The ecological approach is the major ecological approach applied in research of resources and environment oriented policies and laws. The ecological approach (abbreviated as EA), is also known as IEM (Integrated Ecosystem Management), integrated ecological approach (or mode) (IEA), integrated ecosystem management. Or it can be deemed that IEM is the product of applying IEM methodological theory in the field of resources and ecosystem management or the intensive reflection, important embodiment and typical representation of the ecological approach. The Resolution V/6 on Ecosystem Modes passed at the 5th member nations congress of the Convention on Biological Diversity held in Nairobi of Kenya in May 2000, and the 12 IEM principles and 5 IEM guidelines proposed by the expert team were accepted and included as an all important practice framework of the treaty. According to the resolution, "IEM is a strategy for integrated management of land, water and biological resources, and the purpose is to promote protection and sustainable development of these resources in a fair approach." To sum up from the perspective of resources and environmental policy, IEM refers to a kind of integrated strategy and approach for management of natural resources and environment, which requires integrated treatment of various components of the ecosystem, integrated consideration of social, economic and natural (including the environment, resources, biology) need and value, integrated application of multidisciplinary knowledge and method, as well as integrated utilization of administrative, market and social adjustment mechanisms for solutions of resources utilization, environmental protection and ecosystem degradation, so as to create and fulfill multiple benefits with the economy, society and environment, and realize harmonious coexistence of the human and the nature.

The essentials of IEM mainly include various IEM principles, rules and guidelines. The description of the ecological approach stated in IEM Resolution V/6 are:

(1) The ecological approach is a strategy for the integrated management of land, water and living resources that promotes conservation and sustainable use in an equitable way. Thus, the application of the ecological approach will help to reach a balance of the three objectives of the Convention: conservation; sustainable use; and the fair and equitable sharing of the benefits arising out of the utilization of genetic resources.

(2) An ecological approach is based on the application of appropriate scientific methodologies focused on levels of biological organization, which encompass the essential structure, processes, functions and interactions among organisms and their environment. It recognizes that humans, with their cultural diversity, are an integral component of many ecosystems.

(3) This focus on structure, processes, functions and interactions is consistent with the definition of "ecosystem" provided in Article 2 of the Convention on Biological Diversity: "'Ecosystem' means a dynamic complex of plant, animal and micro-organism communities and their non-living environment interacting as a functional unit."

(4) The ecological approach requires adaptive management to deal with the complex and dynamic nature of ecosystems and the absence of complete knowledge or understanding of their functioning.

(5) The ecological approach does not preclude other management and conservation approaches, such as biosphere reserves, protected areas, and single-species conservation programmes, as well as other approaches carried out under existing national

policy and legislative frameworks, but could, rather, integrate all these approaches and other methodologies to deal with complex situations.

The 12 principles of the IEM Resolution V/6 are: ① The objectives of management of land, water and living resources are a matter of societal choice; ② Management should be decentralized to the lowest appropriate level; ③ Ecosystem managers should consider the effects (actual or potential) of their activities on adjacent and other ecosystems; ④ Recognizing potential gains from management, there is usually a need to understand and manage the ecosystem in an economic context; ⑤ Conservation of ecosystem structure and functioning, in order to maintain ecosystem services, should be a priority target of the ecological approach; ⑥ Ecosystems must be managed within the limits of their functioning; ⑦ The ecological approach should be undertaken at the appropriate spatial and temporal scales; ⑧ Recognizing the varying temporal scales and lag-effects that characterize ecosystem processes, objectives for ecosystem management should be set for the long term; ⑨ Management must recognize that change is inevitable; ⑩ The ecological approach should seek the appropriate balance between, and integration of, conservation and use of biological diversity; ⑪ The ecological approach should consider all forms of relevant information, including scientific and indigenous and local knowledge, innovations and practices; ⑫ The ecological approach should involve all relevant sectors of society and scientific disciplines. The 12 IEM principles mentioned above are not only theories but also guidelines for practice.

Major characteristics of IEM are: ① Emphasis on basic conceptions, including conception of ecological (environmental) justice, conception of ecological (environmental) fairness (including intra-generation fairness, inter-generation fairness, inter-species fairness, interregional fairness), conception of ecological order (including maintenance of ecological security and balance), conception of restrained ecological basis or bounded environmental carrying capacity, conception of care the nature, respect life-form and environmental protection, principle and conception of harmonious coexistence of the human and the nature and between humans; ② Methodologically, the system analysis method of the ecology is adopted, particularly integrated ecological approach. In addition to systematic scientific theories, such as "three old theories (system theory, cybernetics and information theory, abbreviated as SCI theories)," "three new theories (dissipative structure theory, coordination theory and catastrophe theory, abbreviated as DSC theories)," entropy theory, chaos theory, grey systems theory and game theory, the policy science research method and law research method are attached great importance and focused upon for application in policy-making and law research.

2. General situation regarding IEM practice in policy-making and legislation in foreign countries [3]

The impact and function of ecological approaches, including integrated ecological

---

[3] Helen M. Ingram, Dean E. Mann, Environmental Protection Policy, published in the Encyclopedia of Policy Research (edited by Stewart S. Nagor), Science & Technology Documents Publishing House, July 1990, P534.

approach, in protection of resources and environment and construction of policy and law involving resources and environment, are widely accepted and recognized. The ecological approaches are important theories and practice guidelines for establishment and implementation of resources and environment oriented policy and law. Presently, they're widely applied in research of resources and environment oriented policy and construction of resources and environment oriented law.

Back to 1970, Lynton Caldwell, an American policy analyst, published an article, where he called for the ecosystem to be used as the basis for public land policy.[4] He thought that this demands American government to adopt a new policy mode. By contrast, the environmental protection movement in America in the 1970s'was not powerful enough to drive American government to make such a revolutionary policy reform. In the early 1980s', Helen M. Ingram and Dean E. Mann, scholars of American environmental policy and law, noted in the conclusion of American environmental policy that contemporary environmental protection movement is a process of pattern transition and new social thought, which are densely reflected by ecology and ecological approach (abbreviated as Eco-approach). The authors made it clear that "the environmental protectionists buy the ecological viewpoint; united or 'general" view prevails as regards the standing of ecosystem and human. ······The balance expected by the environmental protectionists does not mean balanced resources utilization; rather, it implies pursuance of coexistence by protection of food chain and species (including the human being) in terms of ecological significance. The environmental protectionists apply scientific knowledge more accurately, so as to maintain the values compatible to theirs."[5]

So far, many policies and laws embodying IEM conception have been established in foreign countries, and IEM experimentations have been carried out in many countries, some of the which have accumulated rich experiences in IEM application. For instance, the Law on Natural Resources Management of Swede enacted in 1987 is an Environment and Resources law reflecting IEM conception, which defines basic guidelines and detailed requirements for sound development and utilization of natural resources, takes the society, economy and ecosystem as an integrated whole, and establishes a new mode integrating different benefits for utilization of water area and land resources. Article 1 of Chapter 1 defines that "development of land, water and all natural environments shall be carried out in the view of society, economy and ecosystem and on the basis of sustainable utilization." In countries like America, Canada, extensive land use and improper policy have caused serious land degradation and environmental damage in arid regions. For unyielding research and practice, they succeed in integrated natural resources and ecosystem management in different ways. For example, the Article 24 (Preservation of Natural Environment) of the Basic Law on Environmental Policy of South

---

[4] Australian Intergovernmental Agreement on the Environment (IGAE) of 1992 is a document of legal effect, which is cosigned between Australia Federal Government and various states, regions and local government associations for coordinated environmental protection movement.

[5] USDA Forest Service. 1992. Ecosystem management of the National Forests and Grasslands. Memorandum 1330- 1. USDA Forest Service, Washington, D.C..

Korea (established on Aug. 1, 1990 and revised respective on Dec. 31, 1991 and June 11, 1993) defines that "considering that preservation of natural environment and ecosystem is the basis for human existence and living, the nation and the subjects shall make the best of efforts to maintain and preserve natural order and balance." For instance, the Environmental Agreements between the Governments of Australia (1992)[6] states that "preservation of biological diversity and integrity should become the most fundamental consideration." US Forest Service published a new policy on IEM for multiple utilizations in 1992. To adapt to new IEM conception, US Forest Service reformed the "resources based" management policy applied in the past.[7] In 1995, 14 departments of America, including United States Commission on Environmental Quality, United States Department of Agriculture, US Army Department, Department of Defense of the United States of America, US Department of Energy, Department of Housing and Urban Development, Department of Interior, US Department of Justice, US Department of Labor, the US State Department, US Department of Transportation, US Environmental Protection Agency and US Science and Technology Policy Bureau, cosigned the Memorandum of Understanding on Encouraging Ecological Approach. Then Federal Interagency Ecosystem Management Task Force was set up, so as to promote the understanding of ecosystem method for relevant department of the federal government. The Canadian Environmental Protection Act passed in 1999 stated in detail in the foreword that "Canadian government recognizes the significance of ecosystem method." The Article 2 emphasizes that in practicing this act, Canadian government shall observe Canadian constitution and laws as well as "practice ecosystem method in considering specific and basic characteristics of the ecosystem." The Russian Federation Forest Law (enacted in 1997 and revised in 2007) also includes or embodies new conception and thought such as "the ecological approach," and this law applies a modern and artistic ecological approach to forest management by prescribing preservation of biological diversity as well as protection and development of forest resources and environment and social function. For instance, this law states a forest economy planning that is consistent in process, inclusive (EA and multiple targets of forest management), compulsory, investigation based and funded by the national finance. This law defines in the principle of forest management that "forest management must be sustainable, so as to maintain biological diversity and follow the EA requirements," "multiple targets and sustainable forest management shall be implemented for the good of the social majority," "the Measures for forest protection must embody major land condition and ecological characteristics of the forest."

III. General situation regarding IEM application in policy-making and legislation in China

Although the ecological approaches including EA are lately applied in the field of policy

---

[6] Wang Ya, Establishing Scientific Environmental Protection Conception, China Environment News, Mar 25, 2003;

[7] This article was originally published in *Natural Resources Forum*, Issue 32, 2008. Some small changes have been made.

and law regarding resources and environment in China, the evidence indicates that Chinese government and researchers of policy and law are paying more attention and consideration to these ecological approaches.

The ecological approach and IEM conception were introduced in China from the early 21st century for the sake of exploring and establishing an integrated management framework for sustainable development between administrative agents, industries and districts. On Mar. 28, 2002, the State Environmental Protection Administration (SEPA) of the People's Republic of China published the National Environmental Protection Program for the "10th Five-Year Plan," with which "the management thinking of ecosystem mode" is regarded as an important guideline for environmental protection in China; the "guideline" of the program highlights "the management thinking of ecosystem mode. And it advocates establishing macro-system and macro-environment conception, reinforces integrated management of multiple elements and systems in the regions and valleys as well as maintenance of ecological structure and function during proper protection of single element." In Jan. 2003, the State Council published the Program of Action for Sustainable Development in China in the Early 21st Century, where it defines "environmental protection and construction" in an overall manner. In 2003, Yang Chaofei, then director of the Natural Protection Department of SEPA, emphasized "to persist in the management thinking of "ecosystem mode," namely, the guideline of IEM that is recognized and popular in the international arena in recent years, which calls for environmental protection with the aim of rationality of ecological structure, preferable function and integrity of ecological process, transformation from singular element management to management of complex elements, reform of administrative region to valley system management, united management of organic and inorganic systems, scientific management based upon ecological monitoring and scientific research, coordinated management of human activity included in the ecosystem; the environment is a complete and organically correlated system, and various environmental elements shall be managed as an integrated whole in terms of the system perspective, and oneness and integrity of ecosystem management system shall be attached more importance.[8] With strong support from GEF and under the guidance by the Ministry of Finance of the People's Republic of China, for collective efforts be relevant governmental departments, ADB and six Provinces (Regions) in western China, the PRC-GEF Partnership on Land Degradation in Dryland Ecosystems was officially launched in July 2004, and the planned investment within 10 years reaches to 1.5 billion USD for integrated control of land degradation in western China. This is the first time of cooperative long term planning on the environmental field between Chinese government and GEF, and the IEM conception is introduced for control of land degradation in western China. The main purpose is to cre-

---

[8] Zhou Ke is Professor in the School of Law, Renmin Univ. of China, Beijing, PR China; Cao Xia is Professor in the School of Law, Shanxi Univ. of Finance & Economics, Taiyuan, Shanxi, PR China. E-mail: caoxia_502@163.com; Tan Baiping is a lecturer from the Section of Legal Studies, the College of Economics and Management, Beijing Univ. of Technology, Beijing, PR China. The authors declare herein that the paper is for the very Symposium Only.

ate an integrated management framework for sustainable natural resources between industries, regions and fields. Presently, many governmental departments and research institutes are engaged in the project of prevention and control of land degradation in western China, including the Legislation Committee of National People's Congress, National Development & Reform Commission, the Ministry of Science & Technology of the People's Republic of China, the Ministry of Finance of the People's Republic of China, the Ministry of Land & Resources of the People's Republic of China, the Ministry of Water Resources of the People's Republic of China, the Ministry of Agriculture of the People's Republic of China, the Ministry of Environmental Protection of the People's Republic of China, the State Forestry Administration of PRC, Legal systems Office of the State Council, CAS. The Decision on Implementing the Concept of Scientific Development and Strengthening Environmental Protection by the State Council (Published in Dec. 2005) defines in the policy on "Improving ecosystem management system" that "gradually divide the departmental responsibility according to the regional ecosystem management approach, so that strengthen coordinated and integrated environmental supervision." This is the first time when the document in the form of administrative rule of the State Council confirms and emphasizes the application of ecosystem management approach in the resources and environmental protection management, policies and laws in China.

IV. Proposals on application and popularization of IEM in policy-making and legislation in China

So far, several policy documents and legal regulatory documents regarding the ecological approaches (including IEM) have been established in China; nevertheless, we do not have law and administrative regulation in this field. As a matter of fact, some documents of policy and regulation include rich content of the ecological approach (including IEM), but they fail to define in detail of these respects. For instance, the Decision of the Central Committee of the Communist Party of China and the State Council about Speeding up Forestry Development (published on June 25, 2003) is a complex policy and legal document representing the ecological approach thoroughly. The document affirms the standing of "the forest as the main body of land ecosystem," brings forth the guidelines "to establish the sustainable forestry development course by giving priority to ecological construction, to establish national soil ecological security system prevailing by forest vegetation and combination of forest and grassland, construct ecologically civilized society with picturesque landscapes, make great efforts for protection, breeding and sound utilization of forest resources, realize leaping development of the forestry, so as to make the forestry more beneficial to national economy and social progress;" and the document states clearly the basic policy such as "persist in forestry control by law," as well as method, measure and system for one-package forest IEM. Still, this document fails to state application of the ecological approach (including IEM). To strengthen the application and Popularization of the ecological approach (including IEM), the author puts forth the following proposals:

Firstly, the prospective basic law on Environment and Resources, such as Basic Law on National Environment and Resources, should define in detail the application of the ecological approach (including IEM).

Secondly, revision of the Forest Law of the People's Republic of China should define in detail the application of the ecological approach (Including IEM), and transform the IEM guidelines, principles and measures into operable forest rules and systems.

Thirdly, establishment of relevant standard documentation and other policy oriented documents as regards resources and environment, such as laws, administrative rules, local rules, departmental regulations, regulations of local government in the fields of forest, grassland, soil, land water, ocean, wildlife and biological diversity, the ecological methods including IEM method should be applied and popularized to the maximum, as well as implementation in light of the real situation in China.

Fourthly, it's recommended to establish specific IEM legal regulatory documents. Due to the existing ecological management system, it seems impossible to establish a nationally united legal regulatory document applicable to all natural ecosystems. However, it's advisable for relevant national authorities to establish governmental rules and regulations, such as Measures for Forest IEM, Measures for Wetland IEM, Measures for Grassland IEM, Measures for Valley IEM, Measures for Ocean IEM, Measures for Desert IEM, Measures for High Mountain IEM. And the details of these departmental rules and regulations should be established in accordance with the IEM principles, guidelines, guidance and in light of the real situation of diversified ecosystems in China.

CHAPTER V
Climate Change and Land Degradation

38. Brief Introduction on Climate Change Impact and China's Action

Gao Yun

Director, Division of Climate Change, Science & Technology Department, China Meteorological Administration

Abstract: From 1906 to 2005, the average global surface temperature has increased by 0.74 °C. Of the warmest 12 years of the recent 150 years, 11 of those years occurred from 1995 to 2006. In the 20th century, the global sea level has increased by about 0.17 meters, with the yearly rate being 1.8 mm from 1961 to 2003, and 3.1 mm per year from 1993 to 2003. Since the third assessment report of 2001, the forecast on precipitation distribution has been improved. The precipitation in high latitude regions may increase while decreasing in most of the subtropical continental areas. The frequency of high temperature, heat wave and intensive precipitation weather may continuously increase, wind speed of typhoons and hurricanes will increase, precipitation will be more intense, and their destructive forces will be more serious. Furthermore, the flood recurrence interval of "once per thousand years" may decrease to "once per hundred years," and a flood that came once per hundred years may now strike once per fifty years or even shorter. In some regions, extreme unprecedented meteorological events may occur. Economic losses caused by these meteorological disasters have been increasing. The total population suffering from such meteorological disasters is about 600 million and the disaster-affected farmlands can be up to 500 million mu (1 ha = 15 mu) each year. The risks related to safety operation of important projects are increasing. The trend of China's coastal sea level rise will certainly continue unabated. Compared with 2000, the sea level will increase by 0.13 to 0.22 meters by 2050, which will affect the estuary ecosystem and coastal regions economy. Coastal ecosystems like mangroves and coral reefs will degrade. Sea level rise and extreme meteorological events will aggravate oceanic disasters, such as storm surges, red tides, salt sea water intrusion and salinization. As to population centers, large coastal cities like Shanghai and Guangzhou will be directly threatened by sea level rise, which will in turn affect the estuary ecosystem and coastal regions economy. In China, seventy percent of large cities, half of the total population and sixty percent of the national economic production lie in low-lying coastal areas. Thus, the estuary ecosystem and marine biological resources will be severely affected. Finally, this paper describes China's principles, system construction, overall objectives, capacity building and action strategies etc. in counteracting climate changes.

1. China's Climate Change and Future Trends

Climate change refers to climatic changes caused by the change of the earth's atmospheric components due to direct or indirect human activities, except for natural variation in climate observed in a similar period. From 1906 to 2005, the average global surface temperature has increased by 0.74 °C. Of the warmest 12 years of the recent 150 years, 11 of those years occurred from 1995 to 2006. In the 20th century, the global sea level has increased by about 0.17 meters, with the yearly rate being 1.8 mm from 1961 to 2003, and 3.1 mm per year from 1993 to 2003. In most areas of the world, drift snows are melting, especially in spring and summer. The Arctic Sea glacial area is receding and the thickness of sea ice in the spring season is shrinking by 40%. Furthermore, the Northern Hemisphere permafrost active layer is undergoing significant melting.

Under the circumstances of heavy greenhouse gas emissions, the global mean temperature is predicted to rise by 1.1 - 6.4 °C by the end of this century. For situations of low emissions (scenario B1), global temperature is predicted to rise by 1.1 - 2.9 °C, and for situations of high emissions (scenario A1FI), the global temperature is predicted to rise by 2.4 - 6.4 °C. Presently, the estimations on warming and other regional-level characteristics appear to be more credible. The temperature increase in continents and most of the high latitude areas of the Northern Hemisphere is most remarkable. In contrast, the Southern Ocean and North Atlantic Ocean have seen the least amount of temperature increases. Since the third assessment report of 2001, the forecast on precipitation distribution has been improved. The precipitation in high latitude regions may increase while decreasing in most of the subtropical continental areas. The frequency of high temperature, heat wave and intensive precipitation weather may continuously increase, wind speed of typhoons and hurricanes will increase, precipitation will be more intense, and their destructive forces will be more serious. Furthermore, the flood recurrence interval of "once per thousand years" may decrease to "once per hundred years," and a flood that came once per hundred years may now strike once per fifty years or even shorter. In some regions, extreme unprecedented meteorological events may occur.

In the last 100 years, there has been a significant ascending trend in China's observed temperature. Temperature in most areas of China has unabatedly risen in recent fifty years, particularly most evident in northern China. In 2007, China's average temperature was 10.1 °C, which was 1.3 °C higher than the average year and set a historical record. It was the warmest year on record and also the 11th successive year when the temperature was higher than the corresponding period of the average year. On average, the precipitation in the recent 100 years has slightly decreased. However, during the recent fifty years, the "flood south with drought north" phenomenon has frequently occurred in eastern China, while precipitation in western China has increased by 15% to 50%. More interestingly, precipitation in southern China has increased by 5% to 10%, in contrast to a decrease by 10% to 30% in most part of the northern and northeast China. Before 2040, there will not be obvious differences in the warming trends in different regions of China under various scenarios, but the difference will be significant after 2050. Under the low emissions model, the temperature will increase slowly, and it will not exceed 1 - 3 °C. But, temperature will increase 3 — 6 °C when high emissions occur. With regard to different emission scenarios, future precipitation in China tends to increase, but the inter-annual variability of precipitation in different simulation models is considerably high. In 2020, national precipitation will slightly increase, with an increase of 2-5% in 2050, and of 6-14% in 2100.

2. Challenges of Climate Change to China's Sustainable Development

The changes in frequency and intensity of China's extreme meteorological events are quite obvious. High temperatures and heat waves in summer have increased. Especially after 1998, the number of continuous hot weather days above 35 °C has been significantly higher than that of the average year, with the highest number of 9.4 days coming in 2006. However, the frequency of extreme meteorological events has also decreased, cold wave and frost notably. In the recent 50 years, the intensity of tropical cyclones landing on the mainland has increased. Typhoon "Noguri" landed on the Chinese mainland in Hainan province on April 18, 2008, the earliest in the year since 1949. The average date of the first tropical cyclone (typhoon) landing on the Chinese mainland was June 29, and the earliest was May 3 (in 1971). This year, "Noguri" shattered the record by 15 days.

River watershed floods in 2007 for the Huai River was only next to that of 1954, and ten flood storage areas were forced to open for flood diversion. The flood influenced 30.37 million people distributed in Anhui, Jiangsu, and Henan provinces. 1.13 million people were urgently relocated, and sixty-one people died. The flood inundated 3.283 million hectares of crops and damaged 130,000 houses, causing direct economic losses of 17.19 billion RMB. Regional droughts have also become more severe. In northern China, droughts have taken place in 8 of the recent 20 years. The frequency, range and economic losses have all set records since 1886. With global warming, "danger" classified forest fires in arid areas have increased. For example, in spring of 2006, little rain was accompanied by droughts in northern China, and the forest fire danger level was set to high. From May 21 to June 2, 2006, forests caught fire in Heihe of Heilongjiang Province, and Elunchunqi and Yakeshi of Inner Mongolia for attacks by thunder, which was the most severe since 1987.

In China, economic losses caused by meteorological disasters have been rising. Meteorological disasters affect about 600 million people annually and the disaster-affected farmlands reach 500 million mu (1 ha = 15 mu). At the beginning of 2008, a combination of low temperature, rain, snow and frost resulted in about 151.6 billion RMB in economic losses. Global warming may affect agriculture in many areas, causing decreases in agricultural production. Instability of China's agriculture production has been increasing, leading to three important ramifications: firstly, yield fluctuation increases have led to a 5-10% reduction in yield; secondly, changes in layout and configuration in plantation systems and varietals; thirdly, cost and investments have increased, as well as the increasing use of fertilizer, pesticides and herbicides. If proper countermeasures are not taken, there will be a 5% to 10% decrease in plantation production in 2030. Furthermore, conflicts between future supply and demand of water resources will be exacerbated. From 1956 to 2000, the average precipitation in the Yellow, Hai, Liao, and Huai Rivers have decreased by 20 to 120 mm per year, and the runoff has also decreased. Among the above rivers mentioned, the Hai River has decreased the most, by a value of approximately 3.66% each year. Days of summer storm and heavy rain increased in the lower and middle reaches of the Yangtze River, and drought frequency has also increased in north China. The underground water resources in the Hexi Corridor reduced by 45% in 1990s compared with that in 1950s, while water resources fell by 41% in the

Hei River, 56% in the Shiyang River and 42% in the Shule River. The risks related to safety operation of important projects are increasing. Severe precipitation may increase in the upper reaches of the Yangtze River, causing geological hazards like landslides and debris flows, which may threaten the Three Gorges Reservoir and the safety of the dam, and may also have negative impacts on reservoir operation, water storage and hydropower generation. In 2050, the winter's lowest temperature in Qinghai-Tibet Plateau will increase by 3.1 to 3.4 °C, and the summer's highest temperature will increase by 1.8 to 3.2 °C, which will threaten the safe operation of the Qinghai-Tibet Highway and Railway. Frost heaving of seasonal frozen ground and thaw settlement of permanently frozen ground are two outstanding issues in western lines of the South-North Water Transmission Project. The coastal economic developed areas are threatened by sea level rise. For the last 30 years, China's coastal sea level has risen by 0.09 meters, which is slightly higher than the global average. Additionally, disasters like frequent typhoon and storms have caused more serious damages. The trend of China's coastal sea level rise will certainly continue unabated. Compared with 2000, the sea level will increase by 0.13 to 0.22 meters by 2050, which will affect the estuary ecosystem and coastal regions economy. Coastal ecosystems like mangroves and coral reefs will degrade. Sea level rise and extreme meteorological events will aggravate oceanic disasters, such as storm surges, red tides, salt sea water intrusion and salinization. As to population centers, large coastal cities like Shanghai and Guangzhou will be directly threatened by sea level rise, which will in turn affect the estuary ecosystem and coastal regions economy. In China, seventy percent of large cities, half of the total population and sixty percent of the national economic production lie in low-lying coastal areas. Thus, the estuary ecosystem and marine biological resources will be severely affected.

3. Introduction to China's Countermeasures to Climate Change
3.1 Principles of China's Countermeasures to Climate Change

China attaches great importance to counteracting climate changes. The 17th National Congress of CPC advocated that capacity building on counteracting climate changes should be enhanced, and new contributions to global climate protection should be made. China sticks to the following principles in counteracting climate changes:

(1) Counteract climate changes under the sustainable development framework;
(2) Take common but distinguishing responsibilities;
(3) Equally emphasize mitigation and adaptation;
(4) View convention and protocol as the main paths to counteract climate change;
(5) Depend on technology innovation and transfer;
(6) Depend on participation of the entire population and extensive international cooperation;

3.2 Building Institutions to Counteract Climate Change

(1) The Chinese government established a relevant institution on counteracting climate change in 1990;
(2) A group for coordination of countermeasures against climate change was set up in 1998;

(3) The Leading Group on Counteracting Climate Change was set up in 2007, and the premier acts as the group leader who is responsible for formulating important strategies, policies and countermeasures, and also for coordinating and solving big issues in facing climate changes;

(4) The members of the Leading Group on Counteracting Climate Change expanded from 18 to 20 in 2008;

(5) For better scientific decision-making on counteracting climate changes, the Committee of Climate Change Experts was established and has done lots of work to support government's decision-making, enhance international cooperation, and develop civil activities.

4. China's Overall Objectives of 2010 in Counteracting Climate Change

Policies and measures in controlling greenhouse gas emissions come into effect; capabilities enhance in adaptation to climate change; research related to climate change undergoes continuous improvement; new progress made in climate change scientific research; public awareness on climate change raised; and institutions counteracting climate change further strengthened.

Control on Greenhouse Gas Emissions (2010).

• Reduce energy consumption by 20% per unit GNP compared to the level of 2005, and correspondingly mitigate CO_2 emissions;

• Strive for improving renewable energy utilization to reach 10% in primary energy consumption structures and extraction of CBM reaching 10 billion cubic meters;

• Strive for restraining N_2O emissions in industrial processes at the level of 2005;

• Enhance management of animal and fowl waste, liquid and solid waste, enhance utilization of biogas, and control methane emissions;

• Strive to increase forest coverage to 20%, and increase 50 million tons of carbon sinks compared with that of 2005.

5. Capacity Building for Adaptation to Climate Change

(1) To develop monitoring, early-warning and emergency-reaction mechanisms for multi-disasters, decision-making & coordination mechanisms crossing multiple agencies, operation mechanisms with participation by the general public, and to enhance the abilities of monitoring and forecasting meteorological hazards;

(2) To improve infrastructures for rural farmlands, adjust crop systems, screen and breed resistance varieties, and develop biotechnology;

(3) To protect natural forest resources, supervise nature reserves, construct key ecological conservation projects, and set up important eco-function areas to promote natural eco-restoration;

(4) To rationally exploit and optimize distribution of water resources, perfect basic construction mechanisms for farmland water conservancy, and strengthen measures on water-saving and hydrology monitoring;

(5) To enhance the scientific monitoring of sea level change and the supervision of ocean and coastal zone ecosystems.

6. China's Policies and Actions to Actively Mitigate Climate Change
(1) To adjust economic structures and transform developing models;
(2) To save resources and improve resource-use efficiency;
(3) To develop renewable energy and optimize energy structures;
(4) To develop cycling economy and reduce greenhouse gas emissions;
(5) To decrease greenhouse gas emissions of agriculture and rural areas;
(6) To advance reforestation and enhance the capacity of carbon sinks;
(7) To improve research and development and counteract climate changes scientifically;

China has improved its capacity building to monitor and broadcast early warnings for extreme meteorological events, set up corresponding emergency mechanisms for meteorologically derived, as well as secondary, disasters, made great progress in prevention of intensive typhoons, regional storms and floods, and established a comprehensive observation system on climate and climate change.

China has already made achievements through the active implementation of policies and actions in adaptation to climate change, in agriculture, forest and other natural ecosystems, water resources, and coastal and seashore eco-fragile areas.

China is in a key period of building a well-off society as well as in the period of accelerating industrialization and urbanization. Developing the economy and improving people's living standards is a huge task. Therefore, China faces more serious challenges than developed countries in counteracting climate changes.

Climate change is a challenge to the whole international society. Solving climate change issues requires cooperation of all countries in the world and international societies. China is willing to make efforts to cooperate with foreign countries to realize global sustainable development and contribute to protect our world's common climate system.

39. IEM and the Future Opportunities for Growth in China from 2010 onwards; the Ultra Green approach.

Ian R. Swingland
Deputy Chairman, Ultra Green Group, Singapore

Abstract: The Commission of the European Communities has recommended the exclusion of forest carbon credits from the EU-ETS in its next phase. It argues that REDD (reducing emissions from deforestation and ecosystem degradation i.e. avoided deforestation) credits cannot be used reliably because they do not demonstrably represent real, verifiable, additional and permanent reductions in emissions. The Commission argues also that REDD credits would, if allowed, flood the European market, deterring real and permanent improvements in the production and energy infrastructure of the EU.

* Emissions from the loss and degradation of forests, let alone those from conventional agriculture in the developing world are enormous in scale and impact, representing nearly 20% of total greenhouse gas emissions attributable to human activities worldwide. This is more than the worldwide emissions from burning natural gas and the emissions of the entire worldwide transportation sector. Moreover, their impact over the next five years (2008-2012) will easily offset whatever gains are achieved by industrialized countries in that same period under the Kyoto Protocol.

* The loss of forests is occurring at a pace that requires urgent action. At current rates, the environmental services of the world's major forests will collapse long before the last tree has been cut or the last hectare cleared. More than a billion of our fellow human beings are forest-dependent peoples, and the loss of forests is to them an irretrievable catastrophe. What is more, the climate impacts of forest loss go well beyond the global warming effect of higher concentrations of greenhouse gases. (For example, if forests continue to be lost then rainfall patterns, hydrological cycles and soil productivity will be affected in countries that are now major suppliers of rice, grain, sugar, beef and other essential food supplies to the rest of the world.)

* Curbing deforestation is a highly cost-effective way of reducing greenhouse gas emissions. Yet there is currently no mechanism that would compensate countries for the opportunity cost of not deforesting. The World Bank has specifically found that the lack of markets for the national and global environmental services offered by forests has contributed to high rates of deforestation in developing countries.

* We have the scientific and technical tools today to measure and monitor reductions in emissions from deforestation. We know enough to establish historical reference scenarios: since the early 1990s, changes in forest area in developing countries have been measured from space with confidence. Our ability to estimate carbon stocks in particular forests has also improved greatly over the last 10 years. We have designed conservative methods that can be used to ensure that we minimize the risk of over or underestimation of carbon stocks to within a margin of error of +/- 5%. And new technologies

and approaches are being developed that will further reduce uncertainties. The technical challenges for monitoring, verifying and quantifying REDD therefore have been and will continue to be addressed, so that markets can now operate with integrity. Further investment is needed to make these tools readily available to poor countries, but there first needs to be an economic incentive for doing so at the required scale.

* The proposed REDD mechanisms (as foreshadowed by the REDD decision adopted in Bali) will address the problems of leakage and permanence that have plagued the discussion of crediting to date:

** Reductions in emissions from deforestation, if measured relative to a national reference scenario (or as close to it as possible), are by definition net of any in-country leakage, which is the only kind that is normally considered for purposes of the UNFCCC; and

** There is nothing inherently impermanent or "temporary" about REDD, so long as the actual reductions in rates of deforestation are real and the countries involved are held to a reference scenario that requires the long-term conservation of forests as a condition of earning those credits in the first place (i.e., a reference scenario that may well be more restrictive of emissions than a business-as-usual projection).

These issues are not technical or methodological and provide no rationale for excluding REDD from the EU or any other market system.

* There is no empirical support for the "floodgates" argument. Anyone who predicts that REDD credits will quickly overwhelm the European carbon markets greatly underestimates the challenges ahead for developing countries. Major national institutional frameworks are required, readiness mechanisms must be developed, and policies and measures effectively implemented on the ground. Moreover, the UNFCCC Parties have agreed that the "rules of the game" will be negotiated before the reduction targets are set, so the targets will reflect whatever cost-control or other flexibility mechanisms are agreed upon. These will almost certainly provide that only a small proportion, based on historic and predicted deforestation rates, of potential credits will be available in any one year.

In any case, the EU-ETS can simply cap the inclusion of forest carbon, and REDD in particular, to a specific annual volume (or a percentage of the reduction commitments of affected operators) as is being proposed in the most advanced US legislation. This is entirely within the control of the European Parliament and the Council of Ministers.

The active participation of developing countries in an eventual global climate change regime, consistent with the principle of common but differentiated responsibilities, is essential to achieving the ultimate objective of the UNFCCC and has long been the policy of the EU. The exclusion of forest carbon credits, and specifically the failure to even preview the possible inclusion of REDD in the EU-ETS, sends precisely the wrong message. It is critical that every incentive be created now to motivate institutional reforms in developing countries to control and abate deforestation. If we lose the world's forests, we will have lost the fight against climate change.

The IEM approach coupled with Sustainable Forestry Management and the Ultra Green portfolio of technologies and teams of world leading scientists can work with China in growing its position in the carbon sequestration, emissions reduction and global energy market. In partnership, with China's exceptional vision, we can work together to further position China as the world's leader in sustainable development. We have technologies ready to implement now, after years of research, which can deliver our shared vision of global sources of energy being gradually moved to renewable and sustainable sources, harnessing the power of the sun and water while improving the living standards of the population.'

Introduction

If there was unanimity on one issue in Bali November 2007 at COP, it is that tropical forests must play a central and vital role in any realistic effort to mitigate climate change. Another strong message from Bali is that payments for reductions of tropical deforestation and degradation are the key to the developing world's consent to a post-2012 treaty. Bali highlighted that virtually the only way in which most developing nations, and particularly the least-developed ones, can meaningfully participate in and benefit from the carbon markets is through land use change and forestry (LULUCF).[1] The promised benefits of the CDM have, however, largely bypassed almost all tropical and subtropical countries due to its stifling approach to LULUCF.[2] Finally, the important co-benefits of encouraging the conservation and restoration of tropical and sub-tropical forests were recognised, including protecting biodiversity and fresh water sources, and providing the best means of adaptation to climate change for many of the world's most vulnerable people.[3]

After years of controversy, much of it unjustified in my opinion, the essential ecosystem services provided by forestry in mitigating climate change were recognised as a result of the efforts of the Coalition of Rainforest Nations beginning in Montreal at COP 11,[4] the cumulative scientific work of the IPCC and the definitive economics of the Stern Review. Among other things, the Bali decision highlights the enormous damage imposed on the world's most important ecosystems and most vulnerable people by the EU ETS ban on forest carbon credits. The EU ETS is the world's largest operating carbon market[5] and thus the most important, near-term, potential source of the large-scale, long-term investment required to slow and reverse deforestation. Yet it has been made inaccessible for this purpose and to the people most dependent on forests for their survival. This inaccessibility has compounded the market failure caused by the CDM (as to which see below).

The EU ban on LULUCF credits, like the EU policy on biofuels, actually incentivises deforestation and the conversion of rainforest to agricultural use. This generates manifold perverse environmental consequences which are becoming more apparent by the

---

[1] See Decisions COP 13, "Bali Action Plan," and "Reducing emissions from deforestation in developing countries: approaches to stimulate action."
[2] See http://cdm.unfccc.int/Statistics/index.html, passim
[3] op. cit. Decisions COP 13
[4] http://www.rainforestcoalition.org/eng/
[5] Point Carbon, *Global carbon market value grew 80% to €40 billion in 2007* (18 January 2008).

day. Viewed objectively and from the point of view of developing nations, EU environmental policy is clearly designed to encourage continued tropical deforestation. The European Commission, lead by DG Environment, has now adopted two policies, the continued ban of LULUCF credits and its policy on bio-fuels, which not only fail to reflect the Bali consensus, but directly contradict the EU's expressed policies on climate, sustainable development, biodiversity and poverty alleviation. One need not be a cynic to see this as hypocritical. The Commission's recent proposal to amend the legislation governing the EU ETS,[6] in which it proposes a continuing ban on forestry credits beyond 2012, together with a radical curtailment of all other project credits, is simply unjustifiable and explicable only by European myopia and a wilful disregard of the evidence.[7]

There are repeated references in Commission documents to the need for Europe to lead the world in emissions trading; to set an example. This is somewhat reminiscent of the last French administration's promotion of the French economic model; a "model" which no other economy sought to imitate. A continuing EU ETS ban on forestry combined with a phasing out of other carbon credits from the developing world, will not only discourage the participation of the developing world in a post 2012 treaty but it will, in fact, jeopardise the leading position the EU has achieved in the carbon trading markets. All other emerging carbon markets, both compliance and voluntary, in the US, Australia, New Zealand, and the post Kyoto arrangements, propose, or already include, forest carbon credits. The Commission's position therefore not only runs against the tide of the rest of the world, but will also render the EU ETS incompatible with all other trading schemes, directly in contradiction with its expressed desire to link with other schemes.[8] The "linking directive" has, in effect, become the "separation directive."

A further perverse effect of banning forest carbon credits will be to make European heavy industry uncompetitive to the point that, as DG Environment admits, there will be "carbon leakage" of these industries which will be forced, by the manipulated price of EU credits, to relocate to lower, or no, emissions cost countries.[9] The potential impact on employment has recently been acknowledged by leading Member States including France and Germany.[10] Recently floated remedies for this include creating tariff barriers against imported products which do not pay the cost.[11] This is not just the economics of the 1930's that lead to a worldwide depression, such a policy will lead to no net gain to

---

[6] See Commission for the European Communities, "Proposal for a Directive of the European Parliament and of the Council amending Directive 2003/87/EC so as to improve and extend the greenhouse gas emission allowance trading system of the Community," January 23, 2008

[7] Attached as Schedule 1 is a rebuttal to the DG Environment's various rationales for the ban's continuance.

[8] See Official Journal of the European Union, "DIRECTIVE 2004/101/EC OF THE EUROPEAN PARLIAMENT AND OF THE COUNCIL of 27 October 2004 amending Directive 2003/87/EC establishing a scheme for greenhouse gas emission allowance trading within the Community, in respect of the Kyoto Protocol's project mechanisms."

[9] *op. cit.* Commission for the European Communities, January 23, 2008

[10] See "France, Germany Warn EU Climate Plan Risks Jobs," at
http://www.planetark.org/dailynewsstory.cfm?newsid=47167

[11] EU split over plan to levy import tax on polluters, January 8th, 2008. www.timesonline.co.uk

the atmosphere. The losses to the tax base, the inevitable increase in unemployment and associated welfare costs will be exacerbated, in climate terms, by the inevitable loss from the EU of those companies whose management is the most capable and the most willing to focus their efforts on innovation in the very industrial sectors where it is most needed.

Finally, the ban not only contradicts the EU's policies on poverty alleviation, sustainable development and biodiversity, it actually prevents the achievement of the EU's stated goal of climate stabilisation by mid-century. The forecast EU carbon price which will result from the Commission's proposal is as high as €48/tCO2e.[12] This is close to the price used by McKinsey in its analysis of scenarios leading to climate stabilization by 2030.[13] That analysis demonstrates that offsets from the forestry sector, particularly tropical and sub-tropical forestry, must account for a larger share of potential reduction abatement (25%) than any other sector over that time frame to achieve that goal.[14]

The Commission, in its rationale, suggests that its position might change if a successor international treaty is ultimately agreed.[15] In the meantime, it proposes to phase out virtually all project credits, thus severely damaging the market and market participants by artificially reducing demand for CERs.[16] The resulting chilling effect on the market and on investment in emissions abatement everywhere outside of Europe is already apparent. The Commission proposals are already making it clear to those who invest and develop climate change projects outside of Europe that no reliance whatever can be placed on European demand for carbon credits. The Commission also seeks the power to prevent Member States from meeting their Kyoto obligations with credits, such as CERs, which are excluded from the EU ETS and to determine itself whether or not any such credits will be admitted to the EU ETS (or used by Member States) in future whether or not a successor treaty is entered into.[17] The result, of course, will be to raise the price of compliance in the EU and to EU Member States and to reduce the price of compliance to other Annex 1 countries and their industries.

As the Prime Minister and GLOBE 8 have recently suggested, it is time to take regulation of the carbon markets out of the hands of DG Environment and the United Nations and put it in the hands of a regulator with the expertise in financial markets required for an undertaking of this size and importance.[18] If further evidence were required that this

---

[12] See, "Fortis Raises CO2 Forecast, Prices Surge," at http://www.reuters.com/article/rbssFinancialServicesAndRealEstateNews/idUSL2564255920080125

[13] A key difference is that Mckinsey uses the price as a marginal cost whereas the EU Commission sees it as a base from which prices are intended to be manipulated upwards. (cites).

[14] McKinsey & Company, "A Cost Curve for Greenhouse Gas Reductions" *The McKinsey Quarterly*, Volume 1 (2007).

[15] *op. cit.* Commission for the European Communities

[16] See "Trade in Agcert shares suspected, company seeks government protection," Point Carbon 21, February 2008

[17] *op. cit.* Commission for the European Communities

[18] See "UK PM calls for European bank to allocate EUAs post-2012," Point Carbon, 21 February 2008 and "Carbon market control should be taken off UN hands, legislators say," Point Carbon, 25 February 2008

war is too important to be left in the hands of those whose experience is elsewhere, the recent DEFRA proposals to limit "voluntary offset" providers to supplying CERs, which are full compliance instruments, provides it.[19] Why DEFRA thinks it makes any sense for anyone to pay the additional and redundant regulatory costs to it (for its "quality mark" approval) after all of the costs incurred to get CDM approval and then to sell the resulting CERs to voluntary purchasers, as opposed to mandatory purchasers, which are bound to pay a higher price, is beyond most rational observers.[20]

The overwhelming majority of respondents to the clearly perfunctory DEFRA consultation said that the UK voluntary market should include forest credits and credits from other projects which could either not afford CDM costs or did not come within its rules such as avoided deforestation; this has simply been disregarded by DEFRA.[21] To then demand that industry provide a single voluntary standard and prove its efficacy - by essentially creating replica CERs - is disingenuous at best. DEFRA knew that robust standards pre-existed the consultation (Chicago Climate Exchange, WWF Gold Standard and the Climate, Community and Biodiversity Alliance Standards) and that a further comprehensive standard was promulgated, with widespread industry support, during the consultative period (the Voluntary Carbon Standard). Its conclusion that none of them, despite years of wide consultation with all categories of stakeholders, are adequate to protect the UK public is arrogant to say the least. The DEFRA standards amount to no more than a reiteration of CDM rules; this is not "leadership"; it is a retrograde step indicative of a failure to understand or accept market demand and market evolution. This rigid, prescriptive, as opposed to flexible, principles-based, approach to regulation has come to typify the regulatory approach in Europe and under the CDM. It is also an approach which is failing and is destined to fail.

What seems to have been lost on these regulatory bodies, or indeed seen by them as a negative impact, is that the reason that market-based approaches were adopted to deal with climate change is to lower the cost of achieving mankind's climate goals. The persistent effort by regulations such as the EU's explicit ban and the CDM Executive Board's implicit ban, on LULUCF credits, is to rig the market to achieve a higher price than necessary. Whether the result of confusing the role of markets (which is to seek the most efficient use of capital) with that of subsidies (which is to replace profits in early stage innovation) or the result of what one informed observer has called "magical thinking", the clear intention is to command a sudden and radical change in the economy. There have been several such efforts in the past, China's "Great Leap Forward" comes

---

[19] See written and oral submissions to this Committee by SFM on voluntary markets: http://www.publications.parliament.uk/pa/cm200607/cmselect/cmenvaud/331/33102.htm

[20] Even in respect to CERs, which must cross a multitude of hurdles before approval, DEFRA imposes a further layer of regulation prior to distribution to the UK public. See "Draft Code of Best Practice for Carbon Offset Providers Accreditation requirements and procedures – February 2008," at www.defra.gov.uk

[21] Summary of responses to the consultation on establishing a Code of Best Practice for selling offsetting to consumers http://www.defra.gov.uk/environment/climatechange/uk/carbonoffset/pdf/cop-summary¬responses.pdf

to mind and the results have been uniformly destructive. The EU Commission's current effort to rig the price of carbon to administer shock treatment to the European economy is no more likely to succeed. It will certainly do nothing to mitigate climate change as no other country will follow suit in seeking higher rather than lower cost (i.e. more efficient) solutions to climate change .[22]

In the context of global warming a "fortress Europe" approach is futile and counterproductive. The suggestion made by some, that this is just a bargaining position to force the US and China into a successor regime, may sound clever but is, in my view, "too clever by half". It is virtually inconceivable that the United States will accede to UN regulation of its carbon market particularly given the UN's performance as a regulator thus far. It is also unlikely that China will adopt regulations which mandate high carbon prices in an economy at its stage of development. It is unlikely in the extreme that those developing nations which remain dependent on agriculture and forestry will regard being cut off from investment and the carbon market for another 5-10 years as being a signal that they will one day benefit. Their disappointment in the decade of non-delivery by the CDM was palpable at Bali. Given the economic logic of the EU Commission's proposed exclusion of virtually all project credits for the indefinite future, even if a future treaty is negotiated, the likely response of developing nations is obvious: continued response to real, as opposed to imaginary, market signals. Increased timber harvest of native forests, conversion of forests to agriculture and increased production of commodities with real demand (palm oil, beef, soya and sugar) are inevitable; precisely what we all want to avoid.

Finally the suggestion that revenues from the auctioning of permits will be earmarked for various project sectors is, to be polite, disingenuous at best. Finance ministers have already made it clear that they will not accept hypothecation of auction proceeds and even if they did there is no structure for them to allocate such revenues .[23] Projects supported by aid will not suffice. Leaving aside the sorry history of such attempts, the billions of dollars which are required annually and for decades to come to deal just with deforestation will not be forthcoming from the public sector. The private sector, the only realistic source of such funding, will only make large long term investments if it is confident in the long-term stability and predictability of the markets for carbon credits. The persistence of political and regulatory interference, particularly with supply and demand (and hence prices) will only delay and in many cases prevent, such investment. Participants in the carbon market, which is centred in London, are already reconsidering its viability in the face of the EU Commission's proposals and the manifest inefficiencies of the CDM. Once a market collapses it is hard and often impossible to resurrect. Given

---

[22] By way of comparison, pending US Federal legislation proposes to limit carbon prices at US$12 per tonne (Bingaman-Specter, Low Carbon Economy Act, available at www.pewclimate.org) and prices for the Australian system are projected at up to A$20 per tonne.
[23] Council of the European Union, EcoFin 33, Env 51, "Report on the efficiency of economic instruments for energy and climate change," Brussels, 5 February 2008

that the Commission proposes to exclude project-based carbon credits for years to come it is all too likely that other trading centres, such as New York, will take its place.[24]

Solutions

How might mechanisms to tackle emissions from deforestation be developed? How can we ensure that such mechanisms contribute to wider sustainable development aims? Will such mechanism deal with the need to ensure the protection of indigenous people, land use rights and governance? How might forest degradation be dealt with? Are additional mechanisms required to enable the creation of carbon sinks?

The Mechanism

The mechanism for dealing with emissions from deforestation and forest degradation already, at least in principle, exist: the carbon trading market. No new mechanism is required although radically improved regulation is necessary. If forest carbon is made freely available for compliance purposes, as well as in the voluntary sector, of the carbon marketplace the critical problem of tropical forest deforestation and degradation will be successfully addressed and emissions from that source radically reduced. All that is required to accomplish this is the setting of long term targets for emissions reductions and structural change in the regulation of the market as recently proposed by GLOBE 8.[25]

Sustainability

Deforestation and forest degradation is fundamentally about land use. The land will be used for something and the question is what affects the choice of use. Today, in the absence of any market value for forest carbon, the land is and will continue to be used to produce timber and agricultural products. These products, whether on the scale of agri-business or subsistence, have a real value to the landowners, whether public, communal or private. Unless the landowners are offered a price for an alternative land use, such as carbon storage and sequestration, which is at least equal to or higher than what they get for timber, soya, palm oil, beef, sugar or maize, what economists call the "opportunity cost", deforestation and forest degradation will continue. The demand for these products will not abate given rising population and living standards; the question is how to direct such production into sustainability. But there will be no sustainability unless there is long-term commercial sustainability. The moment that payments for carbon storage cease conversion of forest land with its attendant emissions will resume.

The requirement, therefore, is for continuous, predictable payments of the relevant opportunity cost for each forest area for decades to come. No one has yet proposed anything that can be seriously considered capable of accomplishing this aside from crediting carbon stored or sequestered in biomass for use in the compliance markets.

---

[24] See NYMEX Green Exchange http://www.greenfutures.com/
[25] See "Carbon market control should be taken off UN hands, legislators say," Point Carbon, 25 February 2008

The failure to do so for the last 15 years is a tragedy and has compounded the difficulty mankind now faces in dealing with climate change in the limited time left before it goes beyond our control. Fortunately, considerable work has been done and continues to be done to assess the opportunity cost of carbon storage in tropical and sub-tropical forests and it is affordable; that is, it can be done at a price equal to or lower than the cost of technological sources of emissions reduction.[26] The only barrier to these payments being made to landowners and incentivising change in land use is the regulatory structure thus far imposed on the carbon markets particularly by the EU and the CDM.

The opportunity cost of reduced deforestation, was the basis of a study carried out for the Stern Review. This estimated the opportunity cost for eight countries that collectively are responsible for 70% of land-use emissions. If deforestation in these countries were to be reduced by 50%, the opportunity cost would amount to at least $5-10 billion annually (approximately $1-2/tCO$_2$ on average).[27] Although there are various proposals for public sector funding, donor governments and agencies show little sign of being able to contribute funding necessary at that level for the decades required.[28]

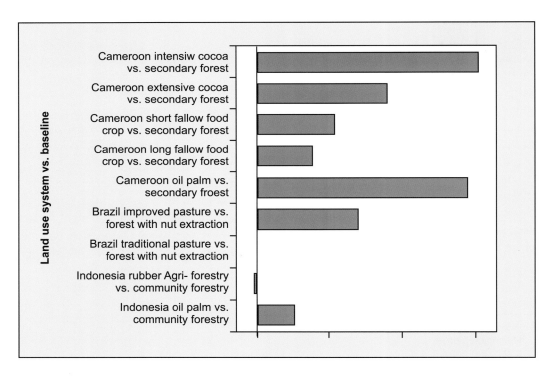

---

[26] Stern, N, 2006, Stern Review: The Economics of Climate and Nepstad et al, 2007, The costs and benefits of reducing carbon emissions from deforestation and forest degradation in the Brazilian Amazon.
[27] Stern, N, 2006, Stern Review: The Economics of Climate
[28] Castro, G. and I. Locker. 2000. Mapping Conservation Investments: An Assessment of Biodiversity Funding in Latin America and the Caribbean. Washington, D.C.: Biodiversity Support Program.

Carbon credits from avoided deforestation allow a real commercial alternative value to be placed on tropical forests if they are integrated into carbon credit trading systems in a fungible and transparent manner. The carbon market can in many cases "tip" the balance of economics in favour of forest conservation. According to the World Bank's most recent study of the subject, tropical forest cleared to pasture is worth between $200-500 per hectare. Based on its average CO_2 storage per hectare of 500 tonnes, its value is between $1500-10,000 per hectare ($3-20/$tCO_2$).[29] Even at the low end range of carbon prices continued deforestation would become unprofitable in many land systems if it is credited in the carbon markets.[30]

The international community, prompted by a proposal from 15 developing countries,[31] now understands this and has been in dialogue over the inclusion of emission reductions from avoided deforestation and degradation in the post-2012 regime for the last two years. At COP 13 in Bali the parties resolved in their decision on '*Reducing emissions from deforestation in developing countries: approaches to stimulate action*' a number of measures to assist in evaluation and implementation of mechanisms to tackle emissions from this source.[32] In relation to specific mechanisms it was resolved 'that policy approaches and positive incentives on issues relating to reducing emissions from deforestation and forest degradation in developing countries will be considered in the context of land use....[33] The role of the carbon market in this effort was recognized in the Bali Roadmap where it is stated that '*various approaches, including opportunities for using markets, to enhance the cost-effectiveness of, and to promote, mitigation actions, bearing in mind different circumstances of developed and developing countries.*'[34] Forests, in short, are finally being recognized as a critical, "cost-effective" key to a future climate change treaty as well as being important for the other environmental services they provide[35] including services which are also essential to adaptation to climate change by the rural poor.[36] It is also now acknowledged that the carbon market is the most appropriate vehicle to deliver these policy goals.

The 26 developing countries in the Coalition of Rainforest Nations[37] have made it clear that either they receive compensation for the carbon sequestration services which their native forests provide to the world or they must continue to exploit them as

---

[29] Chomitz, K, 2007, At Loggerheads? Agricultural Expansion, Poverty Reduction and Environment in the Tropical Forests, The World Bank
[30] *op. cit.* Chomitz
[31] UNFCCC. 2005/CP/L2, "Reducing Emissions from Deforestation in Developing Countries: Approaches To Stimulate Action". 06 December 2005.
[32] See "Decision -/CP.13 Reducing emissions from deforestation in developing countries: approaches to stimulate action"
[33] See "Decision -/CP.13 Reducing emissions from deforestation in developing countries: approaches to stimulate action"
[34] Bali Action Plan http://unfccc.int/documentation/decisions/items/3597.php
[35] Point Carbon, "Japan to make forest management a priority at next G8 summit," 21 February 2008
[36] Adaptation to climate change in agriculture, forestry and fisheries: perspective, framework and priorities, FAO, Rome, 2007
[37] http://www.rainforestcoalition.org/eng/

sources of energy, food and wood products.[38] The implications of the latter are illustrated in that Indonesia is now the third largest emitter of greenhouse gases in the world, almost entirely the result of continued deforestation, and Brazil is fourth for largely the same reason.[39] It is only through linking emission reductions from avoided deforestation and degradation to the carbon market, and in so doing linking forests with the world financial markets, that sufficient capital will be available to ensure that a substantial reduction in deforestation globally and a shift to sustainable sources of supply actually occurs in the time remaining to us.

It is also critically important to understand that the payments to landowners, while necessary, are not sufficient to sustainability. New sustainably managed sources of timber and forest product supply must be created to meet ever increasing demand. This too was acknowledged at Bali.[40] . These too require annual investments measured in the billions of dollars. The continued discouragement of such investment in the developing world by the EU ban on forest credits and the CDM rules on afforestation and reforestation compound the difficulty of reaching the goal of sustainable supply and sustainable development. We cannot just reduce or end the harvest of native forests without creating such alternative supplies; in Kyoto-speak: we cannot have REDD without A/R.

Methods

The methods of linking the land with the capital markets are now well developed and understood. The SBSTA workshop (Cairns, 2007) concluded that *'there is general agreement that methods, tools and data are robust enough to estimate emissions with an acceptable level of certainty and that the IPCC Good Practice Guidance for LULUCF and the 2006 IPCC Guidelines provide a good basis for the estimation of emissions from deforestation and their reductions.'* [41]

Since the early 1990s, changes in forest area have been able to be measured by satellite with confidence .[42] Analysis of remotely sensed data from aircraft and satellites supported by ground based observations is now well developed at the national level.[43] Some developing countries have national level monitoring initiatives in place for the land use sector such as Brazil[44] and India[45]. Other countries are developing these

---

[38] Stilts, Joseph, "Cleaning Up Economic Growth," Project Syndicate, 2005
[39] See Wetlands International: http://www.wetlands.org/ckpp/publication.aspx?ID=1f64f9b5-debc-43f5-8c79- b1280f0d4b9a
[40] Decision -/CP.13 Bali Action Plan
[41] Second workshop on reducing emissions from deforestation in developing countries, Cairns, Australia. March 7-9 2007. Preliminary Chairs' summary
[42] DeFries, R, 2002, Carbon Emissions from tropical deforestation and regrowth based on satellite observations for the 1980s and 1990s, Proceedings of the National Academy of Sciences of the United States of America, Vol.99, No.22 p 14256-14261
[43] Herold, M et al, 2006. Report of the workshop on Monitoring Tropical Deforestation for Compensated Reductions, GOFC-GOLD Symposium on Forest and Land Cover Observations, Jena, Germany, 21-22nd March 2006, GOFC-Gold report series, http://www.fao.org/gtos/gofc-gold/series.html
[44] INPE. 2005. Monitoramento da Floresta Amazonia Brasileira por Satelite, Projeto PRODES
[45] Forest Survey of India.2004. State of Forest Report 2003. Dehra Dun, India

capabilities or have successfully monitored forests with aerial photographs that do not require sophisticated data analysis or computer resources. A variety of methods that are applicable to varying national circumstances regarding forest characteristics, cost constraints, and scientific capabilities are available and adequate for monitoring deforested areas and verifying the accuracy of such measurements. Additionally, the historical remote sensing database is sufficient to develop a baseline of tropical deforestation in the 1990's.[46]

Based on current capabilities, GHG emissions from deforestation can be accurately estimated. These estimates come from changes in the carbon stocks in the aboveground biomass of trees and from other forest carbon pools using models and default data in the IPCC Good Practice Guidelines report.[47] Forest inventories can provide biomass values according to forest type and use, such as mature forest, intensely logged, selectively logged, fallow etc. Many developing countries do not have sufficient data from national forest inventories and they should be assisted in developing this information and the related administrative systems [48]. However, even in their absence, the FAO database provides a default value for national carbon stock with stratification into main ecological zones[49]. Compilation of data from ecological or other permanent sample plots can provide estimates of carbon stocks for different forest types subject to the design of site specific scientific studies.

There are a variety of approaches and potential mechanisms to crediting avoided deforestation and degradation activities which reflect the variety of historical experiences of the countries which are the intended beneficiaries. A flexible combination of approaches is the best way forward whether based on mandatory or non-mandatory emissions targets being adopted by such countries. A sectoral approach based on national boundaries and national administration established under broad principle-based regulation, as opposed to the current CDM approach of prescriptive, project by project assessment, is required both because of its simplicity and its respect for sovereignty and because it eliminates many of the methodological problems, such as leakage and additionality, which have plagued development of the CDM market thus far. National forest sector emission targets adopted by developing countries are by far the most efficient way of encouraging sustainable forest management and reducing deforestation. This approach lays the foundation for the capital markets to operate in a coherent regulatory environment.

---

[46] DeFries, R, et al, 2005, Monitoring tropical deforestation for emerging carbon markets, Tropical Deforestation and Climate Change/ edited by Paulo Moutinho and Stephan Schwartzman, IPAM and Environmental Defense.
[47] Penman, J et al, 2003, Good Practice Guidance for Land Use, Land-Use Change and Forestry. IPCC National Greenhouse Gas Inventories Programme and Institute for Global Environmental Strategies, Kanagawa, Japan.
[48] Second workshop on reducing emissions from deforestation in developing countries, Cairns, Australia. March 7-9 2007. Preliminary Chairs' summary
[49] FAO. 2006. Global Forest Resources Assessment 2000. FAO Forestry paper 147. Food and Agriculture Organization of the UN, Rome.

Indigenous Peoples, Land Use Rights and Governance

The Bali decision on '*Reducing emissions from deforestation in developing countries: approaches to stimulate action*' specifically recognizes '*that the needs of local and indigenous communities should be addressed when action is taken to reduce emissions from deforestation and forest degradation in developing countries.*' One of the key aspects of markets that seem to elude most participants in the debate over forest carbon credits is their requirement that land ownership and entitlement to land use rights are clearly established. Unless the buyer is confident that the carbon credits it purchases have been lawfully created and transferred he will not buy them. Exchanges and clearing houses are specifically designed to ensure that "good delivery" takes place in the commodities which they list and trade. Purchasers of carbon credits in particular, exposed as they are to public and regulatory oversight, do not want to find that the credits they buy are sourced through abuses of human rights or from stolen land. It is critical to all markets that ownership be clearly established and the market for land use rights, like carbon storage and sequestration is no exception.

Markets quickly punish and exclude those who fail to deliver what was bargained for. The carbon markets have appropriate standards for "good delivery" of forest carbon; these standards have already been developed and are being adopted by exchanges, in over-the counter transactions and in the voluntary sector.[50] By the same token good delivery requires lawful delivery requiring compliance with both domestic and international law such as ILO 169 .[51] In short, good governance is integral to well-regulated markets and is demanded by them. The opportunity to benefit from the payments such markets offer is itself a powerful incentive to improved governance including acceptable standards of land tenure and registration and the adoption of the safeguards for vulnerable communities required by long term investors of capital.

Forest Degradation

Although carbon emissions from forest degradation may not be as large per unit area as the complete removal of forest through deforestation, forest degradation occurs over large areas and can contribute significantly to overall emissions from forest loss.[52] Differences between forest and degraded forest are more subtle than in the case of deforestation, and degradation patches are generally small compared with clearings. Monitoring degradation is technically more difficult than monitoring deforestation but is now practicable. Techniques have been developed and are being steadily improved. A team at the Carnegie Institution of Washington, for example, has developed techniques for automated remote-sensing analysis of selective logging utilising Landsat satellite imag-

---

[50] Chicago Climate Exchange (CCX) http://www.chicagoclimateexchange.com/, NYMEX http://www.greenfutures.com/, Climate, Community and Biodiversity Appliance, CCBA: http://www.climate-standards.org/, Voluntary Carbon Standard: VCS: http://www.v-c-s.org
[51] Convention (No. 169) concerning Indigenous and Tribal Peoples in Independent Countries http://www.unhchr.ch/html/menu3/b/62.htm
[52] Asner et al, 2005, Selective logging in the Brazilian Amazon. Science 310: 480-482

ery combined with extensive fieldwork. Their work highlighted that exclusion of selective logging from a monitoring system would miss a substantial source of emitted carbon and that such activity can be monitored remotely.[53] Forest degradation therefore must and can be included in the calculation of biomass subject to emissions reduction targets and for crediting in the carbon marketplace.

Our recommendation is that countries that have the capacity and funding to measure and monitor degradation should be encouraged to do so by the opportunity to sell the carbon credits generated from reduced forest degradation in the same manner as those generated from avoided deforestation activities. We recommend that countries be able to choose the level of carbon accounting for their country (with periodic review). This would allow

countries which have the funding and capacity to generate emissions reductions from reduced degradation to generate carbon credits immediately while those countries who have not reached a stage to implement this level of technological capacity are assisted to do so.

Additional Mechanisms for Carbon Sinks

As will be apparent from the remarks above, some developing countries, although by no means all, lack the infrastructure for full realisation of the markets' potential. The key is to provide them with the capacity to reap those benefits. Some need assistance to measure, monitor and verify their carbon stocks. Some have weak systems of land tenure and registration; others require assistance with administration and public accountability. Some need assistance with law enforcement. These are essential elements to a functioning market. It is here, in capacity building, that the public sector and multinational institutions have a key role to play and a role in which they have expertise and institutionalised experience. If the available resources from the public sector were utilised for this purpose carbon sinks everywhere can be fully valued by the market for the benefit of all concerned. If there is a role for a "fund" (which was the original intention behind the creation of the CDM) this is surely it.

Are the Clean Development and Joint Implementation Mechanisms functioning effectively? How might they be improved? How might they better be used in relation to forestry or other land use emission reduction projects? Should CDM and JI projects play a greater role in sustainable development more widely? To what extent should credits such as those from the CDM and JI be permitted to be used in emissions trading schemes, or contribute to emissions reductions targets?

It is now evident that insofar as land use and forestry is concerned both the CDM and JI have failed to make any contribution. This was, in many ways, failure by design. Despite the fact that all forestry and land use is accounted for in Annex 1 countries, at the outset of the debate over the role of developing countries, the CDM excluded

---

[53] Asner et al, 2005, Selective logging in the Brazilian Amazon. Science 310: 480-482

deforestation entirely.[54] Then, in 2001 at COP7 in Marrakech, it adopted rules (the "Marrakech Accords") which by their terms and in subsequent interpretations by the CDM Executive Board, made it nearly impossible even for afforestation and reforestation projects to be approved.[55]

Needless to say, investment in the forestry sector through the CDM has been virtually nonexistent, despite a relatively large number of approved methodologies for CDM projects in general. Out of 106 approved methodologies, 10, or 9.4%, relate to afforestation and reforestation. However, out of 945 registered CDM projects, only 1, or 0.1%, is an A/R project.[56] Moreover, this project is projected to generate only 340,000 tCO_2 by the end of 2012,[57] in comparison to the projected 1.17 billion tCO_2 from the other registered projects. As of today, there are no forestry projects approved under JI.[58] The obvious bias against LULUCF is such that no meaningful commercial investment is likely under the present regime.

Despite being initially excluded from the CDM, deforestation in developing countries has finally returned to the agenda. The consensus in Bali on the urgent need to deal with tropical deforestation and forest degradation speaks for itself. The developing nations have also made their position clear: unless they are paid for their carbon sinks they will not accede to a post-2012 treaty. The logic of their position is unassailable. Unless they receive compensation for not converting forests to agriculture they cannot develop their economies sustainably or otherwise. They are the low cost producers of many agricultural products as well as the principal source of timber demanded by the industrial world and they will not give up that competitive advantage for nothing. They also must feed growing populations.[59]

To be understood, deforestation has to be seen primarily as a response to market forces both international and domestic. The world's growing demand for food and forest products will not abate; it can only be directed toward sustainable supplies by the same market forces that create the demand. This is the critical role which the carbon markets can play if, but only if, appropriate regulation is now introduced. There needs to be both structural and regulatory realignment before that market can have its desired effects. We suggest below seven steps which need to be taken urgently if tropical forest loss is to be arrested.

The first step is to rectify the bias against the developing world now codified in the Kyoto Protocol. Any successor treaty must treat North and South alike by extending the same comprehensive crediting of forests and agricultural land allowed to Annex 1 countries to all treaty adherents. There is no longer any justification for and many reasons

---

[54] See COP7, Decision 11, "Land use, land-use change and forestry."
[55] See COP9, Decision 19, "Modalities and procedures for afforestation and reforestation project activities under the clean development mechanism in the first commitment period of the Kyoto Protocol."
[56] See http://cdm.unfccc.int/Statistics/index.html
[57] See http://carbonfinance.org/Router.cfm?Page=BioCF&ft=Projects
[58] See http://ji.unfccc.int/JI_Projects/ProjectInfo.html
[59] Global Environment Outlook, The United Nations Environment Programme in 2007

against, continued discrimination against tropical carbon sinks.[60] The goal should be full carbon accounting, whereby all biomass is accounted for in measuring progress toward achieving each country's emission reduction goals. The scientific and technological techniques for this now exist.[61]

The only requirement for qualifying for such treatment should be demonstration of technical and administrative capacity. Those countries which do not have, or cannot afford, such capacity should receive internationally funded support to achieve it such as that proposed by the World Bank's deforestation initiative.[62]

The second step is to remove from the CDM the authority to approve carbon projects for crediting. No agency in the world has, or can have, the requisite capacity, expertise and resources to make judgements as to every project in over 100 countries spanning everything from agriculture to industrial processes. Instead of the current prescriptive approach of the CDM, the successor treaty should adopt a principle-based approach to regulation such as that now adopted by the Financial Services Authority. Project approvals should be entirely devolved to participating countries whose designated agencies would take on the responsibility to conform their rules to broad principles established by or under the successor treaty. There does not need to be paternalistic second guessing by others provided that the market is then allowed to operate within a stable regime.

A necessary third step is therefore an agreement on long-term emission reduction goals which are not subject to periodic political or regulatory interference. A key structural weakness of both the Kyoto Protocol and the EU ETS are the 5-year compliance periods at the end of which political and regulatory interference is virtually assured. This creates wholly unnecessary uncertainty. Investments which must perform over decades cannot be implemented in this context particularly those dealing with forest and land use. Clear overall targets for emissions reduction must be set and adhered to both in terms of the level of reduction and the period in which they must be accomplished. A realistic time period is at least until 2030 and preferably beyond. Fundamental changes in the world's economy will not take place in any shorter period and neither the capital nor trading markets can operate to facilitate that change if the challenge is compounded by periodic political and regulatory interference.

The fourth step is to increase the participation of the private sector in the regulatory process. Uniquely in recent history, the development of climate change goals, legislation and regulation has taken place without any meaningful consultation of the private sector, the capital markets or even financial market regulators. The Stern Review, the first government sponsored economic analysis was published in 2007, some 15 years

---

[60] See Marrakech Accords Decisions, COP 7 of the UNFCCC, Decision 11/CP.7IPCC, 2000, Special Report of the Intergovernmental Panel on Climate Change: Land Use, Land-Use Change and Forestry, Cambridge University Press
[61] See National Carbon Accounting System, Australia http://www.greenhouse.gov.au/ncas/
[62] See Forest Carbon Partnership Facility
http://carbonfinance. org/Router. cfm?Page=FCPF&FID=34267&ItemID=34267&ft=About

after the Rio Conference which launched the Kyoto process. Until its publication virtually no economic or financial analysis was referred to in the climate debate. To the authors' knowledge no meaningful input has yet been sought, by the UN or by the EU, from those most experienced with capital and commodity markets such as central bankers, market regulators and market practitioners. This sealed box approach has lead to such recent fiascos as the price collapse in the market in Phase One of the EU ETS; the acceleration of tropical deforestation and increase in food prices caused by mandated biofuels standards and most recently the suppression of the market for CERs.[63]

Fifth there must be far more open and efficient decision making whether at international or national levels. The lack of transparency in decision making by the CDM is now legendary and has caused enormous harm particularly to those seeking to work within its rules.[64] Symptomatic of the damage to investment is the time it takes to get a project from the period of public comment to the registration. The CDM requires an average number of 237 days for a project to progress from the start of the public comment period until a request for registration; then there is a further average delay from the request for registration until registration of 84 days. It thus takes well over a year to create a CER and the process is slowing down, not speeding up as should be the case now that the CDM Executive Board has years of experience to draw upon. In June 2005 it took an average of 70 days for a project proposal to proceed from a registration request to registration. By June 2007, this had increased to 110 days. For projects that require a new methodology, it takes an average of 305 days from the point of submission of a new methodology to its approval.[65] In other words a project developer seeking innovative solutions will be required to wait two years or more to find out if he can proceed. We simply don't have time for such a bureaucratic process if the developing world is to make a meaningful contribution to dealing with climate change. There is also substantial anecdotal evidence of political and personal bias in decision making.

The sixth step is to repeal, and certainly not to replicate or extend, the regulations and regulatory interpretations which have stifled investment in the forest and land use sector. The required use of counterfactual scenarios and the adoption of arcane concepts with no meaningful analytical underpinning have created a "parallel universe" of regulatory requirements which are completely detached from commercial reality. The first and key step, as mentioned above, is to extend to developing nations the same scope of forest project credit types as is afforded to Annex 1 countries. This leads to the second reform which is to provide for the inclusion of the full spectrum of land use: avoided deforestation, forest degradation, reforestation (natural and assisted), sustainable forest management, afforestation, low till and no-till agriculture. Third, prescriptive rules such as those created by the CDM for afforestation and reforestation should be rejected. These include such counterproductive measures as the following:

---

[63] *op. cit.* Commission for the European Communities
[64] *op. cit* Trade in Agcert Shares
[65] See UNEP Riso Centre at http://cdmpipeline.org/

i. Capping at 1% of compliance requirement the use of A/R credits by Annex 1 countries

CDM forestry rules cap the use of A/R credits to just 1% of an Annex 1's country's annual compliance requirement over the first commitment period; equivalent to 120MtCO$_2$ annually. The 1% rule has clearly had a "chilling effect" on the market, discouraging investment in A/R projects, which offer the only meaningful alternative to meeting timber and fuel demand by continued deforestation of natural forests. There is no such cap on Annex 1 countries.

ii. A/R pr ojects are limited in location to lands deforested or in agricultural use prior to 1990 and which remain deforested at a project's inception. Restoration of land deforested since 1990 or of degraded land is excluded

The result of this rule has been to exclude from the system any credit for regeneration or replanting of forests destroyed since 1990. As a result between 125-195 million hectares of deforested land is now ineligible for CDM forestry (an area three times the size of France) and the area is growing (not least because of the lack of any crediting of avoided deforestation) by an area the size of Greece every year and it is happening in the most bio-diverse areas and the home to many of the world's last remaining indigenous peoples.

iii. Requiring the replacement of A/R credits after a maximum of 60 years.

Forests are a long-term store of carbon. They have covered vast areas of the earth's surface for millennia, and contain 60% of the carbon stored in terrestrial ecosystems.[66] CDM rules require that A/R forest credits be either temporary ('tCERs') or long term ('lCERs') and that all of them be replaced at specific intervals which are unrelated to the forest harvest cycle, with a maximum duration of 60 years. This rule not only reduces incentives for forest restoration but actually encourages the liquidation of healthy forests after no more than 60 years in order to generate cash to buy replacement credits. No other carbon market in the world finds such a rule necessary.

New rules should be principles-based and should allow for the recognition of the full value of tropical and subtropical forest land carbon storage and sequestration. LULUCF projects can bring multiple benefits, all of which are intricately linked and promote sustainable development. They include ecosystem services such as soil protection, erosion control, water purification, reduced flooding, agricultural pollination, local rainfall and biodiversity protection. They also include benefits to local communities and indigenous people by encouraging the resolution of land tenure issues, increasing resilience to adaptation to climate change, such as drought, storms, wildfires and floods.[67] New rules should encourage payment for these services in addition to carbon storage and sequestration and thus begin to fully value the multiple benefits of tropical and sub-tropical forests.

The seventh and final step is to ensure public accountability both of the regulators themselves and the efficacy of the regulations they promulgate. The CDM Executive Board, for

---

[66] IPCC, Land use, land-use change, and forestry: a special report of the IPCC. (Cambridge & New York. Cambridge University Press, 2000)

[67] Swingland, I, 2002, Capturing Carbon and Conserving Biodiversity: The Market Approach, The Royal Society

example, is appointed by an obscure process in which those most affected by its decisions have no say. There is no requirement that its members have any relevant experience or expertise and its resources are not the subject of any budgetary scrutiny. Regulations are promulgated without any serious attempt to determine their costs or benefits, their impact on climate change or in the case of land use and forestry, their impact on biodiversity, on communities or on other critical resources such as fresh water. Neither the markets nor the general public can have any confidence in such a system of regulation.

Summary

None of the manifold benefits and none of the climate change mitigation potential of the tropical and sub-tropical forests are now being realised precisely because of their exclusion from the carbon markets by misconceived regulation. Markets, contrary to many of the underlying assumptions of these counterproductive regulations, are in fact are very good at ensuring the integrity of the products they buy and sell and in punishing bad deliveries and bad deliverers. Market discipline, supported by appropriate, not manipulative, regulation will always be more efficient than bureaucratic attempts to ensure capital formation and price discovery. Markets are excellent at distinguishing the qualities of competing products by pricing them and their associated risks. There is no need to tell them what to do and efforts to do so, to pick winning technologies or approaches, always fail at huge and unnecessary cost. The EU bio-fuels targets are just one recent example of this.

Structural change is required which removes regulatory authority for the carbon market from the UN and EU Commission and vests it in the hands of experienced and accountable bodies such as central banks and securities markets. The role of the UN should be to establish the emissions targets to be undertaken by each country and to set out the broad principles which all countries must adhere to. No attempt should be made to make special rules which discriminate against the full participation of the developing world in the carbon markets such as is embodied in the CDM. The World Bank and other multinational bodies should assemble and distribute public sector funds to capacity building in the developing world so that all countries can benefit from investment driven by the carbon market.

We are firmly of the view that use of forest and land use credits, broadly defined to include the whole spectrum of rural land use, should be permitted to an unlimited extent in emissions trading schemes and in reaching emission reduction targets. Climate research has shown that to avoid catastrophic changes to the global climate and large scale irreversible systemic disruption temperatures must not increase above a threshold of 2 % those in pre¬industrial times.[68] Achieving this target requires significant emission cuts.

---

[68] European Commission Communication "Limiting Global Climate Change to 2° Celsius: The way ahead for 2020 and beyond.", Stern, N, 2006, Stern Review: The Economics of Climate Change, Meinshausen, Malte. "On the Risk of Overshooting 2°C." *Proceedings from International Symposium on Stabilisation of Greenhouse Gas Concentrations -- Avoiding Dangerous Climate Change*, Exeter, 1-3 February 2005 at www. stabilisation2005. com/programme.html.

The task of reducing and mitigating greenhouse gases on this timescale is enormous and to do so at a minimal cost to the world's economy should be a priority for policy makers. Emission trading schemes are designed to put a price on an industrial pollutant and to yield the lowest cost sources of compliance. Thus, opening the scheme as widely as possible to all sources of credits will allow the market to drive investment toward these low cost solutions. Unrestricted trading of REDD and LULUCF credits will provide a major portion, up to 25%, of the solution, will lower the cost of overall compliance and provide time for industry to implement the balance of the solution. If we do not rectify the fundamental error of excluding such credits from the carbon markets we will fail in the attempt to stabilise the climate.

Is there adequate support for developing countries to adapt to climate change?

No and there never will be without large scale private sector investment in rural land use in the developing world. Adaptation to climate change is costly, and to date it is unclear where the necessary funding will come from. According to the Stern Review in OECD countries the cost of making new infrastructure resilient to climate change could range from $15-$150 billion each year (0.05-0.5% of GDP), with their costs reflecting the prospect of higher temperatures in future. The Stern Review highlights that while there are few credible estimates of the costs of adaptation in developing countries it estimates the additional costs of adaptation alone in the developing world are $4-37 billion each year. This includes only the cost of adapting investments to protect them from climate-change risks, and it is important to remember that there will be major impacts that are sure to occur even with efforts at adaptation.[69] Needless to say, many of the key elements of adaptation for the rural poor of the developing world are integral to sustainable management of the tropical and sub¬tropical forests as well as sustainable use of agricultural lands.[70] In the absence of such investment and the stabilisation of their local environments, particularly soils and sources of fresh water, tens if not hundreds of millions of the 1.4 billion people dependent on forest resources for their survival will become involuntary environmental migrants with profound negative effects on their societies and on ours.[71]

---

[69] Stern, N, 2006, Stern Review: The Economics of Climate Change
[70] Adaptation to climate change in agriculture, forestry and fisheries: perspective, framework and priorities, FAO, Rome, 2007
[71] Global Environment Outlook, The United Nations Environment Programme in 2007

40. Climate Change and Carbon Forestry

Liu Shirong, Jiang Youxu and Shi Zuomin
Chinese Academy of Forestry, Beijing, 100091

Abstract: Forest is the main body in the terrestrial ecosystem, which absorbs and stores a great deal of carbon dioxide (CO_2) in the atmosphere. Carbon storage of the global forest is about 77 percent of the global plants. Carbon storage in the forest soil is approximate 39 percent in the global soil. Therefore, deforestation increases CO_2 emissions in the greenhouse gas, which is the next important emission source of CO_2 emissions compared with fossil fuel and cement. Forest ecosystem plays a very important and irreplaceable role and position in the global carbon circle and balance.

The direction of forestry development has been adjusted and transferred for the meet of mitigating the speed of global climate change. Carbon Forestry has been developed in the global since recent years. Carbon forestry includes the following 4 items:

(1) to increase forestry activities for absorbing carbon, which refer to forestation, reforestation, recovery degraded ecosystem, and building agro-forestry system for improving the storing carbon ability of terrestrial plants and soil.

(2) To keep and maintain forest carbon pool, protect carbon storage in the forest ecosystem, reduce carbon emissions to the atmosphere..

(3) To strength forest sustainable management, take a series of management measures, decrease carbon emissions, increase carbon sequestration, and get a maximum benefit of carbon storage.

(4) To take carbon substitute measures: durable wood products replace energy collective materials, regeneration wood fuel(energy plantation) and logging surplus are used to act as fuel.

Clean Development Mechanism was established for mitigating global climate change, building market operational mechanism of forestry carbon sequestration, realizing eco-asset changed into industrial money, in other word, realizing carbon trading. China has a great wide foreground in developing carbon forestry and carrying out carbon sequestration projects of Clean Development Mechanism. Developing carbon forestry plays a great active role in speeding for forestry ecological construction, improving regional ecological environment, reducing poor population, developing forestry bio-material energy. Developing carbon forestry needs to carry out some following researches: forestry carbon sequestration management policy, dynamics of forest carbon source and sink, methodology of carbon inventory, carbon trade. These results will supply experiences, techniques, experiments and models.

Keywords: forest, carbon forestry, carbon sequestration, carbon trading

1. Introduction

IPCC fourth evaluation (2007) report pointed out that global mean
temperature had increased 0.74± 0.18°C in the past 100 years. The increasing number

of temperature in the Northern hemisphere in 20th century is the highest in the past 1000 years. The distribution of precipitation has also been changed. Precipitation increased in the territorial area, especially the area in the middle and high latitude, but it decreased in the Africa. Scientific evidence suggests that the projected climate changes is likely to lead to increased water scarcity, droughts and high rainfall events, loss of biodiversity, shifts in forest types and reduction in food production in dry tropics with increased risk of hunger and flooding due to sea level rise. According to the IPCC fourth report(2007), a global warming by 1.8-4.0 °C is projected by 2100 with land surface warmer than oceans. Sea level is predicted to be up 18-59cm.

The trend of China temperature change is the same as global. In recent 100 years, temperature in China has increased 0.4-0.5 °C, lower than 0.6 °C in the global mean temperature. From regional distribution point of view, the most obvious climate changeable regions in China are western-north, hua-north, and eastern-north. The warming degree in the western region is higher than mean degree of whole country. According to China climate scenarios, mean temperature in the whole nation is predicted to be raised 1.7 °C from 2020 to 2030, 2.2 °C by 2050(Dahe Qin , et al, 2006).

A great deal of various observations and research results show that the concentration of CO2 and other greenhouse gases increase continuously so that global climate is changeable. According to the IPCC fourth report, global temperature rise in the past 50 years is possible to relative with greenhouse gases augment from petroleum and other fossil fuel use, which means human activities heavily result in the concentration of greenhouse gases rise in the atmosphere.

Climate change and its manifestations, particularly rising temperatures, changing precipitation patterns and sea level rise, are of global environmental concern and have the potential to impact most natural ecosystem(such as sea level rise , glacier reduction, frozen soil melt, plants growth reason extend and et.al,) and social economic systems (such as human settlements, health, and economical social sustainable development) in all countries. Global awareness of climate change issues has increased on unprecedented scale, which becomes a very important issue among with international politics, economics, diplomacy and national safety.

Forest is a main body in the terrestrial eco-system, accounts for nearly one third of the terrestrial area. Annual output of photosynthesis is about two third of the terrestrial ecosystem. Forest has double functions in the CO_2 increase and decrease in the atmosphere. Forest heavily absorbs and fixes CO_2 in the atmosphere, which acts as carbon storage pools and buffers. At the same time it also emits CO_2 with deforestation, which is a emission source. Forests play a very important role in the stabilization and adjustment of global carbon circles and carbon balance in the terrestrial ecological system. Therefore, forests act as a very important measure and way for reducing emissions and increasing carbon sinks in the forest problems of Intergovernmental Panel on Climate Change and other relative acts. For the adaption and mitigation global climate change, forest management and forestry development are facing new challenges and opportunities now.

2. An important carbon pools and sequestration——forest eco-system

Forest occupies a huge areas and tremendous biomass. Globally, forest area account for 3.869 billion hectares, approximately 30 percent of terrestrial area. According to IPCC(2001) estimation, the quantity of carbon stored in the terrestrial ecosystem is about 2,477 GtC, soil accounts for approximately 81% of it and vegetation is 19% at the global level. Forest biomass carbon storage is about 77% of vegetation biomass in the global. Forest soil carbon storage accounts for 77% of the soil in the global. The carbon storage per hectare in the forest ecosystem is 1.9-5 times of it in agricultural land. Therefore, forest ecosystem is the largest carbon pool in the terrestrial ecosystem. Its quantity of increase and decrease would heavily impact on CO_2 in the atmosphere (Table 1).

Table 1 Global carbon stocks in vegetation and top 1m of soil
(modified from WBGU 1998 and Watson et al. 2000)

Biome	areas($10^6 hm^2$)	Carbon stocks(GtC)		
		Vegetation	Soils	Total
Tropical forests	1760	212	216	428
Temperate forests	1040	59	100	159
Boreal forests	1370	88	471	559
Tropical savannas	2250	66	264	330
Temperate grasslands	1250	9	295	304
Deserts and semi-deserts	4550	8	191	199
Tundra	950	6	121	127
Wetlands	350	15	225	240
Croplands	1600	3	128	131
Total	15120	466 (19%)	2,011 (81%)	2,477 (100%)

The carbon stocks of Forest vegetation and soil in China are respectively estimated about 5GtC and 15GtC(Xiaoquan Zhang, et al. 2005). Forest carbon sequestration in China has been strength continuously in recent years. According to *National Program for Addressing Climate change* of China, China forest in 2004 absorbed net 0.45 $GtCO_2$-eq, which is about 8% net emissions of greenhouse gases in whole country. Forestation carbon sinks in china will increase continuously in the future. Forest coverage rate is estimated to be 20% by 2010, the quantity of carbon sequestration will increase 0.05GtC compared with it in 2005. It is estimated that the capacity of forest absorbing net carbon by 2050 would increase 90.4% in 1990. The quantity of carbon sequestration produced by the forestation and reforestation activities by 2010, 2030, 2050, is 0.026GtC, 0.124GtC, 0.191GtC.

Compared with croplands and grasslands, forest has more net primary production (NPP) and huge accumulation biomass in the long time, and larger capacity of fixing carbon than grasses and crops. The atmosphere inventory and simulation research showed that the quantity of carbon sequestration stored by terrestrial land in 1980s is 0.2+1.0 $GtC.a^{-1}$, which is the result that 1.9+1.3 $GtC.a^{-1}$ absorbed by terrestrial land minuses 0.2+ 1.0 $GtC.a^{-1}$ emission produced by land use change. In 1990s, the quantity of carbon sequestration reached to 0.7±1.0 GtC.a^{-1}, which is the result that 2.3±1.3 GtC.a^{-1} absorbed by terrestrial land minuses 1.6±0.8GtC.a^{-1} emission produced by land use change(Ciais et al.). Carbon measurement results and model researches in the north America indicated that forest vegetation in the middle and high latitude is a very important carbon sequestration, which played a key function on reducing non-balance of carbon inputs and outputs (Brown et al.,1999; Schimel et al., 2000). From the analysis of measurement results of carbon dioxide flux, northern, temperate and tropical forests have the functions of carbon sequestration. Forest types, climate environmental change, natural and human interferences influences on the capacity of carbon sequestration (Zijun Mao, 2002).

Deforestation or forest destroyed and degeneration could lead carbon emissions of the forest biomass and in the soil, thus reduce potential capacity in carbon stocks and sequestrations. After forest lands changed into croplands, Loss of soil organic carbon reaches 75%. Soil organic carbon would reduce average 30±2.4% after 10 years. If the influence of soil volume-weight was omitted, soil organic carbon reduced 22.1±4.1% (Murty et al., 2002).

According to IPCC estimation, net cumulative global carbon dioxide emissions from land-use change during 1850-1998 are at 136±55GtC, out of this, 87% emissions are from forest areas, 13% emissions from grasslands reclamation, at the same time, emission from fossil fuel and cement is 270± 30 GtC (Ciais et al.). Deforestation, forest destroyed and degeneration is the second carbon dioxide emissions source in the atmosphere next to fossil fuels.

3. Carbon forestry, reducing emissions and increasing sequestrations

Forestry is adjusting and changing its development and management direction for mitigating global climate change, does its best to take an important and irreplaceable function. At present, carbon forestry has been developed in the world. Carbon forestry includes the following issues:

(1) The quantity of forest vegetations and soil carbon stocks is increased by the means of afforestation, reforestation, recovery of degenerated ecological system, and building agro-forestry system and other measures, which strengthens the functions of forest absorbing carbon capacity. The utilization land area for afforestation, reforestation and agro-forestry in the global is approximately 0.345 billion hectares. If afforestation, reforestation and agro-forestry would be finished to be realized on these land areas, the potential of affforestation carbon sequestration would reach 28GtC, the quantity of agro-forestry carbon sequestration would be 7GtC (Reed et al.,2001). The vegetation recovery of 0.217 billion hectares of degenerated land in the tropical areas would strengthen the capacity for fixing carbon, which was 11.5-28.7GtC (FAO,2001b). It is predicted that China will absorb 0.667GtC of net carbon through afforestation and reforestation on the large areas during the first promised term, 2008-2012. By 2050, the capacity of forest absorbing net carbon per year will increase 90.4% in 1990 (Zhang & Xu, 2003). Acted as carbon forestry, agro-forestry in China has a great development space.

(2) to keep and maintain forest carbon pools, in other word, to keep present carbon stocks in the forest ecological system, reducing its emissions to the atmosphere. The main

measures refer to reduce deforestation, to improve forest management ways, to increase the efficiency of wood utilizations and control effectively forest disasters (forest fire, flooding, wind destroy, diseases and insects). In addition, control measures taken are useful for reducing trees and soil interferences to produce carbon emissions, which not only increase carbon stocks continuously in the forest ecological system in the long time, but also realize the goals for keeping biodiversity and developing the functions of ecological system services. These measures are mostly suitable to forest stands, which grow slowly and the quality of dried wood is low, and the regions where the chances of utilization of logging timber are fewer. Since deforestation directly leads to carbon stored in the forest ecological system that would emit to the atmosphere in a lot of years, reducing deforestation speed is a most direct way to mitigate the step of carbon dioxide concentration increase in the atmosphere compared with afforestation and reforestation. The potential capacity for increasing carbon sequestration would reach 14GtC through reducing deforestation in the future 50 years (Reed et al.,2001).

(3) to realize the goals for reducing emissions and increasing carbon sequestrations by the means of forest sustainable management and other carbon management (Figure 2). Reducing interferences of afforestation, sivilculture, and logging influencing on trees and soil is a very important way to keep present forest carbon stocks. Traditional logging operations greatly destroy forest stands. The degree for destroying forest stand decrease 50% through improving the ways of logging operation measures (Sist et al.,1998) , which reduces carbon emissions as forest logging. In addition, the speed of decomposition and

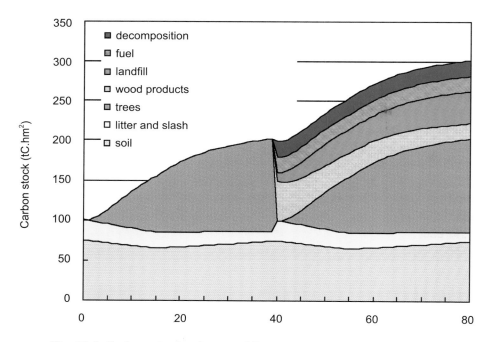

Fig. 43-1 Carbon stocks change of Forest in two logging periods

carbon emissions could be reduced through improving wood utilization rate. Carbon stored in forest stand emission rate to the atmosphere would be decreased through increasing wood production life. Unuseful and old wood products are filled in the land, that extend the time of their carbon emissions, some of carbon will be kept in the long time.

(4)From the regeneration resources point of view, if energy collective materials (cement, steel, plastic, brick) would be replaced by endurable wood product, that not only increases carbon stocks in terrestrial land but also reduce greenhouse gases emissions of fossil fuels, therefore carbon circular procession in the forest ecological system and traditional forest wood production direction should be adjusted. Although carbons stored in wood products return to the atmosphere at last through their decomposition, the regeneration forest resources would absorb these carbons, and avoid unreturnable net carbon emissions that are produced by fossil fuels. Regeneration wood fuels replace fossil fuels so that carbon emissions as human activities could be reduced. Bio-energy includes wood fuels (energy plantation) and logging surplus (Figure 43-1). According to IPCC estimation, bio-energy plants in the global will replace 20-73GtC from 2000 to 2050 (Watson et al.,1996). Carbon substitute measures are suitable for high yield even plantation and short period shrubs management, but the interferences of forest management impacts on soil should be decreased to the lowest degree.

4. Carbon trade and potential

In order to mitigate global climate change and realize the goals of United Nations Framework Convention on Climate Change (UNFCCC), the *Kyoto Protocol* was reached in 1997. The *Kyoto protocol* formulates that greenhouse gases emissions in the industrial countries reduced 5% at least compared with the emission level of 1990 during the promised period from 2008 to 2012. Industrial countries could get carbon credit for counteracting the limited emissions indicator of their promised period through afforestation, reforestation, reducing deforestation, forest management and other activities or carrying out afforestation projects under the Clean Development Mechanism. Marrakech agreement was reached in 2001, to realize the agreements of land use change relative to *Kyoto Protocol*, the definition of forestry activities and relative implementing rules. The ninth signatory states meeting held in December 2003, set up detailed international rules for implementing ways and programs of Clean Development Mechanism(CDM) afforestation project (UNFCCC, 2003). Clean Development Mechanism established double wins mechanism of both benefits for developed countries and developing countries, and also provided new development opportunity for carbon forestry.

Because each country only has its carbon sequestration sums of CDM afforestation and reforestation projects formulated by international organizer, in other word, the quantity of its emissions can not exceed 1% of the emissions on the basal year plus 5(UNFCCC,2001), the countries approved by *Kyoto Protocol* permit to produce carbon sequestration by means of CDM, which is about 0.32 GtC per year. If carbon sequestration per hectare accounts for 30-50 tC, total land area occupied by CDM afforestation projects is approximately 3.2-5.3 million hectares. The scale of CDM afforestation project in the global is lower than the number of afforestation areas per year in China, which is about 6-10% of plan afforestation areas from 2001 to 2020. Therefore, China has a great potential in carrying out CDM afforestation projects.

At present, because CDM projects limited to afforestation, reforestation projects, forestry CDM rules provide marketing operation mechanism for afforestation, reforestation and other forestry activities. Acted as a kind of operating model of marketing activities, the enterprise whose emissions quantity exceeds its formulating number might buy carbon credit to supplement exceeded carbon dioxide emissions, in other word, this operation realized carbon trading that eco-asset could be transferred into industrial money. There is 0.04-0.1 Gt CO_2 eq. per year of marketing potential of CDM afforestation, reforestation projects to be used in the first promised period. China is about 6.7-37% of them, and has 20% marketing numbers in the first promised period, that potential is 0.033GtC, which is equal to 0.7-1.1million hectares of afforestation area. According to carbon sequestration price set by bio-carbon funds of World Bank (about $3-4 US of carbon dioxide per ton), that is equal to $ 0.363-0.484 billion US (Xiaoquan Zhang, et al., 2005).

Generally, CDM afforestation and reforestation projects in developing countries not only produce carbon sequestrations for developed countries to counteract their carbon emissions, but also are helpful for poor population to get carbon benefits, reduce the poor, increase jobs, develop economy and improve environment. Therefore, carbon trading produced by CDM will supply social, economic and environmental benefits for thousands and ten thousands poor people in the some developing countries. If carbon trading would be taken in the global, investments to be supported by some developed countries for some of the poor developing countries would be $ 0.3 billion US.

Developing carbon forestry and implement CDM carbon sequestration projects are accord with forestry development strategies in China, and play an improving role on speeding the step of forestry ecological construction, improving regional ecological system, creating bio-energy and reducing the poor, which also provide a chance for China forestry development. Therefore, it is feasible for China to develop carbon forestry and carry out CDM afforestation projects. At present, Guangxi Zhuang autonomous region in China has successfully implemented the first afforestation and reforestation project in the world. On the base of it, marketing operations of CDM in China should be extend, some relative policies and standards should be set up, its methodology should be researched, experiment models should be established, and much larger space of carbon trading market should be created.

5. Contents of carbon forestry researches

With the development of scientific researches and the continuous procession of international negotiating on UNFCCC and *Kyoto Protocol*, people gradually improve to understand forest roles and functions in mitigating climate change. But this identification still has great uncertain, especially how to optimize utilization of forest and wood production to counteract greenhouse gases led by human activities. If quantity analyses would be researched on carbon and greenhouse gases pools relative to forest and timber production, we would easily understand the optimizing choice of mitigating climate change by forestry ways. At present, some countries and organizations are concern to fix carbon and reduce carbon dioxide in the atmosphere through forestry because the potential value of forest fixing carbon has some uncertain and carbon inventory program is complex. In addition, the carbon that forest fixed is limited, some of them are emitted with the procession of logging, forest fire, or occurrence of disease and insects destroy. Therefore, we should carry out forest sustainable management, and take a development goal of forest fixing carbon benefit and carbon forestry sustainable development, which is one of forest multi-ecological service functions.

Developing carbon forestry should strengthen relative researches on forest carbon management, and forest carbon stocks and wood products carbon stocks in different regions and various types. Therefore, state forest carbon inventory system should be established and state forest carbon and greenhouse gases emissions inventory should be edited, at the same, the quantity role of wood products and bio-energy construction greenhouse gases emissions should be evaluated, which provide scientific base for managing forest carbon and greenhouse gases change effectively. Forest carbon management researches include the following items:

(1) to establish state forest carbon accounting system and greenhouse gases emissions inventory.

(2) to develop long terms measurement on forest carbon and greenhouse gases flux change and evaluate balance of carbon.

(3) to explore carbon and greenhouse gases accounting models and ways.

(4) to evaluate the functions of substitute (wood products and wood) in the state carbon income and output.

(5) to forecast carbon balance of bio-energy(fuel forest and bio-energy forest).

(6) to research the impacts of forest management measures and climate and soil factors on forest carbon, carbon balance and circles.

(7) to research and formulate forest management strategies for mitigating and adapting climate change, and put forward evaluation methods for forest management risks.

References

1. Christine L. Goodale, Michael J. Apps, Richard A. Birdsey, Christopher B. Field, Linda S. Heath, Richard A. Houghton, Jennifer C. Jenkins, Gundolf H. Kohlmaier, Werner Kurz, Shirong Liu, Gert-Jan Nabuurs, Sten Nilsson, and Anatoly Z. Shvidenko, 2002, Forest carbon sinks in the northern hemisphere, Ecological Applications, 12(3): 891-899

2. Fang, J.Y., A.G. Chen, C.H. Peng, S.Q. Zhao & L.J. Ci, 2001, Changes in forest biomass carbon storage in China between 1949 and 1998, Science, 292: 2320-2322

3. Mao Zijun. 2002. Estimation Methods and their researching processes for Carbon ballance in Forest Eco-system. plant Ecology. 26(6): 731-738

4. United Nations Framework Convention on Climate Change. 1992. http://unfccc.int/resource/docs/convkp/conveng.pdf.

5. UNFCCC. Kyoto Protocol to the United Nations Framework Convention on Climate Change. 1997. FCCC/CP/1997/7/Add.1. http://unfccc.int/resource/docs/convkp/kpeng.pdf

6. UNFCCC. Modalities and procedures for a clean development mechanism, as defined in Article 12 of the Kyoto Protocol. In: Report of the conference of the parties on its seventh session, held at Marrakesh from 29 October to 10 November 2001, Addendum: Part two: Action taken by

CHAPTER V: Climate Change and Land Degradation

the conference of the parties, Volume II. 2001. FCCC/CP/2001/13/Add.2, 20~49, http://unfccc.int/resource/docs/cop7/13a02.pdf

7. UNFCCC. Modalities and procedures for afforestation and reforestation project activities under the clean development mechanism in the first commitment period of the Kyoto Protocol. In: Report of the conference of the parties on its ninth session, held at Milan from 1 to 12 December 2003, Addendum: Part two: Action taken by the conference of the parties at its ninth session. 2003. FCCC/CP/2003/6/Add.2, 13~31, http://unfccc.int/resource/docs/cop9/06a02.pdf

8. UNFCCC. Simplified modalities and procedures for small-scale afforestation and reforestation project activities under the clean development mechanism in the first commitment period of the Kyoto Protocol and measures to facilitate their implementation. Buenos Aires, 6–17 December 2004. 2004. FCCC/CP/2004/L.1

9. IPCC (2001). Climate Change 2001 (The Third Assessment Report)

10. IPCC (2007). Climate Change 2007 (The Fourth Assessment Report)

11. Schimel D, Melillo J, Tian H Q, McGuire A D, Kickleghter D, Kittel T, Rosenbloom N, Running S, Thornton P, Ojima D, Parton W, Kelly R, Sykes M, Neilson R, Rizzo B, 2000, Contribution of increasing CO_2 and climate to carbon storage by ecosystems in the United States, Science, 287: 2004-2006

12. Brown S L, Schroeder P E, 1999, Spatial patterns of aboveground production and mortality of woody biomass for eastern U.S. forests, Ecological Applications, 9: 968-980

13. The Royal Society working group on land carbon sinks, 2001, The role of land carbon sinks in mitigating global climate change, Policy document 10/01 of the Royal Society, 18

14. Zhang Xiaoquan, Li Nuyun, Wu Shuhong. 2005. Feasible and potential in China Implementing Afforestation and Reforestation Project of Clean Development Mechanism. Forestry Science

15. Zhang Xiaoquan, Wu Shuhong, Hou Zhenhong, He Ying. 2005. Forest, Forestry Activities and Greenhouse Gases. Forestry Science, 41(6):150-156

16. Qin Dahe. et al., Evaluation Report on China Climate Change. 2006, Science Press

41. Mitigating Climate Change and Better Ensuring Agriculture's Adaptation for impending Climate Change through Conservation Agriculture

Des McGarry
Land and Environment Consultant, Brisbane, Australia

Abstract: Global climate change (CC), as assessed by the 4th Assessment Report of the IPCC, is already occurring and the potential impacts are regarded to be ever-worsening. Agriculture, though occupying 40-50% of the Earth's land surface, contributes "only" 10-12% of total global anthropogenic greenhouse gas (GHG) emissions (5.1 to 6.1 GtCO$_2$-eq/yr in 2005). However, agriculture contributes 47% and 58% of total anthropogenic emissions of N$_2$O and CH$_4$, respectively. CO$_2$ has large annual exchanges between the atmosphere and agricultural lands but the net flux is estimated to be approximately balanced, so accounts for less than 1% of global anthropogenic CO$_2$ emissions. Agricultural CH$_4$ and N$_2$O emissions have increased by nearly 17% from 1990 to 2005; 32% of the increase from Non-Annex I ("least developed countries") countries who were responsible for 75% of all agricultural emissions. The Annex I countries, collectively showed a decrease of 12% in GHG emissions.

Several agencies (eg IPCC) already recognise that the agriculture sector (globally) has developed several documented and tested strategies with strong potential to impact on CC through reduced GHG emissions (mitigation), and achieve readiness of the agricultural sector for CC (adaptation), with concurrent benefits (co-benefits) of attaining more sustainable land management practices, generally, and food and water security, and rural poverty mitigation, specifically. The common term for these strategies is Conservation Agriculture (sometimes termed "low input agriculture") that though non-prescriptive and requiring local fine-tuning to ensure practicality and widespread success, does have three core themes: (a) maintaining a permanent organic cover over the soil (b) minimising soil disturbance (no till) and (c) practising crop rotations for organic matter and biodiversity enrichment. Further, inter-related practices to further enhance CC mitigation and achievement of sustainable systems include biogas (CH$_4$) production from animal wastes, inter-cropping, the use of biofuels and site specific nutrient management for balanced fertiliser usage.

Required for wide and successful implementation of these initiatives will be the close linking of CA and related actions to the on-going development activities of existing International and National CC and sustainable development initiatives, such as the content of IPCC reports, the targets of National Development Strategies and Millennium Development Goals, the Marrakech Process, the National Communications (NC) and National Adaptation Programs of Action (NAPA) of several Non-Annex I countries that already mention such actions in brief. Additionally, placement of these initiatives within higher level policy implementation will help achieve attractiveness of the revised sector management initiatives. Concurrently, implementation of bottom-up, country driven, regionally networked, cross-sectoral approaches will ensure multi-level stakeholder benefits for land, water and biodiversity protection within several key agricultural systems, as well as increased farm profitability towards rural poverty reduction, through lessened input:output ratios.

Selected examples of current CA and related strategies and their role in CC mitigation will be presented. A current aim is the creation of a "framework" of good practices and their role in mitigating CC. Foreseen is that this framework will continue to be populated and enriched with time, to ensure multiple examples of good practice across a wide variety of environments.

Introduction

The purpose of this Paper is to commence discussion of a range of inter-linked, practical strategies already known and practised within some sections of the agriculture sector [1], (mainly Conservation Agriculture and related practices), to aid the mitigation of climate change (CC) while decreasing vulnerability of the agriculture sector to impending future changes, and increasing the adaptation to CC of natural resources generally and agriculture and rural livelihoods specifically. Concurrent co-benefits are also achievable, eg increased economic sustainability (via reduced input:output ratios) towards poverty reduction, and food and water security. The implementation and attainment of these multi-level wins is most guaranteed through holistic, multi-level scenarios, based on sound and acceptable (to end users) practices that are inter-agency and cross-cutting, firmly research (science) based/tested/monitored and supported by necessary policy changes. The desired outcomes are widespread farm-level uptake of these strategies for significant CC gains and co-benefits of improved yields, food and water security, sustainable land management, and reduced rural:urban socio-economic inequities.

This paper will focus on the concepts and practicalities of two inter-related questions: "What can agriculture do to lessen CC?" and "What can agriculture do to ensure adaptation to impending CC" As such, these questions are different to most considerations of CC being addressed in today's literature. These have tended to focus more on the predicted future impacts (direct and indirect) of CC on the agriculture sector, or, in other words the risk to agriculture of CC in terms of cause, magnitude, nature and geographic variability of the predicted changes in climate.

By necessity, this script tends to be academic. However, from the outset it will be stated that the aim or outcome of these deliberations is the implementation of the themes and actions addressed herein. In this way, it is hoped the content of this paper can be used support and develop actions in both loan-funding scenarios (eg World Bank, Asian Development Bank) and grant-funding applications (eg GEF). To these ends, this script will present the bigger, global picture of agriculture and CC, but the final aim will firmly be the addressing of CC issues locally; this being the most

---

[1] See: http://esa.un.org/marrakechprocess/roadmapcsd.shtml

implementable scenario. The goal is to achieve workable and executable scenarios; ones that certainly improve the "lot" of rural communities and individual farmers and concurrently aid/support higher level scenarios (Governmental and International aid agencies) in such areas as food and water security, land degradation and urban expansion, in the ever changing scenario of population growth, rising oil and fertiliser prices, dietary changes and CC.

At this early stage, this paper does not claim to provide an exhaustive breakdown of every component of the current subject matter. Rather, a prime aim, herein, is the creation of a framework that establishes the main components of the debate on "what agriculture can do for CC and adaptation of agriculture for CC", towards the (more important) pragmatic implementation of the parts. To these ends, the framework will consist of several, sequential and inter-related topics, towards achieving several, inter-related aims: (i) a critical review of the literature - to aid clarification (for persons wishing to further investigate this theme) of the diverse content of the myriad of publications on CC currently being produced, in order to separate out those themes relevant to the prime topic addressed here; (ii) review of governmental, internationalagency, NGOs and individuals currently investigating or funding work on CC, again to separate out the relevant works to the current theme; (iii) utilising these two reviews (that by the nature of CC work will be active and ongoing for some years) highlight both the concepts (eg scientific writing on the scales of the problem and expected influences of altered practices on CC) as well as seek out and present already-implemented field-based examples of the major mitigating strategies across a wide range of scales and environments, while concurrently (iv) seeking interested parties to form partnerships of inter-agency, -country, -disciplines etc for information sharing (successes and failures), implementation scenarios, adoption pathways, inter-institute support and training packages, and costs/benefits/barriers at farm, regional and global scales with the aim of achieving upscaling of the relevant agriculture strategies for widespread benefits.

Points (ii) to (iv), above show cognisance that success in these endeavours will be far more assured through linkages with existing strategies both directly on the current theme as well as cross-cutting with related subject areas.

These same points also show that there has been, to date, a certain amount of discussion and some on the ground activities on the themes being addressed here. Written materials (reviewed below) include IPCC and UNFCCC documents, national development plans (e.g. poverty reduction strategy papers), national sustainable development strategies, Millennium goals, country National Communications (NC) and National Adaptation Programs of Action (NAPA) of several Non-Annex I countries. However, these documents mostly provide only general information on Agriculture's role in mitigating CC and fall short in presenting practical, farm-level interventions, or implementable policies, or required capacity building/training etc formulated towards actually achieving the on the ground, CC mitigation/adaptation activities. The specific "on the ground" activities that have been conducted for some decades tend to be specific to individual farmers, farmer associations, and crop-oriented growers' groups (eg Australian cotton and grains groups). Many of these conduct agriculture practices that have strong potential to be CC mitigating strategies, termed Conservation Agriculture (CA). However, at present these are conducted for reason of cost- and input-reduction, and drought and erosion proofing, rather than CC mitigation and preparedness.

Required, therefore, and as such defines the focal point of this paper and the works that emanate from it, is a set of definitive advisory statements of strategies (practical, farm-level interventions) to achieve the mitigation/adaptation to CC. And the link to and support gained from associated supportive policies, as well as the requirements for capacity building/training, networking etc - again formulated to achieve farm level activities. Subsequent steps include pathways for upscaling and replication of these strategies to achieve widespread impacts on both CC and concurrent sustainable use of the environment.

Previous and current works

A start will be made to review current literature and activities in the subject area. The aim is to seek out relevant texts and activities, towards supplementing and enriching the information they provide. Eight of the more major ones will be reviewed here. The content of this list is dynamic and it is expected that others will come to light, for incorporation in future drafts of this paper.

The Pew report [2] (Paustian et al 2006) stated that though the agriculture sector (including land use change) contributes about $1/3^{rd}$ of the total human-induced global warming effect (the remainder mainly being CO_2 emissions from fossil fuel combustion), "….agriculture is unique in that it can bring several, inter-related and synergetic strategies to the mitigation of global warming that also provide multiple co-benefits, that by themselves justify the new practices and provide means whereby agriculture can adapt or improve readiness against CC". In such ways, the agriculture sector has the potential to reduce its own emissions and concurrently both offset and reduce emissions from other sectors through removing CO_2 from the atmosphere via photosynthesis and storing carbon in plants and soils (every tonne of carbon added to and stored in plants or soils removes 3.6 tonnes of CO_2 from the atmosphere), as well as providing biofuels to displace fossil fuel use, and through the adoption of agricultural best management practices (that in themselves aid global, environmental condition and aid adaptation of the sector to CC) reduce emissions from agricultural soils (N_2O), from livestock production and manure (CH_4), and from on-farm energy use (CO_2).

The Marrakech Process began in 2003 to build political support for the implementation of sustainable consumption and production (SCP) and to prepare input for negotiations at CSD 18-19 [3]. The Marrakech Process is a global initiative to support the elaboration of a 10-Year Framework of Programs on SCP, as called for by the WSSD Johannesburg Plan of Action. Its goal is to assist countries in their efforts towards sustainability, to green their economies, to help corporations develop sustainable business models, and to encourage consumers to adopt more sustainable lifestyles. The more

---

[2] Download from: http://esa.un.org/marrakechprocess/issuesagricultureandrural.shtml
[3] Adjunct Professor in Science Communication at the University of Technology, Sydney, Australia; see: http://www.atse.org.au/index.php?sectionid=1207 and http://www.sciencealert.com.au/jca.html

specific report within the Marrakech Process on "SCP in Agriculture and Rural Development" [4] discusses achieving more sustainable consumption and production in the agricultural sector that gives a less input intensive and more resource efficient agriculture as a means to strengthen the competitiveness of the agricultural sector. The report urges the promotion of good environmental practices in agricultural production for sustainable agricultural development and poverty reduction, and rehabilitating degraded, over-used lands through CA practices. The report also notes that biofuel production can assist the agricultural sector contribute to mitigating GHG emissions.

Of direct relevance to the subject matter herein, are the writings of Professor Julian Cribb; a strong advocate of Agriculture's role in positively influencing current global problems of food security and rising prices with impending CC. He argues the important role that the world's farmers can have in aiding the rescuing of civilisation from the multiple, additive problems currently effecting the globe, including: the lowest levels of world food security in 50 years, rising food prices (eg rice has risen from $400 to $1000 a tonne), world population growth (projected 9.1 billion by 2050), increased demand for protein food (especially in China and India), total world food demand forecast to rise 110% by 2050, a global water crisis (cities now consuming half the water once used to grow food while groundwater levels are falling in every country where it is used for agriculture), reduced area of good arable land (from urban expansion, erosion and degradation), massive inflation in the prices of fuel, fertiliser and chemicals, biofuels fast-expanding into historic food production areas (that by 2020 are projected to consume 400 million tonnes of grain annually; the entire world rice harvest) all under the spectre of climate change (up to half the Earth projected to be in regular drought by the end of the century). His discourse is that world authorities need to recognise that "Agriculture Policy is Defence Policy", including and crosscutting as it does with policies in the subject areas of refugees, immigration, environment, health, food and economics. Required is a doubling of world food output, cognisant that there is no "silver bullet solution". Resolution must be achieved using less land (increasingly lower quality land), far less water, far fewer nutrients with projections of ever-increasing drought. The requirements therefore are for integrated, cross-cutting actions on a global scale and by every human, government and international agency. Target areas include increasing water use efficiency (irrigation and rain) in all crops, the implementation of organic and low-input farming systems, raising vegetable production and consumption, replacing protein and carbohydrate based foods with lower input cost and "more direct" foods such as pulses and grains, the commitment to recycle all nutrients - on the farm (eg slurry from bio-digesters), in the food chain or in sewage works (utilising urban sewage for methane production) - and the large scale introduction of 'green cities' that address the environmental impacts of urban development. Recognised is that these challenges are far from trivial. However, just as humanity overcame two previous global food crises with the first agricultural revolution and the green revolution, it is now called on to do so again, with

---

[4] http://www.ipcc.ch/pdf/assessment-report/ar4/wg3/ar4-wg3-chapter8.pdf

the "sustainable food revolution".

The third volume of the Fourth Assessment Report of the Intergovernmental Panel on Climate Change (IPCC, 2007), goes further (in terms of IPCC reporting of CC) than the previous two volumes that considered "only" the physical science basis of climate change and the expected consequences for natural and human systems. More pertinent to the subject matter of this paper, the 3rd volume provides analysis of the technologies, costs, benefits and policy interventions required of different approaches to mitigating and avoiding climate change. Additionally, and again of direct relevance to the current paper, the third volume also analyses "how can climate mitigation practices and policy be aligned with sustainable development practices and policies?" There is a specific chapter (Chapter 8) on "Agriculture" [5] that recognises that "a variety of options exits for (the) mitigation of GHG emissions in agriculture...(including) improved agronomic practices (nutrient use, tillage, residue management), restoration of organic soils (drained) and degraded lands for crop production, improved water and rice management, set-aside of land, land use change (eg conversion of cropland to grassland), agro-forestry, and improved livestock and manure management. Recognised, too is that many of these potential mitigation opportunities use current technologies so could be implemented immediately, but future technological development will be required for efficacy of additional mitigation measures.

An earlier report from the Pew Centre (Burton et al 2006) introduced the concept (again of direct relevance to the current paper) of "integrating CC adaptation considerations across the full range of Sustainable Development (SD)". The report considered this as the most direct and effective means of discouraging investments that heighten climate vulnerability and promoting those that strengthen climate resilience. A development-centred strategy could closely complement the Convention-based approach, helping to ensure that national adaptation strategies prepared with Convention support are implemented, and could over time leverage far more resources than likely would be forthcoming under the climate regime. Proposed investments could be assessed for their own vulnerability to climate variability and climate change and for any broader effect on climate vulnerability within the host country. As with the environmental impact assessments now performed routinely by multilateral lenders, this would provide critical information to decision-makers. Projects in SD that substantially reduce climate vulnerability, or are identified as priorities in national communication and adaptation programs (see next two items) might be given preferential treatment.

National Communications (NC) are the reports that parties to the UNFCC must submit on implementation of the Convention to the Conference of the Parties (COP) [6]. The core elements of NCs for both Annex I and non-Annex I Parties are information on emissions and removals of GHG and details of the activities a Party has undertaken to implement the Convention. NCs usually contain information on national circumstances,

---

[5] http://unfccc.int/national_reports/napa/items/2719.php
[6] http://www.gefweb.org/projects/focal_areas/climate/documents/GEF_Support_for_Adaptation_to_Climate_Change.pdf

vulnerability assessment, financial resources and transfer of technology, and education, training and public awareness; but the ones from Annex I Parties additionally contain information on policies and measures. Annex I Parties that have ratified the Kyoto Protocol must include supplementary information in their national communications and their annual inventories of emissions and removals of GHGs to demonstrate compliance with the Protocol's commitments.

More than 40 least developed countries (LDCs) have received funding under the UNFCCC to prepare National Adaptation Programmes of Action (NAPAs) that draw on existing information to address urgent needs in terms of CC variability, risk and adaptation, and identify priority actions with regard to adaptation to climate change [7]. The rationale for NAPAs rests on the limited ability of LDCs to adapt to the adverse effects of CC. In order to address their urgent adaptation needs with focus on enhancing adaptive capacity to climate variability. A NAPA takes into account existing coping strategies at the grassroots level, and builds upon that to identify priority activities with prominence given to community-level input, recognizing that grassroots communities are the main stakeholders. About 40 NAPAs have been produced to date.

The Global Environment Facility (GEF) administers CC adaptation funding under the UNFCC and allocates funds for implementation projects. As it has become more apparent that global warming is occurring, the Conference of the Parties' (COP) guidance to the GEF has emphasized the need to move from preparation to implementation. The GEF has responded by initiating four different paths to support adaptation activities: the Strategic Priority on Adaptation (SPA), the Least Developed Country Fund (LDCF), the Special Climate Change Fund (SCCF) and the Adaptation Fund (AF) [8]. Consistent with UNFCCC guidance, projects supported by these four avenues will seek to integrate adaptation policies and measures in all sectors of development including water, agriculture, energy, health, and vulnerable ecosystems. SPA aims to reduce vulnerability and to increase adaptive capacity to the adverse effects of climate change through supporting pilot and demonstration projects that address local adaptation needs and generate global environmental benefits in all GEF focal areas. LDCF addresses the extreme vulnerability and limited adaptive capacity of LDCs. LDCF initially supported preparation of National Adaptation Programmes of Action (NAPAs), as detailed in (vii) above. The NAPAs conclude with a list of prioritized project profiles to be subsequently implemented with support from the LDCF. The SCCF, established in response to guidance from the Conference of the Parties to the UNFCCC, was originally aimed at supporting activities in the following areas: adaptation, technology transfer, energy, transport, industry, agriculture, forestry, and waste management, and economic diversification The AF will be financed from the share of proceeds on the clean development mechanism (CDM). Consequently, with the entry into force of the Kyoto Protocol, 2% of the share of the proceeds from CDM projects will be directed to an adaptation fund. Many of the projects

---

[7] IPCC (2007) http://www.ipcc.ch/pdf/assessment-report/ar4/wg3/ar4-wg3-chapter8.pdf
[8] From: From: http://www.fao.org/newsroom/en/news/2008/1000923/index.html

funded under these GEF initiatives are quite recent and are being conducted, now. Of interest will be the outcomes of these projects in terms of the subject matter of this current paper; to be investigated as the GEF projects develop and produce outcomes.

The problem(s)

Why is there a need for Agriculture to aid in the mitigation of CC and concurrently better adapt itself for change? Some of the drivers have already been mentioned, as reviewed by Cribb (2008). The following list adds to those.

(i) A first consideration is Agriculture's contribution to CC. Agriculture, though occupying 40-50% of the Earth's land surface, contributes "only" 10-12% of total global anthropogenic GHG emissions (5.1 to 6.1 GtCO$_2$-eq/yr in 2005) [9]. However, agriculture contributes 47% and 58% of total anthropogenic emissions of N$_2$O and CH$_4$, respectively; particularly important as it is known is that CH$_4$ and N$_2$O have 21 and 310 times the "global warming potential" of CO$_2$. CO$_2$ has large annual exchanges between the atmosphere and agricultural lands but the net flux is estimated to be approximately balanced, so accounts for less than 1% of global anthropogenic CO$_2$ emissions. Agricultural CH$_4$ and N$_2$O emissions have increased by nearly 17% from 1990 to 2005; 32% of the increase from Non-Annex I ("least developed countries") countries who were responsible for 75% of all agricultural emissions. The Annex I countries, collectively showed a decrease of 12% in GHG emissions.

(ii) In terms of food security, FAO estimates put the number of people suffering from chronic hunger worldwide in 2003-5 at 848 million, an increase of 6 million from the 842 million in 1990-2 [10]. Soaring food, fuel and fertilizer prices have exacerbated the problem. Food prices rose 52% between 2007-8, and fertilizer prices have nearly doubled over the past year.

(iii) The diets of large sections of the world's population are changing, particularly in developing countries (Delgado 2003, FAO 2008a) [11] where there has been a pronounced shift away from staples such as cereals, tubers and pulses towards more livestock products, vegetable oils, fruits and vegetables. Total meat production in developing countries increased 5-fold (27 million tonnes to 147 million tonnes) between 1970 and 2005, and, although the pace of growth is slowing down, global meat demand is expected to increase by more than 50% by 2030. One report [12] states that by 2020, developing countries will consume 107 million metric tons (mmt) more meat and 177 mmt more milk than in 1996-8, dwarfing developed-country increases of 19 mmt for meat and 32 mmt for milk in the same period. These increases require more feed (coarse grains and oilseed meals). One projection sees that this increase in livestock production will require annual feed consumption of cereals to rise by nearly 300 mmt by 2020 with

---

[9] FAO - "Current world fertilizer trends and outlook to 2011/12"ftp://ftp.fao.org/agl/agll/docs/cwfto11.pdf
[10] http://jn.nutrition.org/cgi/content/full/133/11/3907S
[11] FAO - "Current world fertilizer trends and outlook to 2011/12"ftp://ftp.fao.org/agl/agll/docs/cwfto11.pdf
[12] The Guardian newspaper (August 2008); http://www.guardian.co.uk/environment/2008/aug/12/biofuels.food

concurrent increased demand for fertilisers. Conversion of grain areas to vegetable and fruit production will also translate into greater fertilizer demand as average application rates for the latter is about double those for grain crops.

(iv) The recent GLADA report (Bai et al 2008) stated that land degradation is increasing in severity and extent in many parts of the world, with more than 20% of all cultivated areas, 30% of forests and 10% of grasslands undergoing degradation. An estimated 1.5 billion people, or a ¼ of the world's population, depend directly on land that is being degraded. This land degradation reduces productivity, impacts negatively on migration, food insecurity, damage to basic resources and ecosystems, and leads to loss of biodiversity. Additionally, land degradation has important implications for CC mitigation and adaptation, as the loss of biomass and soil organic matter releases carbon into the atmosphere and affects the quality of soil and its ability to hold water and nutrients. Importantly, the GLADA study shows land degradation since 1991 has affected new areas and some historically degraded areas were so severely affected that they are now stable having been abandoned or managed at low levels of productivity; all impacting on non-mitigation of CC through reduced soil organic carbon (SOC), decreased vegetative cover and often increased albedo effects, causing increased warming form increased reflectance.

(v) Compounding the ever growing area of degraded land, the world is running out of farm land (Cribb 2008). The area of land where food is grown has declined from 0.45 ha per person in the 1960s to 0.23 ha currently and will continue to fall as population rises. This creates a need to increase output from smaller land areas. Or, increase productivity of what are currently considered marginal lands but only if conducted with sustainable practices or the situation (marginality / degradation) will worsen.

(vi) Oil costs. From the mid-1980s to September 2003, the inflation adjusted price of a barrel of crude oil was <$25/barrel, rising to $60 by August 2005, and then $146 in June 2008, before falling back to $110 by the start of September 2008. The price of oil impacts greatly on many areas of the agriculture sector; fuel for farm machinery is the most direct, but fertiliser price and haulage of farm products are also affected.

(vii) Fertiliser prices. A multitude of recent developments have led to a dramatic increase in world fertiliser prices over the last 18 months [13], causing fertiliser prices to rise more than oil or any other commodities in that period [14]. The world price of diammonium phosphate (DAP) in January 2007 was $335 per tonne; in 14 months this price increased to $1110 per tonne. Over the same period the retail price increased from $610 per tonne to $1220 per tonne. The world price for urea was relatively stable at around $200 per tonne until mid-2004. Over the last six months of 2004 the world price for urea increased to around $325 per tonne. It then fluctuated in the range between $325 per tonne to $400 per tonne until January 2008, after which it increased to above $600 per tonne by May 2008.

---

[13] FAO - "Current world fertilizer trends and outlook to 2011/12"ftp://ftp.fao.org/agl/agll/docs/cwfto11.pdf
[14] FAO - "Current world fertilizer trends and outlook to 2011/12"ftp://ftp.fao.org/agl/agll/docs/cwfto11.pdf

(viii) Ever growing population. Already the urban : rural ratio of land is increasing with some of the best agricultural land being taken for urban development From footnote 16 - In spite of world population growth slowing from 1.26% (1996-2005) to 1.10% (projected 2006-15), absolute annual increments continue to be large. It is anticipated that between 50 and 70 million people will be added annually to the world population until the mid 2030s. Almost all of this increase is expected to take place in developing countries especially the group of 50 LDCs. More food and fibre will be required to feed and cloth these additional people and to increase the daily food uptake of the still 830 million undernourished world wide (2002-04). There is thus significant scope for further increases in demand for food even as population growth slows down.

(ix) Biofuels. High oil prices and the potential for future decline in mineral oil stocks are creating new markets for agricultural commodities that can be used as feedstock for the production of bio-fuels [15]. Bio-fuels are being promoted as contributing to a wide range of policy objectives, most notably as providing greater energy security with regard to liquid fuels, increasing rural incomes, lowering greenhouse gas emissions and providing economic opportunities for developing countries. Currently there is 14 million hectares or 1% of arable land planted with biofuel crops which provide some 1% of transport fuels. Some predictions see this area doubling to 35 million hectares by 2030. This massive growth is causing alarm, in terms of possible negative impact on the food security of millions of people across the world, if biofuels are to be either utilising crops currently used as food (eg maize and sugarcane) or are grown on land currently used for food crops. Balance is required. Additionally, foreseen is the potential to grow certain biofuel crops (the more hardy and drought resistant shrub types) on what is currently deemed "marginal" land. With improved agricultural practices (detailed below) especially no till, stubble mulches and inter- and understorey cropping, these lands could be both made productive, be protected from degradation and achieve long term sustainability.

Linkages are known to exist between rising oil prices, increased fertiliser demand and biofuel production [16]. High oil prices have contributed to price increases for most agricultural crops by both raising input costs, and by boosting demand for agricultural crops used as feedstock in the production of alternative energy sources (biofuels). The combination of high oil prices and the desire to deal with environmental issues is driving the rapid expansion of the biofuels sector. This is likely to boost the demand for feed stocks such as maize, sugar, rapeseed, soybean, palm oil and wheat for many years to come. However, much will also depend on the supply and demand fundamentals of the biofuel sector itself. High oil prices could depress the use of oil-based fertilizers which have been behind much of the increase in farm production during the past half century.

(x) Lessened water. Rural water usage is increasing, competition with industry and ever growing urban areas is increasing, and CC seems to forecast increasing drought in certain areas. Required are cropping practices that are far more water use efficient; of both rain and irrigation waters.

---

[15] From http://www.scidev.net/en/news/climate-cropland-changes-raising-temperatures-in-e.html
[16] http://www.unep.org/pdf/UNEP_Planning_for_change_2008.pdf

(xi) Changing land use with CC. Some studies are predicting that with change in global climatic zones, population growth, dietary changes and biofuel demands vast areas of what are currently natural ecosystems will be converted into croplands. The example is given of vast areas of grasslands in East Africa, expected to be converted to ploughed fields over the next 40 years, as wetter conditions caused by climate change attract crop farmers to grazing grounds [17].

Towards solutions
- General
In the light of the above multitude of inter-linked and in many cases synergetic "pressures" on land, the environment and the agricultural sector, there recently have been many calls for "a change to the norm" in order to offset short term economic pain and long term "doom" of the world's population. A major goal of each of these three initiatives is to guide and inspire many government and civil society personnel who though they may realise that action is needed action was needed, are not exactly sure what steps to take. Three examples will be given.

(i) A recent publication by UNEP on "Planning for Change" (Matthew 2008) [18] urged the global community, in the context of impending CC, to adopt more sustainable lifestyles to both reduce the use of natural resources and CO_2 emissions. Coming out of discussions at the Summit on Sustainable Development (Johannesburg 2002), The UNEP report stated that it is becoming increasingly clear that the world cannot achieve sustainable economic growth with old fashioned consumption and production patterns. In accordance with the "Marrakech process" (a ten-year framework of national and regional initiatives on how to achieve SCP), these guidelines have been developed for governments and other stakeholders to establish national programs on "Sustainable Consumption and Production (SCP). Provided are 10 steps on how to plan, develop, implement and monitor a national SCP program. Discussed also are cross-cutting steps, aimed at linking the program to existing strategies such as national development plans (e.g. poverty reduction strategy papers) and national sustainable development strategies. For monitoring purposes a special focus has been made on the development and application of indicators to measure progress toward SCP. In addition, nine country case studies and other examples of good practice illustrating how governments are implementing SCP programs around the world are provided highlighting lessons learned.

(ii) Smith et al (2007), in Chapter 8 "Agriculture" in the 4th Assessment Report of the IPCC, state that opportunities for mitigating GHGs in agriculture fall into three broad categories, based on the underlying mechanism: (1) reducing GHG emissions by more efficient management of carbon and nitrogen flows in agricultural ecosystems, (2) enhancing removals by correcting SOC losses through improved management, thereby

---

[17] From the "Science and Development Network": http://www.scidev.net/Features/index.cfm?fuseaction=readFeatures&itemid=576&language=1
[18] From the www site of Rolf Derpsch (world no-till advocate): http://www.rolf-derpsch.com/

withdrawing atmospheric CO_2, includes reduced crop residue burning and erosion for increased carbon sequestration, and practices such as agro-forestry, perennial plantings and no-till with crop residue retention to increase SOC; (3) avoiding (or displacing) emissions: eg using biofuels rather than fossil fuels as the carbon is of recent atmospheric origin.

More specific agriculture practices that may mitigate GHGs are:

(a) cropland management (better yields, perennial crops, inter-cropping, legumes and crop rotations, with careful fertiliser regimes, give more SOC; minimal or no till reduces SOC losses from tillage and erosion, and builds SOC and soil fauna); improved irrigation practice can increase yields, hence SOC; draining wetland rice out of season to reduce CH4 emissions; agro-forestry gives increased carbon sinks, reduced erosion; conversion of arable to grassland or perennial shrubs also aids SOC increase.

(b) grazing land management/pasture improvement; improved species and nutrition lead to better SOC and water use efficiency with reduced erosion; ceasing slash and burn reduces many GHGs and builds SOC.

(c) restoration of degraded lands; re-vegetation increases SOC and improves water use efficiency and erosion control.

(d) livestock management; main aim is CH_4 reduction from enteric fermentation; via feeding and dietary changes, and biogas production from collected animal waste.

(e) manure/bio-solid management; as (d).

(f) bio-energy production; biofuels; balance required of food and fuel production from arable lands.

(iii) Conservation Agriculture has been practised in many countries and agriculture sectors for up to 40 years. One definition of CA is: "a concept (with very practical field-level outcomes) for resource-saving agricultural ecosystem production that strives to achieve acceptable profits together with high and sustained production levels while concurrently conserving the environment as one of the promising ways of implementing SLM" (Unger 2006). CA varies in almost all situations, as it is not prescriptive. However, the common features are: (a) maintaining, a permanent vegetative cover over the soil provided by the leaves and stems of the current crop, including cover crops and inter-crops, plus the organic matter provided by a mulch of retained residues from previous crops: (b) minimising soil disturbance by tillage, preferably eliminating inversion tillage altogether (no-till) and (c) practising crop rotations and combinations which contribute to increased SOC as well as maintaining biodiversity above and in the soil, and may help avoid build-up of pest populations and weeds.

Recently there has been a move to better orient the three CA principles with current challenges in the terms of agriculture, food security and CC (FAO 2008b). The workshop sought answers to the question: "Can plough–based farming be replaced with more sustainable systems in order to safeguard the world's future food supplies?" Recognised is that the world's food supplies will increasingly depend on raising production per unit area of farmed land. The need now, therefore, is for farmers to take up more sustainable, productive and profitable ways of production that do not damage the soil, land and environment. The workshop focused its attention principally upon CA based farming systems with their potential to be applied on a global scale (currently there are

100 million hectares of arable crops, grown annually without tillage) to ensure adequacy and security of the world's food supplies while improving farmers' livelihoods. Challenges recognised were how to accelerate the participatory adaptation and large-scale uptake of CA practices, wherever appropriate, and in forms fitted to the diversity of local conditions and constraints.

- More specific

Currently, there are several specific initiatives that appear to promise potential in mitigating CC, adapting agriculture for CC, and concurrently aid sustainable land management for food security and improved rural socio-economics. This will be particularly true with joint determination of these initiatives. Required for each of these is investigation for specific agricultural sectors, countries, regions etc of the requirements to implement and then rapidly upscale (widen the uptake of) these initiatives to maximise local and global benefits, towards future (preferably) spontaneous uptake as widely as possible. Also worthy of investigation is the potential of developing Regional Networks, to accelerate the sharing of training and capacity building in the new techniques, sharing of ideas and technology breakthroughs, equipment, trained staff, laboratory facilities and results form monitoring sites, etc; towards developing a modality of the several initiatives, to aid transferability to other countries, regions, etc. Requirements of creating or developing associated, supporting policy development and capacity building (training) also will aid the wider and successful adoption of these initiatives.

Six examples will be presented here. This list is designed for future enriching and enlarging by others, to address specific needs for as wide an impact as possible.

(1) No till (NT), also termed direct drilling or zero tillage, is an agricultural practice widely recognised as an important contributor to fixing carbon in the soil, thereby reducing the amount of CO_2 released into the air [19] as well as being a generally accepted "best management strategy" for reduced inputs, and land and environmental protection. Retention of plant residues with NT prevents soil erosion (reduced by up to 90%), builds soil biodiversity, decreases use of Nitrogen-based fertilizers (hence release of atmospheric N_2O), improves water infiltration (up to 60%) for future crop use, and requires less cultivation and animal draft (hence fossil fuel use and manure production with less CO_2 and CH_4 emissions). John Landers, who has promoted the technology in Brazil since the 1970s, states: "With the best NT systems you have >1 ton of carbon sequestered (in the soil) per hectare per year." At present there is an estimated 100 million ha of NT, practised worldwide [20], showing the very large potential of this practice to remove atmospheric CO_2.

The aim is to increase the uptake of NT (or reduced till) in the agriculture sector of many countries. Constraints and barriers to this strategy, however, are already well known. Rattan Lal (Lal 2007) has stated that "the adoption of NT farming in …south

---

[19] http://www.i-sis.org.uk/BiogasChina.php
[20] This paper is one of the research fruits of the author for the research project "TA 4357(G)/China/ Capacity Building to Combat Land Degradation" (PRC-GEF Partnership on Land Degradation in Dryland Ecosystems).

east Asia...is practically negligible" with the predominance of NT uptake (and associated multilevel gains and spillovers) being in the Americas, Europe and the Indo-Gangetic plain. Resource poor and small size landholders of south east Asia have limited access to required inputs to NT (herbicides, seeding equipment) as well as having competing demands on crop residues (animal feed and fuel) that NT requires to be left in the field.

(2) Biogas (Methane, CH_4) production (ie the controlled production of CH_4 in anaerobic digesters) presents a powerful, and apparently readily implementable and upscalable use of animal manures and other (waste) farm residues, that otherwise would either be kept in open ponds or spread on farms as raw materials (both recognized as strong emitters of CH_4). The aim is to gain far wider utilization of biogas production and use, particularly in rural and peri-urban areas of selected countries. And in this way follow the lead of the Peoples' Republic of China (particularly over the last 20 years) where there are currently 1.3 million people providing daily CH_4 inputs into 22 million digesters (end 2006 data) with annual production of 6.5 billion m^3 of biogas, with the projection of 50 million people using biogas technology by 2010, producing 5.5 million kW at that time (DuByne 2008) and [21]. China is aiming for 25 billion m^3 of biogas by 2020, providing energy to 25% of households in rural areas. Among the many benefits gained from biogas production, the reduction in CH_4 emissions is only one. Biogas usage also reduces the use of trees and crop residues for firewood, addressing not only smoky (unclean) air pollution which is seen as a particulate "GHG" (at all scales from individuals in family homes to country-wide emissions), but also impacts on the negative cycle of land denudation from tree clearing (for firewood) leading to reduced SOC and soil erosion and reduced-level farm production. Another positive is the use of digester residues (routinely pumped out) as bio-fertilisers for increased land productivity without the need for mineral fertiliser inputs.

Certain countries (eg Indonesia) have commenced investigation into the use of crop residues (in particular rice straw) that is in excess of requirements (particularly in two or three crops a year production cycles, where the mass of rice straw residues slow the next crop planting). The residues are mixed with a starter-catalyst (such as Urea or animal manure) to produce biogas. Prototype biodigesters have been produced, with great potential for upscaling, to reduce straw burning hence GHG production, and provide clean cooking fuel for rural dwellers. Investigations are also underway on the potential of utilising other urban wastes (particularly large amounts of waste vegetable matter from vegetable markets) for CH_4 production and urban energy provision. Waste paper is also being considered with co-benefits of greatly reducing landfill requirements or GHG from burning the paper.

(3) Balanced fertilizer(s) usage is central to CC mitigation (principally N_2O reduction), as well as the achievement of more sustainable development, agriculture adaptation

---

[21] Prof. Cai Shouqiu is the Chairman of China Law Society - Environment and Resources Law Society, professor & doctoral advisor of Wuhan University and Huazhong University of Science & Technology, Domestic Legal Training Specialist of PRC-GEF Partnership on Land Degradation in Dryland Ecosystems.

to CC (through reduced dependence on expensive inputs to maintain yields with more hostile climate) and the reduction of rural poverty. This practice is sometimes referred to as "site specific nutrient management" (SSNM) and has been well-researched from the mid 1990s in south east Asia. SSNM provides a field-specific approach for dynamically applying nutrients to crops as and when needed. This approach advocates optimal use of indigenous nutrients originating from soil, plant residues, manures, and irrigation water. Fertilizers are then applied in a timely fashion to overcome the deficit in nutrients between the total demand by rice to achieve a yield target and the supply from indigenous sources. Research in ASEAN countries show that more balanced fertiliser use leads to reduced losses ex-site, and increased yields with equal or reduced N_2O emissions. Investigation is required on the potential and requirements of applying the SSNM approach to other crops (including upland crops and agro-forestry).

In association with SSNM and to support its use at the farm level, several ASEAN institutes have developed and tested field (farmer usable) soil fertility testing kits and fertilizer field kits (the latter to test the quality of farmer-purchased fertilizers); further ensuring reduced fertilizer usage. The more widespread development of these field kits to a wide range of crops is required as too are the resources required (staff, laboratories, field trials, training, farmer field schools, etc) to develop and extend these.

(5) Rotations. There are multiple, synergetic benefits of introducing diversified crop rotations (particularly with a legume phase), married together with reduced tillage and especially NT. Positive outcomes include increased soil productivity (from increased SOC and N status), improved soil aggregation (leading to increased water use efficiency of rain and irrigation from increased water entry, storage and release to plants) and increased soil micro- and macro- biology (in particular earthworms) with improved soil porosity and mixing of SOC to deeper soil layers. Fertiliser (mineral) inputs are also reduced with direct impacts on GHG emissions both from field applications and N2O release, as well as GHGs in production of the fertilisers. Input costs, particularly with spiralling oil and fertiliser costs are reduced. Sequencing low residue crops like peas, lentils, mustard, and canola with greater residue cereals to reduce the trash loading is attractive in wetter climates. Including a deep-rooted legume like alfalfa or lucerne can help increase the rate of nitrogen cycling and help break through plough layer compaction. Possible also to grow alelopathic crops that cause adjoining plants to die or grow more slowly, that with careful selection can greatly assist weed control, hence reduce the need for (oil based) weedicides and the fuel required to apply them to the field. Smother crops, too, can be grown to achieve similar weed-kill or reduction, with the added bonus that these crops reduce erosion, enhance soil fertility and SOC when they naturally break down, and again help feed the soil biota.

(6) Biofuels. Many possible scenarios remain be investigated to ensure the correct balance of biofuel and food production, and the sustainable use of current arable land and (even more importantly) of land currently classed as "marginal". If conducted correctly (that is firmly based on the CA principles of no-till, in conjunction with cover and rotation crops, with under and alley cropping and balanced fertiliser usage) the use of low fertility land to produce biofuels is a possible example of "win win win" technology. Currently underutilised or degraded land, that is commonly steep and erodible, can be

used for fossil fuel replacement materials, with subsequent increase in local area employment and salary earning, as well as increasing the land quality and long term sustainability by organic matter and leguminous cover/row crops.

Conclusions

Most of the strategies whereby Agriculture can contribute to climate change mitigation and the adaptation of the agriculture sector to the, apparently, inevitable changes are already known. Fortunately, these same initiatives cross over with recognised practices to ensure long term sustainability of land, the agriculture sector and those dependent on its outputs. Stabilising food security in the light of increasing population, dietary changes, spiralling oil, commodity and fertiliser prices only strengthen the need for dramatic reductions in input:output ratios across all components of the Agricultural sector.

The challenge remains, however, of achieving the far wider implementation and "upscaling" of these strategies to ensure both a global impact on CC mitigation, as well as more assured food, water and livelihood security for the world's population through local adaptations to CC. Rapid and more impacting effects will be gained with concurrent application of several of the strategies, rather than in isolation of each other. Countries should strive to work together on these approaches, most likely in comparable ecological/climatic zones, and build regional networks, to aim for commonalities in approach and cross-sharing of technologies, successes, training and capacity building to achieve regional homogeneity of practices, and provide a framework (modality) to carry these over to other regions for more assured global impacts. Immediate emphasis should focus on "what can be initiated now" with understood, in-place technology that has shown success elsewhere and which has been taken up by farming communities as their new "best practice and practice of choice". One aim is to ensure widespread uptake within minimum or no risk scenarios. Benchmarking (pre-change), monitoring and evaluation of the impacts of the altered practices will be required, to demonstrate positives (and negatives) and provide the means to convince others of the need and the positive impacts of change. Economic, environment and social indices all need to be collected, to show the cross cutting nature of the altered practices.

References

Bai ZG, Dent DL, Olsson L and Schaepman ME 2008. Global assessment of land degradation and improvement 1: identification by remote sensing. Report 2008/01, FAO/ISRIC – Rome/Wageningen. pp 59.

Cribb, J. (2008). Tackling the world food challenge. Australian Academy of Technological Sciences and Engineering. Volume 151. August 2008. Available at: http://www.atse.org.au/index.php?sectionid=1207

Delgado, C.L (2003). Rising Consumption of Meat and Milk in Developing Countries Has Created a New Food Revolution. Journal of Nutrition, 133, November 2003.

DuByne, D. (2008). The Biogas Revolution. BBI Bioenergy Australasia. October 2008. p. 26-29.

FAO (2008a). Current world fertilizer trends and outlook to 2011/12, Food and Agriculture Organization of the United Nations, Rome 2008

FAO (2008b). Investing in Sustainable Agricultural Intensification the Role of Conservation

Agriculture - A Framework for Action. Proceeding of a Technical Workshop, FAO (Rome). July 2008. Downloadable from: www.fao.org/ag/ca/

IPCC, 2007: Climate Change 2007: Mitigation. Contribution of Working Group III to the Fourth Assessment Report of the Intergovernmental Panel on Climate Change. B. Metz, O.R. Davidson, P.R. Bosch, R. Dave, L.A. Meyer (eds), Cambridge University Press, Cambridge, United Kingdom and New York, NY, USA.

Lal, R. 2007, 'Carbon management in agricultural soils', Mitigation and Adaptation Strategies for Global Change 12: 303–22.

Matthew, B. (2008). Planning for Change - Guidelines for national programmes on sustainable consumption and production. United Nations [UN] Environment Program. Waterside Press. pp 106.

Smith, P., D. Martino, Z. Cai, D. Gwary, H. Janzen, P. Kumar, B. McCarl, S. Ogle, F. O'Mara, C. Rice, B. Scholes, O. Sirotenko (2007) Chapter 8: Agriculture. In Climate Change 2007: Mitigation. Contribution of Working Group III to the Fourth Assessment Report of the Intergovernmental Panel on Climate Change [B. Metz, O.R. Davidson, P.R. Bosch, R. Dave, L.A. Meyer (eds)], Cambridge University Press, Cambridge, United Kingdom and New York, NY, USA.

Unger, P.W. (2006) Soil and Water Conservation Handbook – policies, practices, conditions and terms. Haworth Food and Agricultural Productes Press. London. p 248.

42. Working Together to Combat Rangeland Degradation

Brant Kirychuk
Manager, Sustainable Agriculture Development Project, Agriculture and Agri-Food Canada

Abstract: Canada went through a severe grassland degradation period beginning in the 1930's. The majority of these grasslands have now recovered but it took several decades. It required a combination of an increased emphasis on research and technology development, a significantly expanded agriculture extension service, and programs to support water development and improved land management. China is facing a devastating land degradation problem. Desertification and erosion are evident in all grassland areas, and grazing livestock production is not close to its potential nor is it comparable to many developed countries. It is possible to recover the valuable Chinese grasslands but it is going to take a significant change in policy and agriculture production practices. A combination of programs and provision of information to the farm sector is required, as was used in other countries to recover from this situation. Farms must be looked upon as a business, while efficient and productive grazing livestock enterprises must be a focus. It should include all aspects of the production system including: grazing systems, livestock management, nutrition, health, record keeping, and marketing. Patience, innovation and a new way of looking at grazing livestock production is required.

Canada, like other areas of the world faced a severe land degradation situation, which became most apparent in the 1930's. A series of social, economic and environmental issues came to a head to create a national disaster. Drastic and immediate measures were implemented and the agriculture landscape in Canada is now considered recovered, and relatively sustainable. It took several decades to reach this state though.

A national immigration policy in the early 20th century for settlement of western Canada had scant appreciation of the fragile nature of the landscape. The result was too many settlers who lacked the appropriate agricultural skills situated on land areas not suitable for cultivated agriculture with the available technology and knowledge. An economic, social and ecological disaster ensued within about two decades. The road to recovery was long and arduous.

Canada's experience in balancing the needs of society, livestock production and range landscapes parallels the challenge in other parts of the world of "too many people, too many animals for too little grass". Appropriate use of rangeland in the prairie region is one of the best expressions of sustainable agriculture in Canada today. Four phases characterize the evolution of sustainable rangeland development: initial development and settlement; social disruption precipitated by economic and environmental crises; period of reclamation followed by an emphasis on range and livestock productivity increases; and, most recently, a broadened institutional approach based on technology and policy adaptations. A major response to the drought and economic impacts of the 1930's was the creation of the Prairie Farm Rehabilitation Administration (PFRA) Community Pastures. This program has evolved into an internationally acknowledged model of sustainable land management integral to the achievement of Canada's biodiversity objectives. The program rehabilitated and conserved nearly one million hectares of severely eroded and drought prone lands representing Canada's largest reclamation efforts of native prairie. Today native vegetation dominates about 85% of the PFRA rangelands that are uniquely valued as some of the largest remaining tracts of contiguous native grasslands.

The health and carrying capacity of western Canadian rangelands (private and public) continues to improve in response to institutional and technological innovations. At the same time, societal demands have increased with respect to the additional public goods arising from healthy rangelands such as biodiversity, research sites, ecosystem conservation, C sequestration, endangered species habitat, and recreation. Rangeland managers are challenged to accommodate such additional and competing societal goals while maintaining farm level profitability. Broadened demands on rangelands have fostered innovations in science technology, institutional partnerships and tools for improved education and awareness to enhance range productivity and biodiversity conservation.

Canada's experience with people, livestock, grasslands and policy innovation has played a direct role in international development cooperation projects. The China-Canada Sustainable Agricultural Development Project seeks to support Global Environment Fund targets for land degradation reduction, biodiversity and climate change. However, in the context of "too many people, too many animals for too little grass", significant institutional innovation will be required to replicate Canadian successes in land reclamation and adaptation of new knowledge. Particular focus will be required on policy and institutional arrangements affecting land tenure, independent farmer associations, effective rural financial services and public investment in extension and awareness systems at the local level. As in Canada, the ultimate success will lie in the ability of farmers and ranchers to realize the complementarities between social, economic and environmental goals for rangeland and biodiversity protection and conservation.

The Sustainable Agriculture Development Project (SADP), initiated by the Chinese and Canadian governments, aims to address the severe degradation of China's rangeland and cultivated lands. With China's Ministry of Agriculture and Agriculture and Agri-Food Canada as delivery agents, the focus is on developing capacity to address this challenge in the rural western regions of China with the objectives of:

(1) adapting of land resource management systems for sustainable agriculture;
(2) enhancing sustainable agriculture extension systems;
(3) improving enabling environment for sustainable land resource management.

In northern China the grasslands are overall extremely degraded, and the grazing livestock production system is on the verge of collapse. The rangelands are not producing to near their potential (Figure 1). Where most grazing livestock production systems in developed countries are seeing an increase in per animal output, most of the grazing

CHAPTER V: Climate Change and Land Degradation

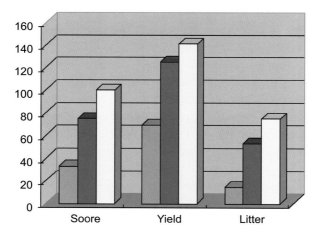

Figure 1: Range health score, biomass yield and litter on 3 range sites in Ewanke Banner, Inner Mongolia

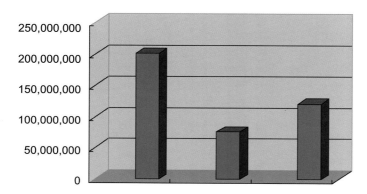

Figure 2: More livestock grazing at Chenbarhu Summer Grazing Site than feed that is available.

livestock systems of China are not seeing any increase, and in fact in many cases it is decreasing. A number of factors are driving this: a grazing livestock population significantly beyond what the grassland can support (Figure 2); keeping of insurance livestock; some livestock, (dairy cows), which are not matched to the grassland environment; exporting hay off the grassland as a cash crop; and a lack of water resources for livestock resulting in reduced productivity. Adding to the challenge is that arid grasslands have a long recovery period, which is going to require patience and significant resources from governments at all levels in China.

The technology is available in China to recover these grasslands, and the grazing and livestock specialists have good knowledge in these fields. The gap appears to be predominantly related to an extension system that is not functioning and not getting information to farmers necessary to have productive, profitable and sustainable grazing livestock production systems. Thus SADP is focussing on developing model extension systems in this field that if successful can be copied else where, and providing extension skills training to specialists so they have the skills and comfort level in working with farmers. The model uses a needs based extension approach, which focuses on the farmer, and addressing what their critical knowledge and information needs are. Secondly the emphasis has been placed on a full farm approach, which addresses livestock nutrition, health, and management, grazing management, as well as marketing and economics. The goal is to take a business and science based approach, thus getting more productivity and economic return from fewer animals.

The following is a description of the areas where the project has had success by adapting methods which were successful in Canada to the situation in China.

Extension Skills Training

The agencies responsible for getting information directly to farmers, notable the Agriculture and Animal Husbandry Bureaus have knowledgeable and for the most part well trained staff in the various technical fields. There is a significant gap in these professionals getting information and advice to the farm level though. Part of the problem rests in the fact that the people tasked to deliver extension programs to farmers, do not have training in extension nor comfort working with farmers. SADP delivered training programs to leading professionals to foster these skills and allow them to train others. There is an outstanding institutional issue which needs to be addressed regarding a lack of funding to do extension, and also a bias as many extension agents are selling inputs. This will need to be addressed to have a credible and functional extension system.

Farmer Field Schools

Over the last few years hundreds of individuals have been trained in delivery of Farmer Field Schools (FFS). Initially the project supported the delivery of FFS by these trained individuals. FFS are generally short one day sessions in local villages on topics of interest or identified need to the local farmer. They are hands on and participatory. In many villages a whole sequence of FFS will be held through the year to address much of the production system. In the last year they have taken on a life of their own and many of the trained personnel are delivering FFS outside of project support. This is a real success of the project.

Demonstration Farms using the Full Farm Approach

Demonstrations have been used for farmers to get a first hand view of practices not commonly used in their area, and also as a tool for training programs. Demonstrations require a lot of resources and on-going follow up to be successful. The project has aimed to do a limited amount of quality demonstrations. The goal has been to demonstrate practices in an area that would have positive production, environmental and economic benefits. In the grazing livestock sector demonstrations included: grazing systems, stored feed, livestock nutrition, and health, record keeping, and marketing. There were 2 different approaches used depending on the province. One method was to create a demonstration farm which showed all practices in one operation. This is very effective to demonstrate the impact of all the practices on a farm basis. The challenge is that it requires a very committed and progressive farmer to implement this number of required changes over a short period of time and on-going follow up and consultation by project staff. The second method was to demonstrate a suite of practices but over a series of farms in an area. Thus a farm would be only demonstrating one or 2 practices. It is much easier to find co-operators who can willingly and effectively manage one new practice on their farm, and generally have been done quite well. The challenge is that a lot of time and travel is required to follow up on each demonstration due to the number of different locations. Further it is quite difficult to present the benefits of the combination of these practices.

Simplified Range Health Models and Pilot Range Health Guide

A number of countries have now developed systems of range health evaluation that are to be used at the farm/ranch level to monitor rangelands, and determine impacts of management. At this point in time there is no system for practical use for farm level range management planning in China. Thus there is little knowledge on what specific tracts of grazing land should be stocked at, as there are no thorough set of tables identifying potential carrying capacities, and what adjustments to stocking rates should be made relevant to condition in the wide variety of eco-sites throughout China. The project conducted wide spread training and discussions with professionals based on simplified approaches to range health applicable at the farm level, and how these could be adapted to China. There was very positive feedback on these systems, and consensus that they could be adapted to be applicable to China. In order to facilitate implementation of a range health model, SADP partnered with the Inner Mongolian Grassland Survey and Design Institute and Inner Mongolia Agriculture University to develop a pilot range health guide for Xilinguole League where there was a wealth of data on the grasslands. This guide will be used to test and demonstrate the applicability of such a model to China.

Publications Targeted at Farmers and Herders

It was quite clear during the delivery of the project that farmers had a real thirst for information both through extension events and written material. Any successful extension program requires written reference material both to support training programs, and also for farmers to use on their own. It appears that there is a very limited amount of unbiased, up to date reference material available to farmers on production, management and marketing. This is a very important need if sustainable agriculture is to advance in China. Thus the project developed publications in both Mandarin as well as some minority languages. This material was made available to farmers at training events, through partners, and at project offices. All reference material is also posted on the project website.
  http://www.ccag.com.cn/english/index_sadp_eng.htm
  http://www.ccag.com.cn/chinese/index_sadp_cn.htm
  Any success of the project can be attributed to starting small and building on successes,

and having a true partnership between the Chinese and Canadian delivery agents who plan and implement the project together, thus building towards long term sustainability of these initiatives.

Based on experience in Canada and other countries it is possible to recover from a devastating land degradation situation. It does require significant commitment, innovation, resources and patience on behalf of governments at all levels, the agriculture sector and farmers. It will take time though and likely production will have to be sacrificed for awhile before the rangeland livestock sector can begin to move forward again.

References

Luciuk, G, M. Boyle, G. Brown, B. Kirychuk, and B. Sonntag. 2008. Too many people, too many livestock, too little grass – a Canadian perspective. Proceedings International Grassland/Rangeland Congress, Hohhot, PRC.

图书在版编目（CIP）数据

综合生态系统管理理论与实践 = Integrated Ecosystem Management Approach and Application / 江泽慧 主编．—北京：中国林业出版社，2009.11
ISBN 978-7-5038-5523-8

I. 综… II. 江… III. ①生态系统－系统管理－世界－文集－英文②土地退化－防治－世界－文集－英文 IV. X321-53 F313-53

中国版本图书馆 CIP 数据核字（2009）第 200902 号

© China Forestry Publishing House 2009

All rights reserved. No part of this publication may be reproduced, stored in a retrieval system, or transmitted in any form or by any means, electronic, mechanical, photocopying, recording or otherwise without the prior permission of the publisher.

Printed in the People's Republic of China
Chinese Publications Number of Archives Library:
ISBN 978-7-5038-5523-8

Integrated Ecosystem Management Approach and Application: Jiang Zehui
I. Integrated… II. Jiang… III. Ecosystem – System Management – International Workshop Proceedings – English IV. X321-53 F313-53

Editors: Wu Jinyou, Li Shun. Designer: Fu Xiaobin
China Forestry Publishing House
No. 7 Liuhai Hutong, Xicheng District
Beijing, P.R. China, 100009
E-mail: cfphz@public.bta.net.cn
Tel: 83286967
Price: RMB 140.00